Key Advances in Clinical Informatics

D0756951

Key Advances in Clinical Informatics

Transforming Health Care through Health Information Technology

Edited by

Aziz Sheikh
University of Edinburgh, Edinburgh, United Kingdom; Brigham and Women's Hospital/Harvard Medical School, Boston, MA, United States

Kathrin M. Cresswell
University of Edinburgh, Edinburgh, United Kingdom

Adam Wright
Brigham and Women's Hospital/Harvard Medical School, Boston, MA, United States

David W. Bates
Brigham and Women's Hospital/Harvard Medical School, Boston, MA, United States

Foreword by Sir John Savill

ACADEMIC PRESS

An imprint of Elsevier

Academic Press is an imprint of Elsevier
125 London Wall, London EC2Y 5AS, United Kingdom
525 B Street, Suite 1800, San Diego, CA 92101-4495, United States
50 Hampshire Street, 5th Floor, Cambridge, MA 02139, United States
The Boulevard, Langford Lane, Kidlington, Oxford OX5 1GB, United Kingdom

British Library Cataloguing-in-Publication Data
A catalogue record for this book is available from the British Library

Library of Congress Cataloging-in-Publication Data
A catalog record for this book is available from the Library of Congress

ISBN: 978-0-12-809523-2

For Information on all Academic Press publications
visit our website at https://www.elsevier.com/books-and-journals

Working together
to grow libraries in
developing countries

www.elsevier.com • www.bookaid.org

Publisher: Mica Haley
Acquisition Editor: Rafael Teixeira
Editorial Project Manager: Mariana Kuhl
Production Project Manager: Julia Haynes
Cover Designer: Mark Rogers

Typeset by MPS Limited, Chennai, India

Contents

Part I
An Introduction to Clinical Informatics

1. An Overview of Clinical Informatics

Kathrin M. Cresswell, David W. Bates, Adam Wright and Aziz Sheikh

2. Inpatient Clinical Information Systems

Kathrin M. Cresswell and Aziz Sheikh

5.   Interoperability

Mark E. Frisse

6.   Privacy and Security

John D. Halamka

Part II
Improving the Quality, Safety and Efficiency of Care

7. Public Policy and Health Informatics
David W. Bates and Aziz Sheikh

8. Health Information Technology and Value
Blackford Middleton and Ngai T. Cheung

9. Organizational and Behavioral Issues
Joan S. Ash and Nancy M. Lorenzi

10. Medication Management, and Laboratory and Radiology Testing

Sarah P. Slight and David W. Bates

11. Bioinformatics and Precision Medicine

Arjun K. Manrai and Isaac S. Kohane

12. Clinical Decision Support and Knowledge Management

Robert A. Greenes

13. Mobile Health

Karandeep Singh and Adam B. Landman

14. A Sociotechnical Approach to Electronic Health Record Related Safety

Dean F. Sittig and Hardeep Singh

15. Predictive Analytics and Population Health

Peter S. Hall and Andrew Morris

16. An Apps-Based Information Economy in Healthcare

Kenneth D. Mandl, Joshua C. Mandel and Pascal B. Pfiffner

Part III
Future Developments

17. Cloud-Based Computing

Mehrdad A. Mizani

18. Social and Consumer Informatics

Felix Greaves and Ronen Rozenblum

List of Contributors

Joan S. Ash Oregon Health & Science University, Portland, OR, United States

David W. Bates Harvard Medical School; Harvard School of Public Health, Boston, MA, United States; Brigham and Women's Hospital/Harvard Medical School, Boston, MA, United States

Alison Callahan Stanford University, Stanford, CA, United States

Ngai T. Cheung Hospital Authority of Hong Kong, Kowloon, Hong Kong

Kathrin M. Cresswell The University of Edinburgh, Edinburgh, United Kingdom

Mark E. Frisse Vanderbilt University Medical Center, Nashville, TN, United States

Felix Greaves Imperial College London, London, United Kingdom

Robert A. Greenes Arizona State University and Mayo Clinic, Scottsdale, AZ, United States

John D. Halamka Beth Israel Deaconess Medical Center and Harvard Medical School, Boston, MA, United States

Peter S. Hall The University of Edinburgh, Edinburgh, United Kingdom

Isaac S. Kohane Harvard Medical School, Boston, MA, United States

Adam B. Landman Harvard Medical School, Boston, MA, United States

Nancy M. Lorenzi Vanderbilt University Medical Center, Nashville, TN, United States

Joshua C. Mandel Boston Children's Hospital, Boston, MA, United States; Harvard Medical School, Boston, MA, United States

Kenneth D. Mandl Boston Children's Hospital, Boston, MA, United States; Harvard Medical School, Boston, MA, United States

Arjun K. Manrai Harvard Medical School, Boston, MA, United States

Blackford Middleton Apervita, Inc., Chicago, IL, United States; Harvard TH Chan School of Public Health, Boston, MA, United States

Mehrdad A. Mizani Middle East Technical University, Ankara, Turkey

Andrew Morris The University of Edinburgh, Edinburgh, United Kingdom

Pascal B. Pfiffner Boston Children's Hospital, Boston, MA, United States

Ronen Rozenblum Harvard Medical School, Boston, MA, United States

Gordon D. Schiff Harvard Medical School, Boston, MA, United States

Nigam H. Shah Stanford University, Stanford, CA, United States

Aziz Sheikh The University of Edinburgh, Edinburgh, United Kingdom; Brigham and Women's Hospital/Harvard Medical School, Boston, MA, United States

Hardeep Singh Michael E. DeBakey Veterans Affairs Medical Center, Houston, TX, United States; Baylor College of Medicine, Houston, TX, United States

Karandeep Singh University of Michigan Medical School, Ann Arbor, MI, United States

Dean F. Sittig The University of Texas Health Science Center at Houston, Houston, TX, United States

Sarah P. Slight Highlander House, Ovington, Northumberland, United Kingdom

Mary J. Tharayil Brigham and Women's Primary Care of Brookline, Brookline, MA, United States

Adam Wright Harvard Medical School; Partners HealthCare, Boston, MA, United States; Brigham and Women's Hospital/Harvard Medical School, Boston, MA, United States

About the Editors

Aziz Sheikh is Professor of Primary Care Research & Development and Director of the Usher Institute of Population Health Sciences and Informatics at The University of Edinburgh. He also serves as Co-Director of Harvard Medical School's Safety, Quality and Informatics Leadership (SQIL) Program. He is one of the world's most published academicians and has advised governments on health information technology in Europe, North America, and Asia. He holds fellowships from seven learned societies, including the American College of Medical Informatics, the Royal Society of Edinburgh, and the Academy of Medical Sciences. He was made an officer of the British Empire in 2014 for his "Services to medicine and healthcare."

Kathrin Cresswell is a health psychologist and experienced mixed-methods researcher who has worked in the field of clinical informatics for over a decade. She has authored around 80 scientific papers in this area in international peer-reviewed journals. During her career, she has collaborated with international leaders on a number of large projects and consulted for the World Health Organization and Harvard Medical School. She is currently holding a prestigious fellowship from the Scottish Government and a Chancellor's Fellowship from The University of Edinburgh to investigate future strategic directions in health information technology.

Adam Wright is Associate Professor of Medicine at Harvard Medical School and Senior Scientist in the Division of General Internal Medicine at Brigham and Women's Hospital. His research focuses on making electronic health records safer and more effective through better design, improved clinical decision support, and harnessing clinical data to drive a learning healthcare system. In addition to his research, he teaches Harvard's introductory biomedical informatics courses and also teaches first medical students. He holds PhD in biomedical informatics from Oregon Health and Science University and BS in mathematical and computational sciences from Stanford.

David Bates is an internationally renowned expert in health information technology, patient safety, and quality. His work has focused on improving clinical decision-making, patient safety, quality-of-care, and cost-effectiveness. He is Chief of the Division of General Internal Medicine at Brigham and Women's Hospital in Boston, Professor of Medicine at Harvard Medical School, and Professor of Health Policy and Management at the Harvard School of Public Health, where he co-directs the Program in Clinical Effectiveness. He has been elected to the Institute of Medicine, and has been the past chairman of the Board of the American Medical Informatics Association.

Foreword

I qualified in Medicine in 1981, just as medical practice began to change from experience-based to evidence-based. Seemingly in the blink of an eye, we are now experiencing the development of data-driven medicine, where the application of new analytic technologies to data derived from linked electronic medical records will revolutionize medicine. Patients will benefit from care protocols distilled from experience and evidence analyzed at population scale, orders of magnitude larger than even the biggest clinical trial. Wherever the reader is in a healthcare career, or whatever the reader's interests (discovery science, precision medicine, public health, clinical trials, citizen-driven health, learning health systems) this valuable book will inform and educate. Indeed, its text is likely being mined and analyzed as you read!

Sir John Savill

Part I

An Introduction to Clinical Informatics

Chapter 1

An Overview of Clinical Informatics

Kathrin M. Cresswell[1], David W. Bates[2], Adam Wright[3] and Aziz Sheikh[1,4]

[1]*The University of Edinburgh, Edinburgh, United Kingdom,* [2]*Harvard Medical School; Harvard School of Public Health, Boston, MA, United States,* [3]*Harvard Medical School; Partners HealthCare, Boston, MA, United States,* [4]*Brigham and Women's Hospital/Harvard Medical School, Boston, MA, United States*

INTRODUCTION: THE EVOLVING AND EXPANDING ROLE OF INFORMATION TECHNOLOGY

Information technology (IT) is increasingly pervading everything we do. For instance, worldwide statistics show that more people have access to a mobile phone than to working toilets (Times Magazine, 2013), and while the existence of genome editing may have seemed unimaginable just a few years ago, tools have now equipped scientists with the ability to genetically modify human embryos (Liang et al., 2015). Such developments are occurring exponentially with an increasing array of technological features, designs, and data generated from applications impacting on all aspects of human life including food, health, energy, and the environment. The first personal computer was released in 1974, and within a mere seven decades, by 2045 computing power is expected to exceed that of all human brains combined. This "technological singularity" is a widely debated hypothetical moment in time where artificial intelligence (AI) will surpass our cognitive limitations (Kurzweil, 2005).

Not surprisingly, most industries have drawn heavily on technological developments to transform their services, and although somewhat lagging behind, healthcare is following suit, driven by increasing pressures on health systems to improve quality, reduce errors, and increase efficiency (Travis et al., 2004). This chapter will provide an overview of past, present, and future developments in the area of clinical informatics. It will introduce the most important concepts and definitions, provide a high-level perspective of the existing empirical evidence base in relation to the effectiveness of health IT (HIT), and provide a contextual overview of the chapters in this book.

Key Advances in Clinical Informatics. DOI: http://dx.doi.org/10.1016/B978-0-12-809523-2.00001-7

The following three sections serve as an overall structure: Section 1 introduces clinical informatics as a discipline, outlining key terms and tensions; Section 2 tackles issues surrounding the impact of clinical informatics applications on quality, safety, and efficiency of care; and Section 3 delves deeper into future developments that are likely to dominate the sector in the foreseeable future, though undoubtedly many other developments will occur that cannot yet be predicted.

A BRIEF HISTORY OF THE FIELD OF CLINICAL INFORMATICS

The first use of HIT in a clinical setting can be traced back to 1952, when Dr. Arthur Rappoport reported his experiences with using the McBee Manual Punch Card in a pathology laboratory setting (Porth and Lübke, 1996). This was followed by the emergence of hospital information systems in the 1960s, with the Latter Day Saints Hospital in Utah (USA, now Intermountain) being the first to implement this in 1967. Others, including The COmputer STored Ambulatory Record and the Regenstrief Medical Record System, followed. The Health Evaluation through Logical Programming system had the ability to collect demographic and clinical data with decision support features. It is used to the present day, but may be replaced by Cerner in the near future (Gardner et al., 1999; Healthcare IT News, 2015). The development of clinical specialty systems for laboratory, radiology, pathology, radiotherapy, pharmacy, and primary care followed. Integration of these was not possible until the 1980s, when larger integrated medical information systems emerged, facilitated by the development of high-speed communication networks. In 1985, the first patient scheduling software called "Cadence" was launched by Epic Systems, followed by EpicCare in 1992. Subsequent developments in the 21st century have been characterized by growing clinical uses of technologies drawing on an ever increasing array of data sources (including patients and various care settings), mobile applications that allow patients and providers to gather and view data "on the go", and the exploitation of digital data generated for reuse (Cresswell and Sheikh, 2016). An overview of key historical developments is provided in Box 1.1.

WHAT IS CLINICAL INFORMATICS?

Clinical informatics represents a highly interdisciplinary field that involves "analyzing, designing, implementing, and evaluating information and communication systems that enhance individual and population health outcomes, improve patient care, and strengthen the clinician-patient relationship (Gardner et al., 2009)." As such, the field can be positioned at the intersection of clinical care, the health system, and information and communication technology. Its interdisciplinary nature is a core feature, and it includes

---

BOX 1.1 Key Historical Developments in Clinical Informatics[1,2,3]

1949: establishment of the German Society for Medical Documentation, Computer Science and Statistics (first professional informatics organization)

1950s: first time IT applied to the field of medicine in biomedical context

1952: Arthur Rappoport reported on using the McBee Manual Punch Card in a laboratory setting

1960s: first peer-reviewed informatics journals launched

1960s: emergence of hospital information systems that included digital patient information

1967: Latter Day Saints Hospital in Utah first hospital to use an Electronic Health Record (EHR)

1970s: first mention of English term "medical informatics"

1960s/70s: clinical specialty systems were developed for laboratory, radiology, pathology, radiotherapy, pharmacy, and primary care

1970: first Computerized Physician Order Entry system used in El Caminio Hospital, California

1980s: development of local, national, and worldwide high-speed communication networks

1980s: emergence of larger integrated medical information systems

1985: first patient scheduling software launched (Cadence)

1988: creation of the American Medical Informatics Association

1990s: emergence of the internet facilitating exchange of clinical data

1992: EpicCare launched

2000s: clinical users could use IT to view/order tests/medications from various databases

2010s: emergence of cloud networks and integration of data across multiple locations

---

1. Collen, M.F., 2015. A History of Medical Informatics in the United States. Ball, M.J. (Ed.), Springer, New York.
2. Hayes, G.M., Barnett, D.E. (Eds.), 2008. UK Health Computing: Recollections and Reflections. British Computer Society.
3. http://www.healthworkscollective.com/frankie-xavier/162251/long-road-digitization-history-healthcare-informatics

---

clinical providers such as physicians, nurses, and pharmacists, but also medical librarians, information scientists, and communication specialists, to name just a few of the types of professionals involved.

Related terms that are sometimes used interchangeably include health informatics, medical informatics, and eHealth. There are a range of published definitions with varying understandings of the field in the published literature—the exponential development of applications and increasing convergence of functionalities complicates navigating the area further (Boogerd et al., 2015; Oh et al., 2005). Various chapters of this book will delve deeper into specific applications and associated concepts. These begin with overviews of inpatient systems, outpatient systems, and clinical documentation in Chapters 2, 3, and 4, respectively.

Overall, existing applications can broadly be divided into three categories (Black et al., 2011): (1) systems informing and supporting decisions (see Chapter 10 on medication, laboratory, and radiology testing; and Chapter 12 on knowledge management and computerized guidelines); (2) storage and management of data (Chapter 11 on bioinformatics and precision medicine); and (3) delivery of expertise and care at a distance (see Chapter 13 on mobile health).

There has been an increasing emergence of clinical informatics as a discipline (Greenes and Shortliffe, 1990). Associated activity includes the growing demand for organizational capacity in this area, but also the need for academic expertise to develop new educational trajectories and evaluate ongoing implementation, adoption, and optimization activities associated with the increasing range of technologies. Formal accreditation and certification of clinical informatics expertise are closely associated activities that are presently receiving attention (Fridsma, 2015; Gadd et al., 2016; Shortliffe et al., 2016), particularly in the United States (Middleton, 2014), but also many other countries.

EMPIRICAL EVIDENCE SURROUNDING EFFECTIVENESS OF CLINICAL INFORMATICS APPLICATIONS

Clinical informatics applications have been shown to result in a number of benefits including, among others, the prevention of life-threatening allergic reactions to medication through systems facilitating clinical decision making (Bates et al., 1999; Kaushal et al., 2003), reductions in prescribing errors (Avery et al., 2012), and the ability to manage diabetes and high blood pressure remotely (Wild et al., 2016). However, it is often difficult to demonstrate the clinical effectiveness and cost-effectiveness of HIT (Black et al., 2011; Chaudhry et al., 2006; Jones et al., 2014), this at least in part reflecting the need for workflow reconfiguration and systems optimization (Cresswell et al., 2017).

Chapter 8 will explore HIT and value in more detail, while Chapter 14 will examine the impact of technology on safety.

There is increasing understanding of the potential risks associated with the introduction of new technologies in healthcare settings (Black et al., 2011; Buntin et al., 2011). The most commonly examined areas in this respect include privacy, confidentiality, and security (see Chapter 6: Privacy and Security); effects on work practices and interdisciplinary working; and difficulties surrounding accessibility of data (Ash et al., 2004; Barrows and Clayton, 1996; Harrison et al., 2007).

The underlying reasons for this overall lack of evidence may partly be due to difficulties evaluating technologies, as these are often embedded in wider organizational change initiatives resulting in difficulties attributing effects (Campbell et al., 2000; Lilford et al., 2009). There is also a growing literature reexamining traditional evaluation paradigms advocating randomized controlled trials (RCTs) as the "gold standard", toward a more flexible use of various evaluation methods including qualitative, mixed methods, human factors, and engineering-based approaches (Cresswell et al., 2017; Klasnja et al., 2011; Yusof et al., 2008). This is because RCTs tend to be costly and time-consuming (resulting in issues surrounding the applicability of results and major challenges associated with changing software/technology) and may not be appropriate for effectively evaluating the range of different rapidly changing existing applications. Conversely, it can be easy and less costly to conduct clinical trials using EHRs, and in industry for example it is now routine to employ "AB" testing, where when it is not clear which of two options is superior, both are tried for half a user base. Whichever is more effective at achieving the desired outcome (e.g., a digital purchase) is then used as the default.

CLINICAL INFORMATICS IN CONTEXT

Having touched upon the challenges inherent in evaluating clinical informatics applications above, it is important to briefly discuss the importance of appreciating the range of contextual dimensions and various stakeholders that are involved in deploying and adopting technologies in healthcare (Fig. 1.1) (Cresswell and Sheikh, 2009).

Contextual aspects may include technical features (e.g., usability), social contexts (e.g., changes in work practices), organizational strategies (relating to implementation and optimization), and wider sociopolitical dimensions (such as state and federal approaches to implementation and regulatory environments). Stakeholders within these various contexts include patients, academics, providers, vendors, developers, third-sector organizations, and policy makers. All of these have different interests that need to be aligned for initiatives to be successful. Chapter 9 will discuss organizational issues in more detail, while Chapter 7 will examine policy considerations and associated international strategies to promote clinical informatics implementations.

Empirical evaluations that take this range of stakeholders and contextual factors into account are now widely advocated (Catwell and Sheikh, 2009). These should involve a longitudinal component to facilitate tracing developments over time, playing an active role in aligning interests through providing formative feedback to stakeholders in participating healthcare settings, and summative feedback to policy makers (Ammenwerth et al., 2003).

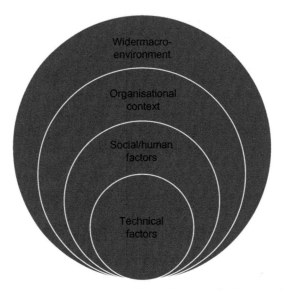

FIGURE 1.1   Overview of contextual factors. Factors important for the successful implementation of EHRs identified in the literature. *Adopted from Cresswell, K., Sheikh, A., 2009. The NHS Care Record Service (NHS CRS): recommendations from the literature on successful implementation and adoption. J. Innov. Health Inform. 17 (3), 153–160.*

VISIONS SURROUNDING FUTURE DEVELOPMENTS IN CLINICAL INFORMATICS

A central theme of this book will be examining state-of-the-art developments in clinical informatics, exploring progress toward realizing the vision of more effective, better quality, and safer care through the application of IT in healthcare settings. Key current developments in this respect are likely to include the creation of integrated health informatics infrastructures where data can be seamlessly shared between settings and applications (see Chapter 5 on interoperability, Chapter 16 on application programming interfaces, and Chapter 17 on cloud-based computing), and the creation of learning health systems that effectively draw on digital data collected in a variety of settings and by a variety of stakeholders to improve performance and services (see Chapter 15 on predictive analytics and population health, Chapter 18 on social/consumer informatics, and Chapter 19 on machine learning and AI).

It is important to place this work in the context of a continuously evolving field, where innovations are created at a rapid pace. HIT has, if appropriately conceptualized, developed and implemented, the potential to continue to have major transformative effects on healthcare and can through so doing help deal with one of the most pressing healthcare challenges facing healthcare worldwide, namely to achieve more in terms of health gain for less and to support patient involvement/enablement/empowerment.

CONCLUSIONS

Significant international policy efforts and investments in clinical informatics are taking place to improve adoption and use of healthcare IT, with the underlying aim of improving healthcare safety, quality, and efficiency. We have outlined some of the past, present, and potential future developments in this domain and provided an overview of definitions and core issues in the field. Subsequent chapters in this book aim to share state-of-the-art developments with nonexpert clinical and academic audiences across the globe.

REFERENCES

Ammenwerth, E., Gräber, S., Herrmann, G., Bürkle, T., König, J., 2003. Evaluation of health information systems—problems and challenges. Int. J. Med. Inform. 71 (2), 125–135.

Ash, J.S., Berg, M., Coiera, E., 2004. Some unintended consequences of information technology in health care: the nature of patient care information system-related errors. J. Am. Med. Inform. Assoc. 11 (2), 104–112.

Barrows, R.C., Clayton, P.D., 1996. Privacy, confidentiality, and electronic medical records. J. Am. Med. Inform. Assoc. 3 (2), 139–148.

Bates, D.W., Teich, J.M., Lee, J., Seger, D., Kuperman, G.J., Ma'Luf, N., et al., 1999. The impact of computerized physician order entry on medication error prevention. J. Am. Med. Inform. Assoc. 6 (4), 313–321.

Black, A.D., Car, J., Pagliari, C., Anandan, C., Cresswell, K., Bokun, T., et al., 2011. The impact of eHealth on the quality and safety of health care: a systematic overview. PLoS Med. 8 (1), e1000387.

Boogerd, E.A., Arts, T., Engelen, L.J., van De Belt, T.H., 2015. "What is eHealth": time for an update? JMIR Res. Protoc. 4 (1), e29.

Buntin, M.B., Burke, M.F., Hoaglin, M.C., Blumenthal, D., 2011. The benefits of health information technology: a review of the recent literature shows predominantly positive results. Health Aff. 30 (3), 464–471.

Campbell, M., Fitzpatrick, R., Haines, A., Kinmonth, A.L., 2000. Framework for design and evaluation of complex interventions to improve health. Br. Med. J. 321 (7262), 694.

Catwell, L., Sheikh, A., 2009. Evaluating eHealth interventions: the need for continuous systemic evaluation. PLoS Med. 6 (8), e1000126.

Chaudhry, B., Wang, J., Wu, S., Maglione, M., Mojica, W., Roth, E., et al., 2006. Systematic review: impact of health information technology on quality, efficiency, and costs of medical care. Ann. Intern. Med. 144 (10), 742–752.

Cresswell, K., Sheikh, A., 2009. The NHS Care Record Service (NHS CRS): recommendations from the literature on successful implementation and adoption. J. Innov. Health Inform. 17 (3), 153–160.

Cresswell, K.M., Sheikh, A., 2016. Key global developments in health information technology. J. R. Soc. Med. 109 (8), 299–302.

Cresswell, K.M., Bates, D.W., Sheikh, A., 2017. Ten key considerations for the successful optimization of large-scale health information technology. J. Am. Med. Inform. Assoc 24 (1), 182–187.

Fridsma, D.B., 2015. Update on informatics-focused certification and accreditation activities. J. Am. Med. Inform. Assoc. 22 (2), 489–490.

Gadd, C.S., Williamson, J.J., Steen, E.B., Fridsma, D.B., 2016. Creating advanced health informatics certification. J. Am. Med. Inform. Assoc. 23 (4), 848−850.

Gardner, R.M., Pryor, T.A., Warner, H.R., 1999. The HELP hospital information system: update 1998. Int. J. Med. Inform. 54 (3), 169−182.

Gardner, R.M., Overhage, J.M., Steen, E.B., Munger, B.S., Holmes, J.H., Williamson, J.J., et al., 2009. Core content for the subspecialty of clinical informatics. J. Am. Med. Inform. Assoc. 16 (2), 153−157.

Greenes, R.A., Shortliffe, E.H., 1990. Medical informatics: an emerging academic discipline and institutional priority. JAMA 263 (8), 1114−1120.

Harrison, M.I., Koppel, R., Bar-Lev, S., 2007. Unintended consequences of information technologies in health care—an interactive sociotechnical analysis. J. Am. Med. Inform. Assoc. 14 (5), 542−549.

Healthcare IT News, 2015. Intermountain live with Cerner EHR. Available from: http://www. healthcareitnews.com/news/intermountain-live-cerner-ehr (last accessed 01.01.17).

Jones, S.S., Rudin, R.S., Perry, T., Shekelle, P.G., 2014. Health information technology: an updated systematic review with a focus on meaningful use. Ann. Intern. Med. 160 (1), 48−54.

Kaushal, R., Shojania, K.G., Bates, D.W., 2003. Effects of computerized physician order entry and clinical decision support systems on medication safety: a systematic review. Arch. Intern. Med. 163 (12), 1409−1416.

Klasnja, P., Consolvo, S., Pratt, W., 2011. How to evaluate technologies for health behavior change in HCI research. In: Proceedings of the SIGCHI Conference on Human Factors in Computing Systems. ACM, pp. 3063−3072.

Kurzweil, R., 2005. The Singularity Is Near: When Humans Transcend Biology. Penguin.

Liang, P., Xu, Y., Zhang, X., Ding, C., Huang, R., Zhang, Z., et al., 2015. CRISPR/Cas9-mediated gene editing in human tripronuclear zygotes. Protein Cell 6, 363−372.

Lilford, R.J., Foster, J., Pringle, M., 2009. Evaluating eHealth: how to make evaluation more methodologically robust. PLoS Med. 6 (11), e1000186.

Middleton, B., 2014. First diplomates board certified in the subspecialty of clinical informatics. J. Am. Med. Inform. Assoc. 21 (2), 384.

Oh, H., Rizo, C., Enkin, M., Jadad, A., 2005. What is eHealth (3): a systematic review of published definitions. J. Med. Internet Res. 7 (1), e1.

Porth, A.J., Lübke, B., 1996. History of computer-assisted data processing in the medical laboratory. Eur. J. Clin. Chem. Clin. Biochem. 34 (3), 215−229.

Shortliffe, E.H., Detmer, D.E., Munger, B.S., 2016. Clinical informatics: emergence of a new profession. In Clinical Informatics Study Guide. Springer International Publishing, pp. 3−21.

Time Magazine, 2013. More people have cell phones than toilets, U.N. Study shows. <http:// newsfeed.time.com/2013/03/25/more-people-have-cell-phones-than-toilets-u-n-study-shows/> (accessed 12.07.16).

Travis, P., Bennett, S., Haines, A., Pang, T., Bhutta, Z., Hyder, A.A., et al., 2004. Overcoming health-systems constraints to achieve the Millennium Development Goals. Lancet 364 (9437), 900−906.

Wild, S.H., Hanley, J., Lewis, S.C., McKnight, J.A., McCloughan, L.B., Padfield, P.L., et al., 2016. Supported telemonitoring and glycemic control in people with type 2 diabetes: the telescot diabetes pragmatic multicenter randomized controlled trial. PLoS Med. 13 (7), e1002098.

Yusof, M.M., Papazafeiropoulou, A., Paul, R.J., Stergioulas, L.K., 2008. Investigating evaluation frameworks for health information systems. Int. J. Med. Inform. 77 (6), 377−385.

RECOMMENDED FURTHER READING

Avery, A.J., Rodgers, S., Cantrill, J.A., Armstrong, S., Cresswell, K., Eden, M., et al., 2012. A pharmacist-led information technology intervention for medication errors (PINCER): a multicentre, cluster randomised, controlled trial and cost-effectiveness analysis. Lancet 379 (9823), 1310–1319.

Black, A.D., Car, J., Pagliari, C., Anandan, C., Cresswell, K., Bokun, T., et al., 2011. The impact of eHealth on the quality and safety of health care: a systematic overview. PLoS Med. 8 (1), e1000387.

Coiera, E., 2015. Guide to Health Informatics. CRC Press.

Collen, M.F., 2015. A History of Medical Informatics in the United States. Springer, New York.

Cresswell, K., Blandford, A., Sheikh, A., Reconsidering paradigms for the evaluation of health information technology. Submitted to JAMIA.

Hovenga, E.J., 2010. Health Informatics: An Overview. IOS Press.

Chapter 2

Inpatient Clinical Information Systems

Kathrin M. Cresswell[1] and Aziz Sheikh[1,2]
[1]The University of Edinburgh, Edinburgh, United Kingdom, [2]Brigham and Women's Hospital/Harvard Medical School, Boston, MA, United States

INTRODUCTION

Hospitals are typically large complex organizational environments, where many different specialties and groups of healthcare professionals work together to provide patient care (Czarniawska, 1997). The introduction of health information technology (HIT) presents significant opportunities to reduce variations in the quality of care, improve safety, and also reduce the currently high and ultimately unsustainable costs of healthcare provision. Opportunities, for instance, include central management of large volumes of electronic data for quality improvement (e.g., helping to identify and target high-risk clinical areas) (Murdoch and Detsky, 2013), sophisticated decision support for different specialties (e.g., to support prescribing decisions in those with impaired kidney function) (Kaushal et al., 2003), and improved streamlining of work practices of healthcare workers (e.g., facilitating ordering of diagnostic tests) (Brandao de Souza, 2009). There are however also important challenges. These may be technical, such as the creation of large-scale infrastructures required to integrate different specialty systems, or social, including workarounds being introduced by intended users (e.g., if the system hinders users fulfilling their tasks such as nurses' inability to any longer request X-rays in the new online image requesting system) (Lapointe and Rivard, 2005).

This chapter will begin by discussing the range of different technical inpatient information system functionality currently used in healthcare settings internationally. This will be followed by considering the complex high-pressure social and organizational environments into which these technologies are implemented and the impact of these contextual factors on adoption and implementation. A discussion of potential risks emerging from these sociotechnical challenges and some suggestions on how to mitigate these then follows. In order to plan for obstacles and minimize risks, we argue that

Key Advances in Clinical Informatics. DOI: http://dx.doi.org/10.1016/B978-0-12-809523-2.00002-9
13

implementation of digital inpatient systems is best conceptualized as a journey, with continuing optimization activities required to ensure ongoing development to suit changing needs of different stakeholder groups. The chapter will conclude with some international examples of sophisticated system implementation and optimization, and a brief discussion on potential future technological developments.

A BRIEF HISTORY OF INPATIENT INFORMATION SYSTEMS

The use of information systems in healthcare contexts is a relatively recent development. The first suggestion that computers could be used to support care was made not until around 60 years ago and the first rudimentary hospital information systems were not introduced until the early 1970s (HealthWorksCollective, 2014). These tended to be "home-grown" systems that were developed by hospitals themselves and extensively customized to suit local needs. Commercially available systems that could be bought and implemented by a wider range of hospitals were not developed until later. The first "home-grown" electronic medical record systems were introduced by the Regenstreif Institute, The Mayo Clinic, Intermountain Healthcare, and the Medical Center Hospital of Vermont (Baker et al., 2008; NASBHC, 2015; Panesar et al., 2014). Although most early inpatient systems were developed in the United States, some notable exceptions include the DIOGENE hospital information system implemented in 1978 by the University Hospital in Geneva Switzerland (Borst et al., 1999), the Danderyd Hospital Computer System in Stockholm Sweden (Abrahamsson, et al., 1970), and the Leyden University Hospital Information System in the Netherlands (Bakker, 1982).

Information technology can support care activities and help to improve safety and efficiency (Bates and Gawande, 2003). Nonetheless, many healthcare professionals continue to struggle with integrating the increased demand for recording of information in suboptimal commercial systems when providing hands-on patient care (NASBHC, 2015; Wachter, 2015), although there are variations among hospitals. This is likely one of the underlying reasons that adoption of electronic systems has been slow in comparison with other sectors such as banking and the retail.

HOW IS HIT USED IN INPATIENT SETTINGS?

The most common type of HIT in inpatient settings are electronic health records (EHRs). These can be seen as digital repositories of health information with different levels of complexity (Fig. 2.1) (Black et al., 2011). They may range from relatively simple Patient Administration Systems that hold patient demographic data to complex systems that allow sharing of data

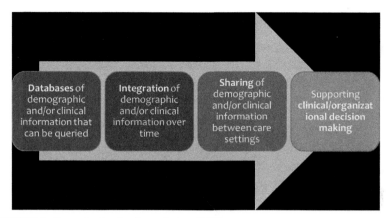

FIGURE 2.1 Increasing levels of complexity of EHRs.

between settings (including primary/social care and the home) and support clinical and organizational decision making.

EHRs may therefore be viewed as a basic information infrastructure to which additional functionality can be added. The potential range of this functionality is vast and rather than catalogue it, we consider some of the most prominent examples. These include computerized provider order entry (CPOE) and clinical decision support (CDS) systems that draw on existing knowledge bases (such as formularies and patient-specific notes) to support healthcare professionals in complex and high-risk tasks such as prescribing. In doing so, they help to improve prescribing performance by suggesting medications and by alerting individuals to potential drug−drug interactions and contraindications (e.g., switching to an alternative if a patient has an allergy to penicillin, Fig. 2.2) (Kaushal et al., 2003; Westbrook et al., 2012).

While CPOE and CDS systems are mainly used for prescribing-related tasks, inpatient settings also have systems tailored the needs of specific professional inpatient groups. These specialty systems may, for instance, include cancer care and associated chemotherapy systems, radiology systems, and pediatrics systems. Common components of radiology specialty software are for instance Picture Archiving and Communication Systems (PACS). These allow X-rays to be accessed and viewed digitally, facilitating sharing and effective real-time communication between healthcare professionals through increased availability and accessibility (Pilling, 2014) (Fig. 2.3).

Electronic data held within the different systems discussed above can be collected, analyzed, and reused. This is commonly referred to as "secondary uses"—drawing on data to review and improve organizational and clinical performance (Safran et al., 2007). Some case study examples illustrating secondary data uses in inpatient settings are provided in Box 2.1.

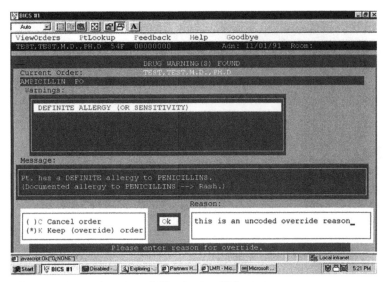

FIGURE 2.2   An illustration of a penicillin alert in a CDS system.

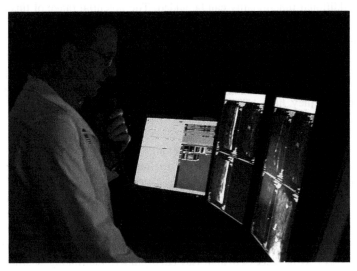

FIGURE 2.3   An illustration of a PACS. *https://commons.wikimedia.org/.*

HIT in inpatient settings not only consists of software. Hardware applications on which software can run may vary widely across settings. Devices can range from computers on wheels and desktop computers, to tablets and other mobile devices used by healthcare professionals and patients. These can help to ensure that patient information is readily accessible when needed.

BOX 2.1 Case Study Examples Illustrating Secondary Data Uses in Healthcare Settings

Kaiser Permanente provides care for 8.6 million members in California and parts of Colorado. In 2002, the organization began working with the Epic Systems Corporation to create and implement an integrated EHR (KP HealthConnect). Although centrally governed, the system is configured in line with the needs of different localities. An in-house analytics team routinely draws on data to improve the quality and efficiency of care provided. This includes data warehousing, and both, operational analytics (clinical and financial) and research analytics (to support research into for example health disparities and comparative effectiveness research).

Geisinger Health System provides care for over 2.6 million people in Pennsylvania. It began implementation of EpicCare, an integrated EHR in 1996. Having collaborated with IBM to create an integrated data warehouse, the system now supports a range of research projects drawing on EHR data. As part of this, the organization is able to simulate reengineering of care processes and allows for testing of new software tools in clinical environments and then commercializing these (xG Health Solutions).

The University Hospital Birmingham NHS Foundation Trust in the United Kingdom has established a Quality and Outcomes Research Unit that is a collaboration between information analysts and clinicians. It uses data analysis to improve care delivery drawing on health information. The unit now oversees the development and implementation of appropriate quality outcome metrics in each of the Trust's specialties.

THE SOCIAL AND ORGANIZATIONAL ENVIRONMENTS INTO WHICH HIT IS DEPLOYED

There is now increasing recognition that the interplay of social and technical factors is important in shaping implementation and adoption of a new system in organizations—commonly known as the sociotechnical perspective (Berg, 1999; Sittig and Singh, 2010). This is characterized by the assumption that technical factors shape social dynamics (such as when a new system results in increasing administrative activity for healthcare professionals), but such social dynamics can also have technical consequences (e.g., when changes to technical design are made to suit local needs, preferences, or practices).

Although sociotechnical systems studies can provide important insights into how adoption of technology can be facilitated, it is important to recognize that the social and organizational environment in inpatient settings varies markedly from organizations in other high-risk industries (such as aviation) and also across different inpatient institutions. For example, inpatient healthcare organizations are characterized by distributed hierarchical structures resulting from the relative autonomy of different healthcare professions on one hand, and managerially led organizational processes with a focus on performance on the other (Kinston, 1983). Tensions resulting from these dynamics can become particularly apparent when implementing organizational changes associated with HIT as leadership roles need to be renegotiated and ways of working need to change (Hersh, 2004). This may be the case when there is an increasing focus on data input activity associated with a new HIT, which may be viewed by healthcare professionals as detracting from providing "hands-on" patient care. Consequently, traditional managerial tools may only be partly applicable to inpatient change initiatives, particularly those involving technology (Edmondson et al., 2001).

Several theoretical models applying change management approaches to healthcare settings have been developed. These bring together different perspectives of health technologies in use (the social context), technical features (the new system), organizational environments, and wider contextual considerations (Box 2.2) (Cornford et al., 1994; Greenhalgh et al., 2004; Hanseth and Lyytinen, 2010; May and Finch, 2009; Williams and Edge, 1996).

Wider contextual considerations are particularly important. However, they are often neglected in HIT implementations (Cresswell et al., 2010). Examples include regional or national health system policies (e.g., those emphasizing progress in certain domains such as antimicrobial stewardship),

---

BOX 2.2 Examples of Theoretical Models That May Be Used to Inform the Implementation and Adoption of Technology in Inpatient Settings

Sociotechnical Approaches
Actor–Network Theory: Helps to conceptualize the interrelated nature between humans and technical systems.
    Cornford et al.'s Evaluation Framework: Practical approach that can help to conceptualize relationships between technical systems, human perspectives, and organizational contexts.

Organizational Change in Health Service Organizations
Greenhalgh et al.'s Diffusion of Innovations in Service Organizations: Comprehensive framework exploring how innovation is introduced and adopted in health service organizations.
    Normalization Process Theory: A theoretical framework that helps to explain how certain healthcare practices become routine over time.

Wider Contextual Considerations
Information Infrastructures: This angle can help to explore how health technology may be viewed as an ever-expanding information infrastructure that is shared between heterogeneous settings and constantly evolving.
    Social Shaping of Technology: This perspective can help to view technology within wider social contexts—it also pays attention to historical and economic factors.

economic considerations (e.g., financial incentives or disincentives), strategies (e.g., those with the aim to promote safety and quality of care), availability of expertise (e.g., in HIT and management), and resources (financial or otherwise). For instance, national health systems will by definition require a different approach to implementing HIT when compared to privately-funded health systems that have very different incentives, resources, and patient populations.

IMPLEMENTATION AND ADOPTION CHALLENGES

In line with sociotechnical approaches, implementation and adoption challenges can be understood to be due to interrelated social and technical challenges that implementing organizations and system users face over time (Berg, 1999; Sittig and Singh, 2010). Innovations can be disruptive to long-established practices and such issues are therefore par for the course (Cresswell et al., 2012). It has, for instance, been found that some healthcare professional groupings feel that using a computer takes away time from "hands-on" patient care (Hsu et al., 2005). Integrating new HIT-related practices to ensure that this is not the case can be an ongoing challenge.

Underlying reasons may be due to a lack of system usability, which is the ease of system use in relation to a specific purpose (Sheikh et al., 2011). This purpose may be to provide patient care and also to ensure efficient organizational functioning. Here, it is important to acknowledge that only a system that is used will derive value and therefore system usability is a serious issue. However, usability can be problematic especially in inpatient settings, as the design of existing tools often fails to account for the various needs of different organizational groupings (e.g., specialty settings, managerial, different healthcare professions). Therefore a careful balancing act is often needed to ensure a degree of streamlining organizational processes through generic tools, while permitting at least some flexibility to allow tailoring to various user needs (Yusof et al., 2008). In this context, it has been found that identifying an underlying need that a system can address (e.g., to improve prescribing safety) can be helpful in engaging different user groups (Martin et al., 2012).

User engagement is an ongoing issue that most implementations need to tackle. If a system is not usable and/or different user groups do not feel that it adds value to their work, it is likely to be used in ways other than intended by implementers or rejected altogether (Cresswell et al., 2011). Using a system in ways other than intended is a very popular way of users to cope with systems that are perceived as cumbersome and/or as interrupting existing practices (Ferneley and Sobreperez, 2006). As a result, users can develop innovative ways to "work around" perceived system limitations. Fig. 2.4 gives an example of such workarounds.

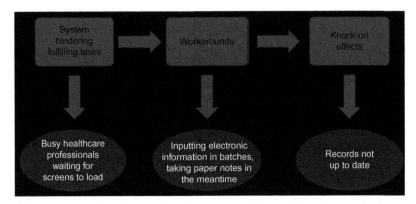

FIGURE 2.4 Example of workarounds employed by system users to address perceived system limitations.

Workarounds by themselves are not necessarily anything to worry about. They may in fact be viewed as innovative practices surrounding system integration within social contexts and may even be used as a source of suggestions for improvements in design (Lalley and Malloch, 2010). But they may also result in unintended consequences and present risks to the safety and efficiency of care. A classic example of such unintended consequences is provided by Koppel et al. (2005) who reported that the implementation of a CPOE system resulted in increased rates of medication errors due to a lack of system usability and integrated presentation of information. Similarly, a phenomenon known as "alert fatigue," where clinical users feel inundated with pop-ups, can lead to important alerts being not acted on and thereby facilitate the potential for errors (Phansalkar et al., 2013).

CONTINUING SYSTEM DEVELOPMENT AND OPTIMIZATION

Many of the challenges outlined above are ongoing, as new technologies are developed and systems are optimized to function within individual user and organizational contexts over time. Implementation, adoption, and system optimization is therefore best conceptualized as an ongoing journey (Cresswell et al., 2013; McDonald et al., 2004). We describe the key phases of this journey in Fig. 2.5.

An important aspect of continuing system development is the progressive integration of information from different sources and settings. This needs to be conceptualized in terms of technical interfacing (of different technologies) and end user/data integration (i.e., the ability seamlessly to draw on systems and data to facilitate the delivery of care) (Cresswell et al., 2017). Information integration may be viewed as an exponential extension from information in discrete settings, toward organizational integration in inpatient

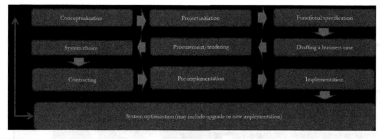

FIGURE 2.5 Illustration of the implementation journey.

settings, and eventually cross-organizational integration covering the entire health system.

Achieving these increasing levels of integration will necessitate continuous technological investment (e.g., infrastructures, hardware, software) and organizational capacity development (e.g., organizational change expertise, technical managers, data analysts). Although benefits realization is in most instances an important driver of HIT implementation, realizing returns on investment can be problematic (Jones et al., 2012). This is in part due to the fact that cash-releasing benefits are often hard to measure and to attribute, as the introduction of technological systems needs to be accompanied with broader organizational change initiatives (Chaudhry et al., 2006). In addition, there are often substantial up-front investments required and the realization of benefits is likely to materialize in the medium- to longer-term as systems are optimized (Cresswell et al., 2016).

In order to ensure that benefits are realized and unintended consequences are minimized, ongoing implementation evaluation is important (Catwell and Sheikh, 2009). This should have formative and summative, as well as qualitative and quantitative elements, in order to explore processes and outcomes associated with the introduction of a new HIT within its social/organizational context. Ideally, such evaluation would begin before the introduction of a new technological system in order to establish baseline measurements against which to assess progress and continue throughout the implementation and adoption lifecycle. An example of an ongoing evaluation is provided in Fig. 2.6.

INTERNATIONAL STRATEGIES AND IMPLEMENTATION PROGRESS

As mentioned above, different sociopolitical and economic environments are likely to have important consequences for local implementations of HIT. These are often characterized by national approaches to achieve increasing information integration. With a focus on high-income settings, where the

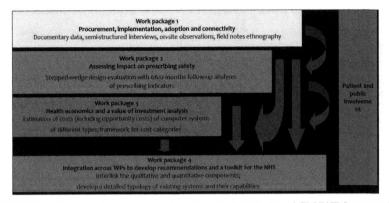

FIGURE 2.6 Example of an ongoing evaluation of the introduction of CPOE/CDS systems for prescribing in English hospitals.

majority of progress has been made to date, we will outline some of the most prominent approaches in the subsequent paragraphs.

The United States is currently one of the most advanced countries in relation to large-scale progress of hospital EHR implementation and adoption with three out of four hospitals now having at least a basic EHR system (The Office of the National Coordinator for Health Information Technology, 2015). The tactical use of national monetary incentives for implementing systems with increasing functionalities, and penalties for missing nationally set targets in policies promoted through the HIT for Economic and Clinical Health (HITECH) Act and Meaningful Use in the United States, has so far proved to be successful. A key factor for the success of HITECH was that it allowed hospitals autonomy with respect to which accredited EHR system to procure, while still setting national goals. Similar approaches based on national interoperability standards and financial incentives have also proven successful in other countries (e.g., Denmark), albeit on a much smaller scale (The Commonwealth Fund, 2011).

However, this approach may not be suited to health systems that are organized and funded differently such as the National Health Service (NHS) in the United Kingdom. Here, an initial "top—down" approach to implement EHRs in hospitals, characterized by the attempted implementation of centrally procured technical systems from a limited number of suppliers, is now gradually being replaced by a nationally-guided approach based on standards (Sheikh et al., 2011). As a result, there is now increased systems choice for individual hospitals and an associated increase in implementation activity. However, central incentives are relatively limited, somewhat inhibiting progress as individual hospitals lack capacity and resources.

Similar issues are faced by Australia, where implementation of hospital EHRs is also guided by nationally determined technical standards. However, different states have considerable autonomy and can thus still choose their own systems and EHR implementation approaches (The Commonwealth Fund, 2015). These disparate efforts increasingly present challenges relating to health information exchange across settings and geographical locations. The focus is therefore currently on building national infrastructures designed to help address this issue (e.g., single national patient identifiers) (The Guardian, 2015).

LOOKING AHEAD—FUTURE TRENDS AND DEVELOPMENTS

In any activity, a vision of a future state is essential for progress. In relation to inpatient settings, there are several examples of how the future facilitated by technological innovation may be conceptualized. Overall, hospitals are likely to become much "smarter"—that is, they will consist of increasingly large networked technologies that synchronously act together with humans to improve quality and efficiency of care (Wachter, 2015). This technological infrastructure across the hospital environment and different specialties should allow the central real-time monitoring of data (e.g., bed occupancy or predicting patients who are likely to deteriorate), and environmental surveillance capabilities should ensure that electronic devices are responding to changing demands in physical environments (e.g., temperature changes). Several such ventures are currently under way internationally, but these require significant monetary resources and have relatively long implementation timescales of around 20 years (Fierce Medical Devices, 2015; Global Construction Review, 2015; Korea.net, 2014; One Healthcare Worldwide, 2014). Recent advances in robotics and artificial intelligence are likely to play important roles in the smart hospital of the future (The Guardian, 2015). An example case study of a smart operating room is provided in Box 2.3.

Part of these networked and smart technologies is also likely to be the increasingly active role of patients. Patient access to electronic records such as in the Open Notes initiative where patients have access to their doctor's notes in primary care settings has already been found to bring clinical benefits with minimal adverse effects on clinician work practices (Delbanco et al., 2012).

There is also now increasing evidence that applications drawing on natural language processing can help in making diagnoses and even outperform humans (Bennett and Hauser, 2013). Real world examples in this respect include drawing on IBM's Watson in the Memorial Sloan-Kettering Cancer Center in New York (IBM, 2015). Watson Oncology, which is now commercially available, has the ability to trawl through extremely large amounts of data and it can interpret clinical information and integrate this with the available literature to make diagnoses and suggest treatment options in next to no time.

BOX 2.3 Example Case Study of a Smart Operating Room

The Florida Hospital Carrollwood has recently signed a contract to build 12 smart operating rooms through the Stryker Complete Integration System. They have already used robotic surgery procedures and now wish to integrate these with the hospital's EHR system to support a range of professional workflows and improve the quality of patient care. It will also draw on data from real-time location services (that identify and track people and/or objects), smart presets (e.g., temperature) and integrate surgical imaging. This will facilitate data storage and support advanced data analytics as information will be held and managed all in one place. Operating information will be shared with patients and an in-build camera will allow recording of procedures that can then be used for teaching or for remote discussion of patients among doctors. The system also includes automatic time-out functionality and integrates information from disparate sources to provide real-time notifications to the operating team. The investment in such systems that integrate with EHR systems can be anywhere between $20,000 and $500,000, but the hospital expects to see returns on investments in around 8 years.

A related line of development concerns the use and reuse of the increasing amounts of electronic healthcare data to improve safety and performance. In hospital care, this is most often referred to in relation to achieving the pinnacle of EHR adoption: the Healthcare Information and Management Systems Society (HIMSS) Level 7 (HIMSS Analytics, 2015). HIMSS Analytics is a global not-for-profit organization that tracks EHR adoption internationally by applying an EMR Adoption Model^{SM} (EMRAM) consisting of eight stages. These range from ancillary electronic functionality (such as laboratory results, radiology and pharmacy systems), to a complete EHR that allows integration of processes (e.g., documentation and results), advanced analytics, and sharing of data across the institution (Fig. 2.7).

As can be seen, technological applications in inpatient settings are increasing exponentially, and advances in interoperability and integration of different systems are likely to result in major changes to the way care is delivered. However, in order to achieve this, it will be essential to allow information to be effectively exchanged across different settings, specialties, and systems through the creation of interfaces. Such work can be promoted through standards that guide software developers in building systems that are interoperable within and across healthcare settings (Liyanage et al., 2015). Opening up of application programming interfaces and underlying protocols of software operation is likely to facilitate the development of future hybrid models of delivering care with HIT by creating opportunities for innovative local developments (Mandl et al., 2015). These may include connecting and/or building on existing functionalities to create tools that facilitate the seamless exchange and exploitation of data and to create new user interfaces tailored to local needs.

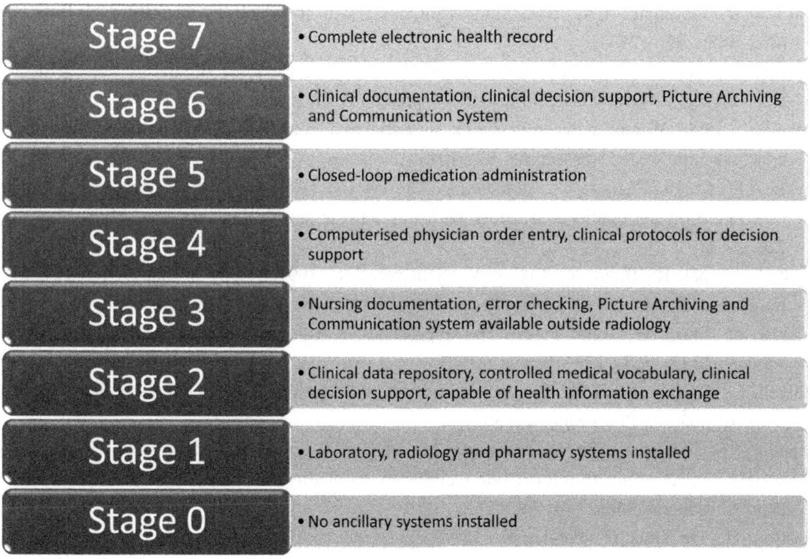

Stage 7	• Complete electronic health record
Stage 6	• Clinical documentation, clinical decision support, Picture Archiving and Communication System
Stage 5	• Closed-loop medication administration
Stage 4	• Computerised physician order entry, clinical protocols for decision support
Stage 3	• Nursing documentation, error checking, Picture Archiving and Communication system available outside radiology
Stage 2	• Clinical data repository, controlled medical vocabulary, clinical decision support, capable of health information exchange
Stage 1	• Laboratory, radiology and pharmacy systems installed
Stage 0	• No ancillary systems installed

FIGURE 2.7 HIMSS Analytics EMRAM. *Adapted from HIMSS Analytics, 2015. HIMSS Analytics Provider Solutions. http://www.himssanalytics.org/provider-solutions. (accessed 12.11.15).*

CONCLUSIONS

An increasing range of new and often very complex technologies are being used in inpatient settings, and the adoption of EHRs is now inevitable. However, implementation and adoption of systems is complicated by the organizational complexity of hospitals and the various social consequences often associated with technological change initiatives. Implementation will likely deliver most benefit when these applications are linked with outpatient systems. Innovative technical developments bring significant opportunities but it is essential to recognize that they will also bring new challenges, including unintended consequences and risks that need to be addressed through careful monitoring and mitigation.

REFERENCES

Abrahamsson, S., Bergström, S., Larsson, K., Tillman, S., 1970. Danderyd Hospital computer system: II. Total regional system for medical care. Comput. Biomed. Res. 3, 30–46.

Baker, G.R., MacIntosh-Murray, A., Porcellato, C., Dionne, L., Stelmacovich, K., Born, K., 2008. Intermountain healthcare. High Performing Healthcare Systems: Delivering Quality by Design. Longwoods Publishing, Toronto, pp. 151–178.

Bakker, A.R., 1982. Organization of a cooperation for further development and implementation of an integrated hospital information system, Medical Informatics Europe, 82. Springer, Berlin, Heidelberg, pp. 14–20.

Bates, D.W., Gawande, A.A., 2003. Improving safety with information technology. N. Engl. J. Med. 348, 2526–2534.

Bennett, C.C., Hauser, K., 2013. Artificial intelligence framework for simulating clinical decision making: a Markov decision process approach. Artif. Intell. Med. 57, 9–19.

Berg, M., 1999. Patient care information systems and health care work: a sociotechnical approach. Int. J. Med. Inform. 55, 87–101.

Black, A.D., Car, J., Pagliari, C., Anandan, C., Cresswell, K., Bokun, T., et al., 2011. The impact of eHealth on the quality and safety of health care: a systematic overview. PLoS Med. 8, 188.

Borst, F., Appel, R., Baud, R., Ligier, Y., Scherrer, J.R., 1999. Happy birthday DIOGENE: a hospital information system born 20 years ago. Int. J. Med. Inform. 54, 157–167.

Brandao de Souza, L., 2009. Trends and approaches in lean healthcare. Leadership in health services 22 (2), 121–139.

Catwell, L., Sheikh, A., 2009. Evaluating eHealth interventions: the need for continuous systemic evaluation. PLoS Med. 6, 830.

Chaudhry, B., Wang, J., Wu, S., Maglione, M., Mojica, W., Roth, E., et al., 2006. Systematic review: impact of health information technology on quality, efficiency, and costs of medical care. Ann. Intern. Med. 144, 742–752.

Cornford, T., Doukidis, G., Forster, D., 1994. Experience with a structure, process and outcome framework for evaluating an information system. Omega Int. J. Manag. Sci. 22, 491–504.

Cresswell, K., Worth, A., Sheikh, A., 2010. Actor-Network Theory and its role in understanding the implementation of information technology developments in healthcare. BMC Med. Inform. Decis. Making 10, 67.

Cresswell, K., Morrison, Z., Crowe, S., Robertson, A., Sheikh, A., 2011. Anything but engaged: user involvement in the context of a national electronic health record implementation. Inform. Prim. Care 19, 191–206.

Cresswell, K., Worth, A., Sheikh, A., 2012. Integration of a nationally procured electronic health record system into user work practices. BMC Med. Inform. Decis. Making 12, 15.

Cresswell, K., Bates, D.W., Sheikh, A., 2013. Ten key considerations for the successful implementation and adoption of large-scale health information technology. J. Am. Med. Inform. Assoc. 20, e9–e13.

Cresswell, K., Bates, D.W., Sheikh, A., 2017. Ten key considerations for the successful optimization of large-scale health information technology. J. Am. Med. Inform. Assoc. 24, 182–187.

Cresswell, K.M., Mozaffar, H., Lee, L., Williams, R., Sheikh, A., 2016. Safety risks associated with the lack of integration and interfacing of hospital health information technologies: a qualitative study of hospital electronic prescribing systems in England. BMJ quality & safety. Available from: http://dx.doi.org/10.1136/bmjqs-2015-004925.

Czarniawska, B., 1997. A four times told tale: combining narrative and scientific knowledge in organization studies. Organization 4, 7–30.

Delbanco, T., Walker, J., Bell, S.K., Darer, J.D., Elmore, J.G., Farag, N., et al., 2012. Inviting patients to read their doctors' notes: a quasi-experimental study and a look ahead. Ann. Intern. Med. 157, 461–470.

Edmondson, A.C., Bohmer, R.M., Pisano, G.P., 2001. Disrupted routines: team learning and new technology implementation in hospitals. Adm. Sci. Q. 46, 685–716.

Ferneley, E.H., Sobreperez, P., 2006. Resist, comply or workaround? An examination of different facets of user engagement with information systems. Eur. J. Inform. Syst. 15, 345–356.

Fierce Medical Devices, 2015. Philips nabs latest 'smart hospital' deal for $225M in Ontario. <http://www.fiercemedicaldevices.com/story/philips-nabs-latest-smart-hospital-deal-225m-ontario/2015-11-10> (accessed 12.11.15).

Global Construction Review, 2015. Habtoor Leighton to build new "smart" hospital in Dubai. <http://www.globalconstructionreview.com/news/habtoor-leig9hton-build-ne2w-sma2rt-hospital-dubai/> (accessed 12.11.15).

Greenhalgh, T., Robert, G., Macfarlane, F., Bate, P., Kyriakidou, O., 2004. Diffusion of innovations in service organizations: systematic review and recommendations. Milbank Q. 82, 581−629.

Hanseth, O., Lyytinen, K., 2010. Design theory for dynamic complexity in information infrastructures: the case of building internet. J. Inform. Technol. 25, 1−19.

HealthWorksCollective, 2014. The long road to digitization: a history of healthcare informatics. <http://www.healthworkscollective.com/frankie-xavier/162251/long-road-digitization-history-healthcare-informatics> (accessed 12.11.15).

Hersh, W., 2004. Health care information technology: progress and barriers. JAMA. 292, 2273−2274.

HIMSS Analytics, 2015. HIMSS Analytics Provider Solutions. <http://www.himssanalytics.org/provider-solutions> (accessed 12.11.15).

Hsu, J., Huang, J., Fung, V., Robertson, N., Jimison, H., Frankel, R., 2005. Health information technology and physician-patient interactions: impact of computers on communication during outpatient primary care visits. J. Am. Med. Inform. Assoc. 12, 474−480.

IBM, 2015. Watson Oncology. <https://www.mskcc.org/about/innovative-collaborations/watson-oncology> (accessed 12.11.15).

Jones, S.S., Heaton, P.S., Rudin, R.S., Schneider, E.C., 2012. Unravelling the IT productivity paradox—lessons for health care. N. Engl. J. Med. 366, 2243−2245.

Kaushal, R., Shojania, K.G., Bates, D.W., 2003. Effects of computerized physician order entry and clinical decision support systems on medication safety: a systematic review. Arch. Intern. Med. 163, 1409−1416.

Kinston, W., 1983. Hospital organisation and structure and its effect on inter-professional behaviour and the delivery of care. Soc. Sci. Med. 17, 1159−1170.

Koppel, R., Metlay, J.P., Cohen, A., Abaluck, B., Localio, A.R., Kimmel, S.E., et al., 2005. Role of computerized physician order entry systems in facilitating medication errors. JAMA 293, 1197−1203.

Korea.net, 2014. Saudi hospital adopts Korean IT. <http://www.korea.net/NewsFocus/Business/view?articleId = 120492> (accessed 12.11.15).

Lalley, C., Malloch, K., 2010. Workarounds: the hidden pathway to excellence. Nurse Leader 8, 29−32.

Lapointe, L., Rivard, S., 2005. A multilevel model of resistance to information technology implementation. MIS Q.461−491.

Liyanage, H., Krause, P., de Lusignan, S., 2015. Using ontologies to improve semantic interoperability in health data. J. Innov. Health Inform. 22, 309−315.

Mandl, K.D., Mandel, J.C., Kohane, I.S., 2015. Driving innovation in health systems through an apps-based information economy. Cell Syst. 1, 8−13.

Martin, J.L., Clark, D.J., Morgan, S.P., Crowe, J.A., Murphy, E., 2012. A user-centred approach to requirements elicitation in medical device development: a case study from an industry perspective. Appl. Ergon. 43, 184−190.

May, C., Finch, T., 2009. Implementing, embedding, and integrating practices: an outline of normalization process theory. Sociology 43, 535−554.

McDonald, C.J., Overhage, J.M., Mamlin, B.W., Dexter, P.D., Tierney, W.M., 2004. Physicians, information technology, and health care systems: a journey, not a destination. J. Am. Med. Inform. Assoc. 11, 121–124.

Murdoch, T.B., Detsky, A.S., 2013. The inevitable application of big data to health care. JAMA 309, 1351–1352.

NASBHC, 2015. <http://www.nasbhc.org/atf/cf/%7BCD9949F2-2761-42FB-BC7A-CEE165C701D9%7D/TA_HIT_history%20of%20EMR.pdf> (accessed 12.11.15).

One Healthcare Worldwide, 2014. Smart hospitals continue to evolve. <http://www.1ohww.org/bundang-smart-hospitals-continue-evolve/> (accessed 12.11.15).

Panesar, S., Carson-Stevens, A., Salvilla, S., Sheikh, A., 2014. Patient Safety and Healthcare Improvement at a Glance. John Wiley & Sons.

Phansalkar, S., Van der Sijs, H., Tucker, A.D., Desai, A.A., Bell, D.S., Teich, J.M., et al., 2013. Drug–drug interactions that should be non-interruptive in order to reduce alert fatigue in electronic health records. J. Am. Med. Inform. Assoc. 20, 489–493.

Pilling, J.R., 2014. Picture archiving and communication systems: the users' view. Br. J. Radiol. 76, 519–524.

Safran, C., Bloomrosen, M., Hammond, W.E., Labkoff, S., Markel-Fox, S., Tang, P.C., et al., 2007. Toward a national framework for the secondary use of health data: an American Medical Informatics Association White Paper. J. Am. Med. Inform. Assoc. 14, 1–9.

Sheikh, A., Cornford, T., Barber, N., Avery, A., Takian, A., Lichtner, V., et al., 2011. Implementation and adoption of nationwide electronic health records in secondary care in England: final qualitative results from prospective national evaluation in "early adopter" hospitals. BMJ. 343.

Sittig, D.F., Singh, H., 2010. A new sociotechnical model for studying health information technology in complex adaptive healthcare systems. Qual. Saf. Health Care 19, i68–i74.

The Commonwealth Fund, 2011. Electronic health records: an international perspective on "meaningful use". <http://www.commonwealthfund.org/~/media/Files/Publications/Issue%20Brief/2011/Nov/1565_Gray_electronic_med_records_meaningful_use_intl_brief.pdf> (accessed 12.11.15).

The Commonwealth Fund, 2015. International profiles of health care systems. <http://www.commonwealthfund.org/~/media/files/publications/fund-report/2015/jan/1802_mossialos_intl_profiles_2014_v7.pdf> (accessed 12.11.15).

The Guardian, 2015. Robot revolution: rise of 'thinking' machines could exacerbate inequality. <http://www.theguardian.com/technology/2015/nov/05/robot-revolution-rise-machines-could-displace-third-of-uk-jobs> (accessed 12.11.15).

The Office of the National Coordinator for Health Information Technology, 2015. Adoption of electronic health record systems among U.S. non-federal acute care hospitals: 2008–2014. <https://www.healthit.gov/sites/default/files/data-brief/2014HospitalAdoptionDataBrief.pdf> (accessed 12.11.15).

Wachter, R.M., 2015. The Digital Doctor: Hope, Hype, and Harm at the Dawn of Medicine's Computer Age. McGraw-Hill, New York.

Westbrook, J.I., Reckmann, M., Li, L., Runciman, W.B., Burke, R., Lo, C., et al., 2012. Effects of two commercial electronic prescribing systems on prescribing error rates in hospital inpatients: a before and after study. PLoS Med. 9, 138.

Williams, R., Edge, D., 1996. The social shaping of technology. Res. Policy 25, 865–899.

Yusof, M., Kuljis, J., Papazafeiropoulou, A., Stergioulas, L., 2008. An evaluation framework for Health Information Systems: human, organization and technology-fit factors (HOT-fit). Int. J. Med. Inform. 77, 386–398.

RECOMMENDED FURTHER READING

Berg, M., 2003. Health Information Management: Integrating Information and Communication Technology in Health Care Work. Routledge.

Berwick, D., 2010. In: Liang, L.L. (Ed.), Connected for Health: Using Electronic Health Records to Transform Care Delivery. John Wiley & Sons.

Cresswell, K., Ali, M., Avery, A., et al., 2011. The Long and Winding Road...An independent evaluation of the implementation and adoption of the National Health Service Care Records Service (NHS CRS) in secondary care in England. <http://www.cphs.mvm.ed.ac.uk/grantdocs/526%20-%20Final%20report%20v31st%20Mar%20FINAL.pdf> (accessed 12.11.15).

Greenhalgh, T., Robert, G., Bate, P., et al., 2004. How to spread good ideas: a systematic review of the literature on diffusion, dissemination and sustainability of innovations in health service delivery and organisation. <http://www.netscc.ac.uk/hsdr/files/project/SDO_FR_08-1201-038_V01.pdf> (accessed 12.11.15).

The e health revolution—easier said than done. <http://www.aph.gov.au/about_parliament/parliamentary_departments/parliamentary_library/pubs/rp/rp1112/12rp03#_ftn118> (accessed 12.11.15).

Chapter 3

Outpatient Clinical Information Systems

Adam Wright and David W. Bates

Brigham and Women's Hospital/Harvard Medical School, Boston, MA, United States

INTRODUCTION

Over the last two decades, healthcare has seen a considerable shift from treating patients on an inpatient basis to outpatient care. Many illnesses that were once only treated in hospitals and procedures once only performed in hospitals can now be accommodated in the outpatient setting, increasing the complexity of outpatient care. Even relatively complex care such as chemotherapy is increasingly being delivered in the outpatient setting.

Using the United States as an example, between 2004 and 2010, the number of hospital admissions for Medicare beneficiaries (mostly older adults and disabled persons) fell 7.8%, while outpatient volume increased 33.6%, with similar trends seen for other payers (Vesely, 2014).

As outpatient care continues to replace inpatient care and the patients seen become more complex, the need for sophisticated, full-featured outpatient clinical information systems, particularly electronic health records (EHRs), has increased. Some EHR systems naturally span inpatient and outpatient settings, but most care providers use separate systems, or separate modules of the same system, across inpatient and outpatient settings, with varying degrees of integration.

This chapter begins with a brief history of outpatient clinical information systems (a broader category that includes standalone e-prescribing systems, specialty-specific software, and other tools), with a focus on outpatient EHRs, followed by a detailed review of common outpatient EHR components. Next, it covers both the benefits and implementation and adoption challenges of outpatient EHRs, before turning to international perspectives and a look to the future. Although this specific chapter focuses on outpatient EHRs, challenges, opportunities, and advances that apply to outpatient EHRs are found throughout the entire book.

Key Advances in Clinical Informatics. DOI: http://dx.doi.org/10.1016/B978-0-12-809523-2.00003-0

A BRIEF HISTORY OF OUTPATIENT CLINICAL INFORMATION SYSTEMS

The use of clinical information systems in the outpatient setting began, in most countries, with administrative systems for scheduling and billing which enabled outpatient practices to function more efficiently than with prior paper systems. Over time, clinical features were added to the systems, eventually leading to complete EHRs, such as EpicCare, a graphical outpatient EHR first released in 1992.

In some cases, outpatient EHRs were developed initially as extensions to inpatient systems (see Chapter 2: Inpatient Clinical Information Systems). For example, the Brigham Integrated Computer System was originally developed primarily as an inpatient system, but an outpatient module, Miniamb, was added in 1989, with a problem list, medication list, allergy list, and visit documentation. Eight years later, Miniamb was replaced by the Longitudinal Medical Record (LMR), a new web-based outpatient EHR which had more advanced documentation capabilities, as well as additional clinical decision support (CDS) (see Chapter 12: Clinical Decision Support and Knowledge Management).

The use of outpatient clinical information systems has dramatically increased in the last 15 years. The pace of this evolution has varied by country. In a study by Jha et al. (2008) looking at EHR adoption by general practitioners in Australia, Canada, Germany, the Netherlands, New Zealand, the United Kingdom, and the United States found significant variation. Australia, New Zealand, the Netherlands, and the United Kingdom led with near-universal adoption of EHRs in the outpatient general practice setting, while the United States and Canada lagged. In the United States, adoption of outpatient EHRs was 18.2% in 2001, but increased to 82.8% by 2014 and is now likely higher (Fig. 3.1 shows the trajectory) (Hsiao and Hing, 2012).

The features of EHRs have also advanced over time, and the US National Center for Health Statistics has defined two levels of EHR use: "any EHR system," which means use of any "electronic medical records or electronic health records (not including billing records)," while a "basic system" "has all of the following functionalities: patient history and demographics, patient problem lists, physician clinical notes, comprehensive list of patients' medications and allergies, computerized orders for prescriptions, and ability to view laboratory and imaging results electronically" (Hsiao and Hing, 2012).

In the United States, the federal meaningful use program has been a major driver of EHR adoption (Blumenthal and Tavenner, 2010; Wright et al., 2013). The program provided financial incentives for eligible healthcare professionals to use outpatient EHRs and to meet a series of increasing requirements for "Meaningful Use" of the system—these requirements are spelled out in a series of core measures (which all providers had to meet)

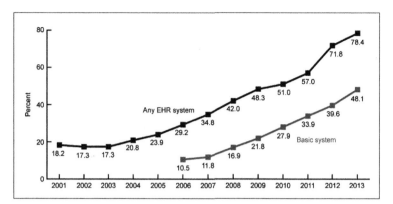

FIGURE 3.1   Use of EHR systems in the outpatient setting in the United States.
Notes: EHR is electronic health record. "Any EHR system" is a medical or health record system that is either all or partially electronic (excluding systems solely for billing). Data for 2001–07 are from in-person National Ambulatory Medical Care Survey (NAMCS) interviews. Data for 2008–10 are from combined files (in-person NAMCS and mail survey). Estimates for 2011–13 data are based on the mail survey only. Estimates for a basic system prior to 2006 could not be computed because some items were not collected in the survey. Data include nonfederal, office-based physicians and exclude radiologists, anesthesiologists, and pathologists. *Adapted from CDC/NCHS, National Ambulatory Medical Care Survey and National Ambulatory Medical Care Survey, Electronic Health Records Survey.*

and menu measures (providers choose from a list of measures). The measures from Stage 1 of the Meaningful Use program are provided in Table 3.1 (Blumenthal and Tavenner, 2010).

The Commonwealth Fund has tracked the development and use of EHRs around the world as part of their program on International Health Care System Profiles (Commonwealth Fund, 2016). Table 3.2 shows the adoption figures from this tracking report, which are from 2015—across the countries, use was very high, with Canada the lowest at 73% and New Zealand achieving 100% adoption.

COMMON OUTPATIENT EHR COMPONENTS

Most outpatient EHRs are composed of various modules, such as medication lists, problem lists, allergy lists, results review, and documentation. Fig. 3.2 shows the layout and display of these modules on the summary screen of the LMR, a representative outpatient EHR used at Partners Healthcare in Boston. In this section, we consider the purpose, capabilities, and special challenges for each of these modules.

TABLE 3.1 Core and Menu Measures for Stage 1 of the United States Meaningful Use Incentive Program for EHR Adoption

Summary Overview of Meaningful Use Objectives[a]

Objective	Measure
Core Set of Objectives to Be Achieved by All Eligible Professionals, Hospitals, and Critical Access Hospitals to Qualify for Incentive Payments	
Record patient demographics (sex, race, ethnicity, date of birth, preferred language, and in the case of hospitals, date and preliminary cause in the event of death)	Over 50% of patients' demographic data recorded as structured data
Record vital signs and chart changes (height, weight, blood pressure, body mass index, growth charts for children)	Over 50% of patients 2 years of age or older have height, weight, and blood pressure recorded as structured data
Maintain up-to-date problem list of current and active diagnoses	Over 80% of patients have at least one entry recorded as structured data
Maintain active medication list	Over 80% of patients have at least one entry recorded as structured data
Maintain active medication allergy list	Over 80% of patients have at least one entry recorded as structured data
Record smoking status for patients 13 years of age or older	Over 50% of patients 13 years of age or older have smoking status recorded as structured data
For individual professionals, provide patients with clinical summaries for each office visit; for hospitals, provide an electronic copy of hospital discharge instructions on request	Clinical summaries provided to patients for over 50% of all office visits within 3 business days; over 50% of all patients who are discharged from the inpatient department or emergency department of an eligible hospital or critical access hospital and who request an electronic copy of their discharge instructions are provided with it
On request, provide patients with an electronic copy of their health information (including diagnostic-test results, problem list, medication lists, medication allergies, and for hospitals, discharge summary and procedures)	Over 50% of requesting patients receive electronic copy within 3 business days
Generate and transmit permissible prescriptions electronically (does not apply to hospitals)	Over 40% are transmitted electronically using certified EHR technology

(Continued)

TABLE 3.1 (Continued)

Summary Overview of Meaningful Use Objectives[a]

Objective	Measure
Computer provider order entry (CPOE) for medication orders	Over 30% of patients with at least one medication in their medication list have at least one medication ordered through CPOE
Implement drug–drug and drug–allergy interaction checks	Functionality is enabled for these checks for the entire reporting period
Implement capability to electronically exchange key clinical information among providers and patient-authorized entities	Perform at least one test of EHR's capacity to electronically exchange information
Implement one CDS rule and ability to track compliance with the rule	One CDS rule implemented
Implement systems to protect privacy and security of patient data in the EHR	Conduct or review a security risk analysis, implement security updates as necessary, and correct identified security deficiencies
Report clinical quality measures to CMS or states	For 2011, provide aggregate numerator and denominator through attestation; for 2012, electronically submit measures
Eligible Professionals, Hospitals, and Critical Access Hospitals May Select Any Five Choices From the Menu Set	
Implement drug formulary checks	Drug formulary check system is implemented and has access to at least one internal or external drug formulary for the entire reporting period
Incorporate clinical laboratory test results into EHRs as structured data	Over 40% of clinical laboratory test results whose results are in positive/negative or numerical format are incorporated into EHRs as structured data
Generate lists of patients by specific conditions to use for quality improvement, reduction of disparities, research, or outreach	Generate at least one listing of patients with a specific condition
Use EHR technology to identify patient-specific education resources and provide those to the patient as appropriate	Over 10% of patients are provided patient-specific education resources
Perform medication reconciliation between care settings	Medication reconciliation is performed for over 50% of transitions of care

(*Continued*)

TABLE 3.1 (Continued)

Summary Overview of Meaningful Use Objectives[a]

Objective	Measure
Provide summary of care record for patients referred or transitioned to another provider or setting	Summary of care record is provided for over 50% of patient transitions or referrals
Submit electronic immunization data to immunization registries or immunization information systems	Perform at least one test of data submission and follow-up submission (where registries can accept electronic submissions)
Submit electronic syndromic surveillance data to public health agencies	Perform at least one test of data submission and follow-up submission (where public health agencies can accept electronic data)
Additional Choices for Hospitals and Critical Access Hospitals	
Record advance directives for patients 65 years of age or older	Over 50% of patients 65 years of age or older have an indication of an advance-directive status recorded
Submit electronic data on reportable laboratory results to public health agencies	Perform at least one test of data submission and follow-up submission (where public health agencies can accept electronic data)
Additional Choices for Eligible Professionals	
Send reminders to patients (per patient preference) for preventive and follow-up care	Over 20% of patients 65 years of age or older or 5 years of age or younger are sent appropriate reminders
Provide patients with timely electronic access to their health information (including laboratory results, problem list, medication lists, medication allergies)	Over 10% of patients are provided electronic access to information within 4 days of its being updated in the EHR

[a]*This overview is meant to provide a reference tool indicating the key elements of meaningful use of health information technology. It does not provide sufficient information for providers to document and demonstrate meaningful use in order to obtain financial incentives from the Centers for Medicare and Medicaid Services (CMS). The regulations and filing requirements that must be fulfilled to qualify for the Health IT financial incentive program are detailed at www.cms.gov.*
Source: Reprinted from Blumenthal, D., Tavenner, M., 2010. The "meaningful use" regulation for electronic health records. N. Engl. J. Med. 363 (6), 501–504.

TABLE 3.2 Primary Care Physicians' Use of EHRs in 2015, by Country

Country	Adoption
Australia	92%
Canada	73%
France	75%
Germany	84%
The Netherlands	98%
New Zealand	100%
Norway	99%
Sweden	99%
Switzerland	54%
United Kingdom	98%
United States	84%

Data from Commonwealth Fund International Health Care System Profiles.

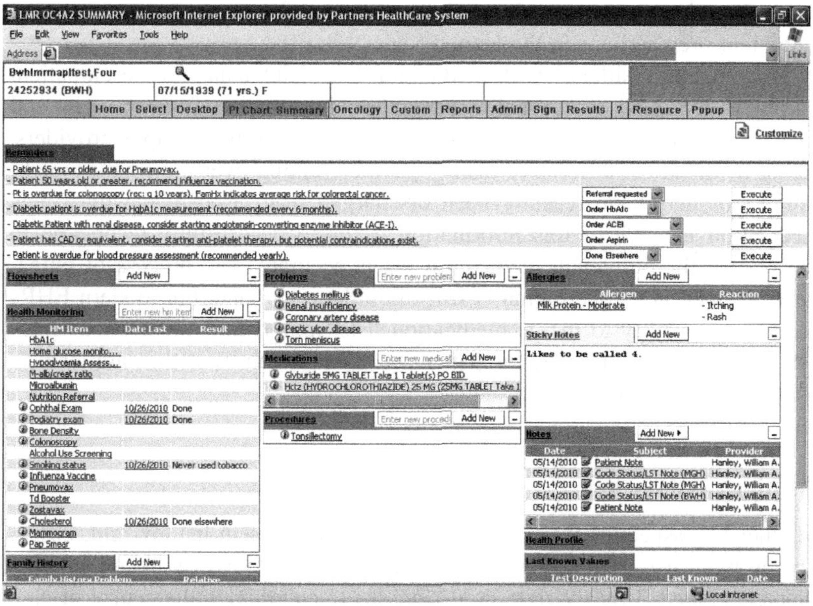

FIGURE 3.2   Summary screen from the LMR, a representative outpatient EHR used at Partners Healthcare in Boston, MA.

Medication Lists and Electronic Prescribing

Before EHRs were widely adopted, many providers used standalone systems for electronic prescribing and medication list management. While some early systems used uncoded medication lists, almost all current outpatient EHRs use a medication dictionary (typically developed and maintained by a third party) that prescribers can use to construct medication orders for CPOE. This dictionary may also contain information on routes, dosing, warnings, etc., which can be used to construct an accurate prescription efficiently.

During the entry of a medication, a range of CDS can be applied to ensure the safety of the proposed medication (see Chapter 12: Clinical Decision Support and Knowledge Management). This CDS is a key driver of the potential value of computerization in the outpatient (or inpatient) setting (Kaushal et al., 2006; Poon et al., 2010; Garrido et al., 2005).

In addition to CDS, another benefit of outpatient EHRs is the ability to electronically transmit prescriptions to pharmacies. This transmission has numerous benefits, including a lower risk of transcription errors, a better experience for the patient (medications may already be ready for pickup when they arrive at the pharmacy) and the possibility of a closed loop, with feedback to the provider on whether a patient has picked up an ordered prescription, and electronic communication of refill requests. The NCPDP SCRIPT standard (Friedman et al., 2009) is almost universally adopted in the United States and enables secure electronic transmission of prescriptions from physicians to pharmacies and is now used for about three-quarters of outpatient prescriptions in the United States (April).

A particular challenge with medication list maintenance in the outpatient setting is the need to reconcile medication information from a variety of sources. A patient may be receiving medications from multiple providers, taking medications differently than prescribed, taking over the counter medications, or may have recently been discharged from a hospital with a new medication list. All of this information needs to be carefully reconciled to ensure that the outpatient medication list is accurate. The process of outpatient medication reconciliation is still often done by hand, but modern EHRs have the capability to merge medication lists, import new medications from external sources, or involve the patient or other members of the care team in the reconciliation process.

Problem Lists

Another key aspect of the outpatient EHR is the problem list (Wright, 2015). The problem list was originally conceived by Weed (cross-ref documentation chapter) (Weed, 1968; Wright et al., 2014b) to summarize the course of care in hospitalized patients, but naturally extends to longitudinal care in the outpatient setting. The problem list is designed to be a single source of information about a patient's active and past health conditions and evidence suggests that accurate problem lists are associated with higher quality care (Samal et al., 2014; Hartung et al., 2005), perhaps due to better provider awareness of a patients' problems, or because problems can drive CDS.

Nearly all modern outpatient EHRs support structured, coded problem lists. The most common terminologies used in problem list maintenance are SNOMED and ICD-10-CM, though a range of subsets, alternatives, and user-friendly terminologies are also available.

Problem lists have many uses. In addition to their obvious clinical purpose, they also drive CDS (Wright et al., 2007), serve as a structure for clinical documentation (see Chapter 4: Electronic Clinical Documentation) (Mehta et al., 2014), and are used in quality measurement and research. Most problem lists are manually maintained, and research (Wright et al., 2011b) suggests that outpatient problem lists are often incomplete—we conducted a 10-site international study which showed that problem list completeness for diabetes ranged from 60.2% to 99.4% (Wright et al., 2015). Based on this study, we identified key strategies for problem list usage, which are detailed in Box 3.1.

BOX 3.1 Success Factors for Problem List Usage, Based on Interviews With Informatics Leaders of Four Top-Performing Sites

1. Financial incentives: Two of the four top-performing sites had financial incentives related to problem list completeness. In one case, the site had a program for chronic diseases, including diabetes. The responder explained, "Our [pay for performance program] effectively incentivizes us to keep accurate problem lists, especially for major morbidities. This is because quality payments are partly driven by the number of patients on any particular morbidity register, e.g. diabetes, hypertension etc. We are not really incentivized to keep problem lists for more minor morbidities, but because [pay for performance] covers quite a lot of morbidities it is easier to record morbidities for everything." The other system had financial contracts that featured risk adjustments based on the chronic diseases a patient had, meaning that greater reimbursement and, in turn, potential physician bonuses, depended on complete documentation of problems, including diabetes.

2. Problem-oriented charting: The top-performing site used a mandatory version of problem-oriented charting stating, "The way in which the electronic records are structured means that we are encouraged to record each of the problems the patient presents with before recording history, examination, medications, investigations, formulation etc." This system provides a strong forcing function to record problems, including diabetes, because otherwise there is no place to enter documentation.

3. Gap reporting: Two of the four sites generated regular reports of patients who appeared to have various chronic conditions, including diabetes, but

(Continued)

BOX 3.1 (Continued)

did not have the condition on their problem list, and share these reports to providers. These reports, which one site called "gap lists" could then be used to update patient problem lists.

4. Shared responsibility: Most sites depended entirely on physicians to maintain the problem list. However, two of the top four sites also had care managers update the problem lists. For example, if a patient is followed by a diabetes care management program, the care manager would ensure that diabetes appeared on his or her problem list. One of the sites also generates reports of patients potentially eligible for care management programs, including patients with high HbA1c scores, combining both the gap reporting and shared responsibility practices.

5. Links to billing codes: Most sites separate the problem list from encounter-based diagnosis coding for billing; however, one of the top sites automatically feeds billing diagnoses to the problem list. This results in a high rate of problem list completeness, as clinicians usually remember to bill patients for diabetes, even if they might not otherwise add it to the problem list. One drawback of this approach is that, if a patient is billed for multiple related ICD-9 codes (e.g., "Diabetes mellitus," "Diabetes mellitus without mention of complication," and "Diabetes mellitus without mention of complication, type II or unspecified type, uncontrolled") over several visits, the problem list can become cluttered with near duplicate terms.

6. Organizational culture: A final and harder-to-characterize practice reported at several of the top sites was simply an organizational culture or practice of assiduous use of the problem list within and across groups. In these organizations, use of the problem list was simply expected and widely practiced. Moreover, at these sites, both primary care providers and specialists considered themselves to have shared responsibility for problem list maintenance. We observed a similar phenomenon in a prior ethnographic study of problem list usage at Brigham and Women's Hospital, where certain practices and specialties had a culture of problem list usage, often due to leadership or peer expectations (Wright et al., 2011a), and others did not.

Source: Wright, A., McCoy, A.B., Hickman, T.T., Hilaire, D.S., Borbolla, D., Bowes, W.A., et al., 2015. Problem list completeness in electronic health records: a multi-site study and assessment of success factors. Int. J. Med. Inform. 84 (10), 784–790.

Another key strategy for improving problem list completeness is problem-list focused CDS. Fig. 3.3 shows a sample problem list CDS intervention, which led to a threefold increase in problem list usage at the Brigham and Women's Hospital in Boston, MA.

Allergy Lists

Because of the centrality of drug allergy prevention, almost all EHRs have a structured field for documenting allergies, which can be consulted during

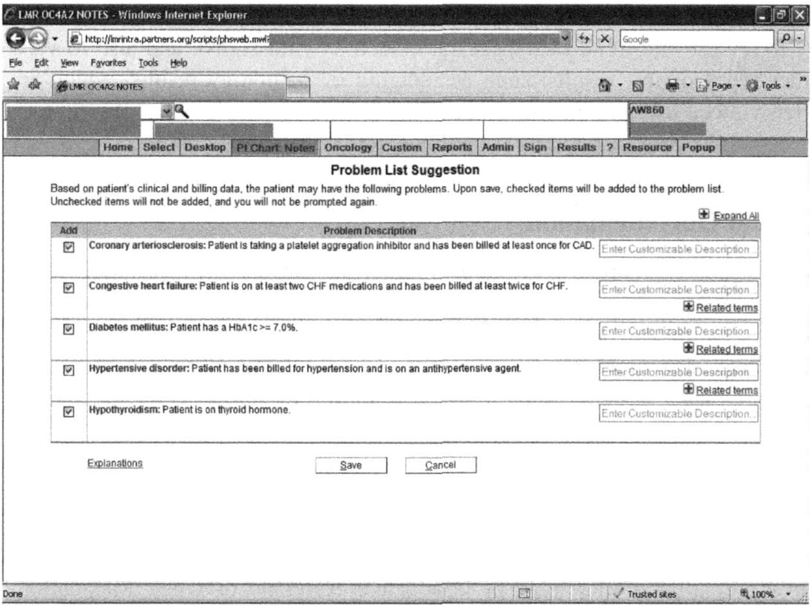

FIGURE 3.3   Problem list CDS intervention in an outpatient EHR.

prescribing. Allergy documentation is deceivingly complex, as patients may report allergies to a specific drug, an ingredient used in drug preparation (such as a color or preservative), a class of drugs, food allergens, or environmental allergens (Goss et al., 2013)—most systems have at least some level of support for all of these types of allergies and also have some capacity to capture the reaction, severity, and certainty of the allergy. Drug intolerances are often collected along with drug allergies, and the difference between allergies and intolerances is not always distinguished in EHRs.

Most EHRs also have some degree of drug allergy alerting. The most basic form is direct matching, where a patient is about to be prescribing a drug that exactly matches a drug on his or her allergy list. More sophisticated drug-allergy alerting systems may alert based on drug classes or known cross-sensitivity risks.

Results Review

Reviewing results of laboratory testing (see Chapter 10: Medication Management, and Laboratory and Radiology Testing) is one of the most basic and also most useful features of outpatient EHRs. Nearly all outpatient EHRs have the ability to review laboratory data in tabular form, and most also have the capability to display the data graphically (Sittig et al., 2015), which can be useful for spotting trends in data over time (Samal et al., 2011).

In addition to laboratory results, outpatient EHRs typically have the ability to review and trend flowsheet data (such as vital signs), and also to view other types of test results, such as imaging or pathology results, which are often viewed using a document paradigm.

Clinical Documentation

Clinical documentation is covered in detail in Chapter 4, Electronic Clinical Documentation. Documentation capability in outpatient EHRs is almost universally available, but the range of documentation modes varies considerably, typically including unstructured, free-text documentation, form- or template-based documentation, and the ability to integrate documentation from outside sources, such as transcribed or scanned documentation.

Clinical Decision Support

Most outpatient EHRs have at least some elements of CDS. Medication-related decision support is the most common type of outpatient CDS, perhaps especially drug—drug interaction checking and drug allergy alerting, and is almost universally available (see Chapter 10: Medication Management, and Laboratory and Radiology Testing) (McEvoy et al., 2017). Reminders for health maintenance and screening items are also widespread and can deliver substantial value. CDS is covered in more detail in Chapter 12, Clinical Decision Support and Knowledge Management.

Patient Portals and Personal Health Records

Most outpatient EHRs in the United States have some capability that allows patients to view data in a standard or unstructured form, although this is unusual in most other countries, with some exceptions including Australia. This technology is often called a patient portal or a tethered personal health record and, at a minimum, might allow patients to view their visit information and medication list. More advanced portals may support secure messaging, allowing a patient to contact his or her care providers electronically and the ability to view laboratory data (Grant et al., 2006; Poon et al., 2007) or problem list data (Wright et al., 2014a). Less common is the ability for patients to edit or add to their record through the personal health record (Grant et al., 2006; Poon et al., 2007), receive and act on preventive care reminders (Wright et al., 2012), or to review clinical notes, such as through OpenNotes (Delbanco et al., 2010) (see Chapters 4 and 18: Electronic Clinical Documentation; Social and Consumer Informatics).

Integration with External Systems

Compared to inpatient EHRs, most outpatient EHRs have relatively few interfaces to external information systems. Common system-to-system interfaces include inbound interfaces for receiving laboratory data and other clinical reports, as well as outbound e-prescribing interfaces—both of these interfaces are used almost universally. In countries with fee-for-service payment systems, external integration with billing systems is also common. Less common interfaces include electronic transmission of laboratory and imaging orders, or inbound interfaces from health information exchanges (see Chapter 5: Interoperability), though integrations with health information exchange are becoming much more prominent.

BENEFITS OF OUTPATIENT EHRs

Outpatient EHRs share many of the benefits of inpatient EHRs, described in Chapter 2, Inpatient Clinical Information Systems, but also have many of their own consideration. It can be challenging to measure the benefits of outpatient EHRs because many patients receive care from a variety of outpatient providers, and benefits (especially around quality) can take a long time to accrue.

Quality

The effects of EHRs on quality in the outpatient setting, at least to date, have been less pronounced than in the inpatient setting. One large review of data from the US NAMCS in 2007 showed no meaningful difference in quality between practices using outpatient EHRs and those using paper systems (Linder et al., 2007). A similar study in 2010 that also used NAMCS data found the same results, even when also considering the use of outpatient CDS (Romano and Stafford, 2011).

However, a systematic review of the effects of outpatient EHRs on quality in 2011 found modest positive results for structure and process measures, but limited evidence for effect on clinical outcomes (Holroyd-Leduc et al., 2011). More targeted studies have also found some specific quality benefits, particularly for reducing medication errors (Kaushal et al., 2010; Abramson et al., 2011, 2013), or when looking at the relationship between specific EHR features and specific clinical quality measures (Poon et al., 2010).

Costs

As with quality, the empiric evidence for the effects of outpatient EHRs on costs is unclear, although some models suggest there should be substantial benefit (Wang et al., 2003). This evidence was considered at length in a report sponsored by the Agency for Healthcare Research and Quality in the

United States entitled Costs and Benefits of Health Information Technology, which found that there should be some expected return on investment for implementing ambulatory EHRs, though the time to break even ranged from 3 to 13 years, depending on the modeling approach (Shekelle et al., 2006).

One particularly effective intervention in the outpatient setting is the display of cost information during laboratory test ordering. An early randomized trial at the Regenstrief Institute showed a 13% reduction in the cost of tests ordered when this information was displayed (Tierney et al., 1990).

EHRs may also enable care to be moved to less expensive settings. A serial cross-sectional study of care in the Northwest and Colorado regions of Kaiser Permanente found a decrease in the number of office visits after implementation of an EHR, with some visits being replaced by telephone contact (Garrido et al., 2005).

Access to Information for Physicians and Patients

One clear win with outpatient EHRs is better access to information. Outpatient EHRs, when well designed and fully integrated, allow multiple providers to view well-organized patient data from a variety of locations, including outside of the office, overcoming the physical limitations of paper charts. This is a clear driver of physician satisfaction with EHRs (physicians are often frustrated with other aspects of their EHR, but still appreciate the improved access), and increased access to information has been associated with better quality (Kern et al., 2008).

Outpatient EHRs also increase patient access to information. A large-scale randomized trial of patient engagement features in a tethered personal health record at Partners Healthcare system showed that increased patient access to data had positive effects on quality and patient engagement, and that various features (such as family history data, enhanced medication lists, and reminders provided directly to patients) had meaningful clinical impacts (Wright et al., 2012; Grant et al., 2006; Poon et al., 2007; Schnipper et al., 2008; Volk et al., 2007), and similar results have been seen in other studies.

IMPLEMENTATION AND ADOPTION CHALLENGES

Most of the implementation and adoption challenges described for inpatient clinical information systems described in Chapter 2, Inpatient Clinical Information Systems apply equally in the outpatient setting. In this chapter, we will outline some additional implementation challenges that are specific to the outpatient setting:

Costs

Although initial costs represent a major issue for EHR adoption in all settings, it can be particularly challenging for small practices, especially single-physician practices. Nonetheless, some data suggest this is not the biggest issue—if providers are paid for delivering high-quality, efficient care other factors dominate the initial costs (Wang et al., 2003). To help small practices deal with the initial costs and ongoing maintenance, vendors have emerged which offer managed, cloud-based application service provider models for EHRs (see Chapter 17: Cloud-Based Computing). The challenges of EHR adoption and maintenance have also caused a wave of consolidation, particularly in the United States (Barkholz, 2016), with smaller practices often joining larger practices or health systems that can provide an EHR.

Availability of Internal Technical Resources for System Implementation

Related to cost, an additional challenge for some outpatient practices is a lack of internal technical resources with sufficient sophistication to implement and manage an outpatient EHR. Although some are very sophisticated, many smaller outpatient practices lack dedicated information technology staff, may not have a computer network, and may not have the internal skill necessary to implement an EHR.

Integration and Interfaces

To achieve their full potential, outpatient EHRs need to interface with a variety of external systems, including external laboratories, imaging centers, health information exchanges, and pharmacies. These interfaces are costly, and outpatient practices sometimes choose not to make complete investments, for example, creating an interface to receive laboratory results but not transmit laboratory orders, or omitting interfaces to specialized laboratories to which they send only a small number of tests.

Specialty Workflows

Outpatient care models vary widely, particularly in certain specialties such as ophthalmology, procedure-based gastroenterology, or psychotherapy. Common outpatient EHRs may not fit these specialty workflows well, requiring extensive customization, adaption of processes, or the adoption of specialty-specific systems. However, regulation and a general trend toward homogenization of EHR software has made the creation and use of specialty-specific systems more complex, particularly in the United States.

Longitudinal Care

Inpatient medicine is typically very intense, with a relatively large number of consultations, tests, medications, and procedures occurring in a condensed period of time, typically within a single health system. In contrast, outpatient care, and primary care in particular, is typically longitudinal, with multiple providers seeing a patient over a long period of time. As such, managing care over time, including preventive care, medication refills, goals of therapy, and long-term clinical plans must be accommodated. Patients may also see multiple providers. In the case of clinically and technically integrated systems (or when interoperability is robust), this may not prove much of an issue but, more typically, if the providers use different EHRs, integrating and reconciling information to form a complete clinical picture can be challenging.

Challenges in Resource-Constrained Environments

International perspectives on EHR adoption in developing nations have been presented throughout this chapter. However, strides are also being made in the use and adoption of outpatient EHRs in the economically developing world. One prominent effort is OpenMRS (Mohammed-Rajput et al., 2011; Mamlin et al., 2006), a project and community led by the Regenstrief Institute that also involves numerous developers, community members, and nongovernmental organizations. The OpenMRS community members have developed an open-source EHR of the same name. OpenMRS has deployed EHRs around the world, particularly in Africa, Haiti, India, China, and the Philippines (OpenMRS, 2016). The system is particularly modular and customizable, and it has been deployed to support a wide range of clinical applications.

LOOKING AHEAD—FUTURE TRENDS AND DEVELOPMENTS

Because outpatient clinical information systems sit at the core of clinical computing, their future is very much aligned with the future of clinical informatics outlined in this book. In particular:

- *Usability and safety* issues are becoming more important, as providers have increased their expectations for EHRs and unintended consequences have come to the fore. These advances are covered in detail in Chapter 14, A Sociotechnical Approach to Electronic Health Record Related Safety.
- *Interoperability* will also increase over time due to regulatory pressure as well as patient and provider demand—advances in the relevant standards and infrastructure are covered in Chapter 5, Interoperability, and the attendant privacy issues in Chapter 6, Privacy and Security.
- *Cloud* models for the delivery of outpatient EHR software are also becoming more prevalent, particularly for smaller provider organizations, and for disaster recovery—Chapter 17, Cloud-Based Computing covers these issues.
- *Better decision support* will be pivotal—this is covered in Chapter 12, Clinical Decision Support and Knowledge Management.

- *Apps and APIs* are also becoming more prevalent—standard and models are covered in Chapter 16, An Apps-Based Information Economy in Healthcare.
- *Big data and machine learning* are also emerging as organizations look to derive value from data and improve the operation of their outpatient EHRs—these advances are covered in Chapters 11 and 19, Bioinformatics and Precision Medicine; Machine Learning in Healthcare.
- *Patient engagement and consumer and social issues* are also becoming increasingly important for outpatient EHRs and are covered in Chapter 18, Social and Consumer Informatics.

CONCLUSIONS

As adoption of outpatient EHRs has dramatically increased, we are at a promising time in the evolution of outpatient systems. Rather than adoption, our challenge is now optimization, improvement, and, most importantly, deriving value. Much has changed in the last 20 years, just as much is likely to change in the next 5−10 years—opportunities to make systems safer, faster, more usable, and, overall better abound, and the road ahead is bound to be an exciting one.

REFERENCES

Abramson, E.L., Barrón, Y., Quaresimo, J., Kaushal, R., 2011. Electronic prescribing within an electronic health record reduces ambulatory prescribing errors. Jt Comm. J. Qual. Patient Saf. 37, 470−478.

Abramson, E.L., Pfoh, E.R., Barrón, Y., Quaresimo, J., Kaushal, R., 2013. The effects of electronic prescribing by community-based providers on ambulatory medication safety. Jt Comm. J. Qual. Patient Saf. 39, 545−552.

April, B. E-Prescribing Trends in the United States.

Barkholz, D., 2016. IT needs are driving the upswing in doc practice mergers. Modern Healthcare.

Blumenthal, D., Tavenner, M., 2010. The "meaningful use" regulation for electronic health records. N. Engl. J. Med. 363, 501−504.

Commonwealth Fund, 2016. International Health Care System Profiles: What is the status of electronic health records? [Online]. http://international.commonwealthfund.org/features/ehrs/.

Delbanco, T., Walker, J., Darer, J.D., Elmore, J.G., Feldman, H.J., Leveille, S.G., et al., 2010. Open notes: doctors and patients signing on. Ann. Intern. Med. 153, 121−125.

Friedman, M.A., Schueth, A., Bell, D.S., 2009. Interoperable electronic prescribing in the United States: a progress report. Health Aff. 28, 393−403.

Garrido, T., Jamieson, L., Zhou, Y., Wiesenthal, A., Liang, L., 2005. Effect of electronic health records in ambulatory care: retrospective, serial, cross sectional study. BMJ 330, 581.

Goss, F.R., Zhou, L., Plasek, J.M., Broverman, C., Robinson, G., Middleton, B., et al., 2013. Evaluating standard terminologies for encoding allergy information. J. Am. Med. Inform. Assoc. 20, 969−979.

Grant, R.W., Wald, J.S., Poon, E.G., Schnipper, J.L., Gandhi, T.K., Volk, L.A., et al., 2006. Design and implementation of a web-based patient portal linked to an ambulatory care electronic health record: patient gateway for diabetes collaborative care. Diabetes Technol. Ther. 8, 576−586.

Hartung, D.M., Hunt, J., Siemienczuk, J., Miller, H., Touchette, D.R., 2005. Clinical implications of an accurate problem list on heart failure treatment. J. Gen. Intern. Med. 20, 143−147.

Holroyd-Leduc, J.M., Lorenzetti, D., Straus, S.E., Sykes, L., Quan, H., 2011. The impact of the electronic medical record on structure, process, and outcomes within primary care: a systematic review of the evidence. J. Am. Med. Inform. Assoc. 18, 732−737.

Hsiao, C.-J., Hing, E., 2012. Use and Characteristics of Electronic Health Record Systems Among Office-Based Physician Practices, United States, 2001−2012. US Department of Health and Human Services, Centers for Disease Control and Prevention, National Center for Health Statistics.

Jha, A.K., Doolan, D., Grandt, D., Scott, T., Bates, D.W., 2008. The use of health information technology in seven nations. Int. J. Med. Inform. 77, 848−854.

Kaushal, R., Jha, A.K., Franz, C., Glaser, J., Shetty, K.D., Jaggi, T., et al., 2006. Return on investment for a computerized physician order entry system. J. Am. Med. Inform. Assoc. 13, 261−266.

Kaushal, R., Kern, L.M., Barrón, Y., Quaresimo, J., Abramson, E.L., 2010. Electronic prescribing improves medication safety in community-based office practices. J. Gen. Intern. Med. 25, 530−536.

Kern, L.M., Barrón, Y., Blair III, A.J., Salkowe, J., Chambers, D., Callahan, M.A., et al., 2008. Electronic result viewing and quality of care in small group practices. J. Gen. Intern. Med. 23, 405−410.

Linder, J.A., Ma, J., Bates, D.W., Middleton, B., Stafford, R.S., 2007. Electronic health record use and the quality of ambulatory care in the United States. Arch. Intern. Med. 167, 1400−1405.

Mamlin, B.W., Biondich, P.G., Wolfe, B.A., Fraser, H., Jazayeri, D., Allen, C., et al., 2006. Cooking up an open source EMR for developing countries: OpenMRS—a recipe for successful collaboration. AMIA Annu. Symp. Proc. 529−533.

McEvoy, D.S., Sittig, D.F., Hickman, T.T., Aaron, S., Ai, A., Amato, M., et al., 2017. Variation in high-priority drug-drug interaction alerts across institutions and electronic health records. J. Am. Med. Inform. Assoc. 24 (2), 331−338.

Mehta, N., Vakharia, N., Wright, A., 2014. EHRs in a web 2.0 world: time to embrace a problem-list Wiki. J. Gen. Intern. Med. 29, 434−436.

Mohammed-Rajput, N.A., Smith, D.C., Mamlin, B., Biondich, P., Doebbeling, B.N., Open, M.R. S.C.I., 2011. OpenMRS, a global medical records system collaborative: factors influencing successful implementation. AMIA Annu. Symp. Proc. 2011, 960−968.

OpenMRS, 2016. OpenMRS Atlas [Online]. http://atlas.openmrs.org/ (accessed 09.01.17).

Poon, E.G., Wald, J., Schnipper, J.L., Grant, R., Gandhi, T.K., Volk, L.A., et al., 2007. Empowering patients to improve the quality of their care: design and implementation of a shared health maintenance module in a US integrated healthcare delivery network. Medinfo 2007: Proceedings of the 12th World Congress on Health (Medical) Informatics; Building Sustainable Health Systems. IOS Press, 1002.

Poon, E.G., Wright, A., Simon, S.R., Jenter, C.A., Kaushal, R., Volk, L.A., et al., 2010. Relationship between use of electronic health record features and health care quality: results of a statewide survey. Med. Care 48, 203−209.

Romano, M.J., Stafford, R.S., 2011. Electronic health records and clinical decision support systems: impact on national ambulatory care quality. Arch. Intern. Med. 171, 897−903.

Samal, L., Wright, A., Wong, B.T., Linder, J.A., Bates, D.W., 2011. Leveraging electronic health records to support chronic disease management: the need for temporal data views. Inform. Prim. Care 19.

Samal, L., Linder, J.A., Bates, D.W., Wright, A., 2014. Electronic problem list documentation of chronic kidney disease and quality of care. BMC Nephrol. 15, 1.

Schnipper, J., Gandhi, T., Wald, J., Grant, R., Poon, E., Volk, L., et al., 2008. Design and implementation of a web-based patient portal linked to an electronic health record designed to improve medication safety: the Patient Gateway medications module. J. Innov. Health Inform. 16, 147−155.

Shekelle, P., Morton, S.C., Keeler, E.B., 2006. Costs and benefits of health information technology. Evid. Rep. Technol. Assess.(132), 1−71.

Sittig, D.F., Murphy, D.R., Smith, M.W., Russo, E., Wright, A., Singh, H., 2015. Graphical display of diagnostic test results in electronic health records: a comparison of 8 systems. J. Am. Med. Inform. Assoc. 22, 900−904.

Tierney, W.M., Miller, M.E., McDonald, C.J., 1990. The effect on test ordering of informing physicians of the charges for outpatient diagnostic tests. N. Engl. J. Med. 322, 1499−1504.

Vesely, R., 2014. The great migration. Hospitals & Health Networks/AHA 88, 22.

Volk, L.A., Staroselsky, M., Newmark, L.P., Pham, H., Tumolo, A., Williams, D.H., et al., 2007. Do physicians take action on high risk family history information provided by patients outside of a clinic visit? Stud. Health Technol. Inform. 129, 13.

Wang, S.J., Middleton, B., Prosser, L.A., Bardon, C.G., Spurr, C.D., Carchidi, P.J., et al., 2003. A cost-benefit analysis of electronic medical records in primary care. Am. J. Med. 114, 397−403.

Weed, L.L., 1968. Special article: medical records that guide and teach. N. Engl. J. Med. 278, 593−600.

Wright, A., 2015. Clinical Problem Lists in the Electronic Health Record. Apple Academic Press, Toronto; New Jersey.

Wright, A., Goldberg, H., Hongsermeier, T., Middleton, B., 2007. A description and functional taxonomy of rule-based decision support content at a large integrated delivery network. J. Am. Med. Inform. Assoc. 14, 489−496.

Wright, A., Maloney, F.L., Feblowitz, J.C., 2011a. Clinician attitudes toward and use of electronic problem lists: a thematic analysis. BMC Med. Inform. Decis. Mak. 11, 36.

Wright, A., Pang, J., Feblowitz, J.C., Maloney, F.L., Wilcox, A.R., Ramelson, H.Z., et al., 2011b. A method and knowledge base for automated inference of patient problems from structured data in an electronic medical record. J. Am. Med. Inform. Assoc. 18, 859−867.

Wright, A., Poon, E.G., Wald, J., Feblowitz, J., Pang, J.E., Schnipper, J.L., et al., 2012. Randomized controlled trial of health maintenance reminders provided directly to patients through an electronic PHR. J. Gen. Intern. Med. 27, 85−92.

Wright, A., Henkin, S., Feblowitz, J., McCoy, A.B., Bates, D.W., Sittig, D.F., 2013. Early results of the meaningful use program for electronic health records. N. Engl. J. Med. 368, 779−780.

Wright, A., Feblowitz, J., Maloney, F., Henkin, S., Ramelson, H., Feltman, J., et al., 2014a. Increasing patient engagement: patients' responses to viewing problem lists online. Appl. Clin. Inform. 5, 930−942.

Wright, A., Sittig, D.F., Mcgowan, J., Ash, J.S., Weed, L.L., 2014b. Bringing science to medicine: an interview with Larry Weed, inventor of the problem-oriented medical record. J. Am. Med. Inform. Assoc. 21, 964−968.

Wright, A., McCoy, A.B., Hickman, T.T., Hilaire, D.S., Borbolla, D., Bowes III, W.A., et al., 2015. Problem list completeness in electronic health records: a multi-site study and assessment of success factors. Int. J. Med. Inform. 84, 784–790.

RECOMMENDED FURTHER READING

Agency for Healthcare Research and Quality, 2016. Electronic Medical Record Systems [cited 2017 Jan 3]. https://healthit.ahrq.gov/key-topics/electronic-medical-record-systems.

Bates, D.W., Bitton, A., 2010. The future of health information technology in the patient-centered medical home. Health Aff. 29 (4), 614–621.

Bates, D.W., Ebell, M., Gotlieb, E., Zapp, J., Mullins, H.C., 2003. A proposal for electronic medical records in US primary care. J. Am. Med. Inform. Assoc. 10 (1), 1–10.

Hartley, C.P., Jones, E.D., 2005. EHR Implementation: A Step-by-Step Guide for the Medical Practice. American Medical Association.

Hillestad, R., Bigelow, J., Bower, A., Girosi, F., Meili, R., Scoville, R., et al., 2005. Can electronic medical record systems transform health care? Potential health benefits, savings, and costs. Health Aff. 24 (5), 1103–1117.

Mourar, M., 2012. EHR Optimization and Operations Guide for Medical Practices. Medical Group Management Association, Englewood, CO.

Williams, T., Samarth, A., 2011. Electronic Health Records for Dummies. Wiley, Hoboken, NJ, xxii, 358 pp.

Chapter 4

Electronic Clinical Documentation

Gordon D. Schiff[1] and Mary J. Tharayil[2]

[1]*Harvard Medical School, Boston, MA, United States,* [2]*Brigham and Women's Primary Care of Brookline, Brookline, MA, United States*

Of all the topics in this book, none is more critical to good clinical care than electronic clinical documentation (ECD). Its structure (how it is designed), processes (how it is used, during and outside the visit), and outcomes (the notes produced and the care they drive) provide the foundation for high-quality patient care, efficient and effective care delivery, and clinician and patient satisfaction (Reiser, 1991a,b; Cusack et al., 2013; Kuhn et al., 2015). However, as Donabedian (1966) pointed out 50 years ago when he set forth this classic structure/process/outcome paradigm, answering the deeper question of "what goes on here" may be even more important than narrow structure/process/outcome metrics in thinking about quality as it relates to clinical notes.

As we discuss in this chapter, all is not well with clinical documentation. A series of widespread "chief complaints" have emerged (Table 4.1) which threaten to compromise and distract from the many potential positives of ECD, and even the overall quality of clinicians work life in a variety of practice settings (Kuhn et al., 2015; Krist et al., 2014; Friedberg et al., 2013; Martin and Sinsky, 2016; Cusack et al., 2013; Toll, 2012; Hanauer et al., 2016). Given our personal background (authors are both practicing primary care physicians in the United States) and the fact that ambulatory primary care comprises the most frequent outpatient encounter, we will mainly discuss the role of and issues surrounding ECD in that setting. While many of the issues discussed are highly generalizable to other settings (inpatient, emergency department, long-term care, home health) and to other clinical roles (physical therapists, social workers, home health aides), we use documentation in the primary care setting as the prototype to better understand ECD and hopefully contribute to the promotion of quality and safety in this domain.

We begin by outlining the historical background, key roles, and goals for ECD and then delve into the complaints and issues that must be addressed if

Key Advances in Clinical Informatics. DOI: http://dx.doi.org/10.1016/B978-0-12-809523-2.00004-2

we are to more fully understand existing practices to enable us to make genuine progress. While often there is not clear "right answer" to the various issues and dilemmas posed, many are ripe for more explicit analysis, data collection, debate, and most importantly, collective action to discover and build better ways for us to create our notes.

WHAT IS ECD?

The American Health Information Management Association and others define clinical notes by enumerating a cluster of *note types* (e.g., Admission History and Physical Examination Report (H & P), Progress Notes, Consultation Reports/Notes, Discharge Summary, and Operative Reports) and then specify various *elements of these notes* (e.g., Chief Complaint, History of Present Illness, Past Medical History, Family History, Social History, Review of Systems, Assessment, Plan). To this, regulatory agencies such as the US Center for Medicare and Medicaid or Joint Commission have added requirements for specific legal and billing elements for acceptable notes (time, date, signature, printed name, etc.) (AHIMA, 2011; System CSRH, 2011; Commission TJ, 2016).

However, rather than narrowly focusing on these traditional elements and requirements, we aim to conceptualize clinical notes in a broader historical, sociotechnical, and more timely clinical context. We will do so by examining bigger picture issues as well as looking more closely at details of clinical notes in everyday practice. However, as we are dealing with evolving technology and underdeveloped and unstandardized processes and practices (i.e., much of this in its infancy), there are enormous opportunities for research and learning in the future.

To bring us to the present in the theory and practice of clinical documentation, we begin by noting a number of key historical landmarks. The first is the elaboration of the concept and adoption of the Problem-Oriented Medical Record, and specifically the SOAP (Subjective, Objective, Assessment, Plan) note format, by Lawrence Weed in the 1960s−80s (Weed, 1968a). Arguing that records needed to be redesigned to function more as longitudinal records of the patient's trajectory rather than as isolated stand-alone entries for each clinical encounter, he advanced the idea of tying notes together via "problem lists." This was revolutionary and transformative, with most practices and practitioners adopting this new format to varying degrees. In addition, Weed was visionary in positing that such records needed to be electronic rather than on paper, something that only decades later would become a widespread reality (Weed, 1997; Wright et al., 2014).

Another landmark event was the 1991 report from the Institute of Medicine (IOM; now the National Academy of Medicine) Committee on the *Computer Based Patient Record* which envisioned new ways in which electronic health records (EHRs) could transform notes and care (Dick et al.,

1997). During the ensuing quarter century, a host of "home grown" and commercial systems implemented various aspects of the EHR. Often starting with ancillary areas such as laboratory results storage/lookup, and electronic ordering of tests and medications, the conversion of notes from paper to electronic eventually rose to the top of the agenda for hospitals and clinics. We have now reached the point where more than 85% of all notes in the United States (with similar trends worldwide) are entirely electronic (Henry et al., 2016). However, the road has been a bumpy one, with various hybrid solutions such as scanning of paper handwritten or dictated notes, and systems that maintained dual systems of paper records that include hard copies of electronic notes/reports, as hybrid solutions. Such notes are not searchable, or even at times (in the case of scanned handwritten notes) legible. And, as the road to full electronic notes is becoming better paved, new potholes have arisen and have led to a series of recent thoughtful reports spotlighting these issues, such as the Policy Position Paper From the American College of Physicians on Clinical Documentation in the 21st century, and a policy report from the American Medical Informatics Association on The Future State Of Clinical Data Capture And Documentation (Kuhn et al., 2015; Payne et al., 2015).

CHIEF COMPLAINTS: CURRENT PROBLEMS WITH ECD

As in a clinical note, a good starting point for understanding the current issues related to ECD is a listing of the "chief complaints" that are being articulated by key stakeholders, in particular the clinicians whose frustrations are painfully palpable (Table 4.1).

GOALS FOR REDESIGNED CLINICAL NOTES—PARTICULARLY TO SUPPORT DIAGNOSIS

Before we discuss these important and quite widespread concerns, it is worth stepping back to consider the goals of clinical notes as we strive to think broadly and creatively envision the full potential of ECD in the EHR of the future.

The authors are guided by a conviction (supported by some data) that good/improved clinical notes are essential for clinicians to be good diagnosticians (Schiff and Bates, 2010). At the simplest level, depending on human memory to recall key details from the history or physical exam is unreliable. Notes are also an essential vehicle for recording our thought processes for ourselves and conveying these with other care team members. Moreover, writing notes should not only support our diagnostic assessments, but also ought not distract us from paying attention to and thinking about the patient. While some have argued that such distractions are inherent in our EHR/clinical note design (Hartzband and Groopman, 2008), we disagree and maintain

TABLE 4.1 Chief Complaints

Note bloat: Notes have become filled with templated/checkbox items and copy/pasted information, often of little relevance to current encounter (often done primarily for billing requirements) or containing inaccurate/out-of-date information. Notes also often include duplicate information that is present in other sections of the electronic medical record (EMR) (e.g., medications, allergies).

Degraded usability: Stemming from the above, note readability has been compromised with difficulties in being able to see at a glance what is happening with the patient and glean a succinct meaningful picture of how the patient's illness is evolving, and the clinician's thinking and management plans. Nonstandardized formatting and design that poorly supports cognitive processing is widespread.

Time burden to write notes: Clinicians report spending many additional hours completing their notes after clinical sessions (often at home), with resulting complaints of unsustainable workloads, burnout, and decreased family time. Clinicians also worry about excessive additional documentation requirements mandated for billing or various quality metrics.

Information overload/overlooked: In earlier eras, lack of information was major problem (unable to find the chart, test results, prior notes). This has now been supplanted (especially within integrated EMRs) by massive numbers of notes that are impossible for any human to review. It has become difficult to ensure that important items that need follow-up are tracked appropriately over time and not overlooked.

Interoperability barriers: We still suffer from lack of access to electronic notes residing in *other* systems. Due to the inability of various systems to communicate with each other, sharing information remains difficult despite much emphasis on sharable formats (such as the Continuity of Care Document based on the HL7 (Health Level 7) CDA architecture), and overcoming inability of various systems to communicate with each other.

Suboptimal note-writing functionality and integration with workflows: Excessive mouse-clicks/checkboxes, poor screen design, and confusing and burdensome navigation do not support or integrate well into clinical workflow. Information often has to be entered multiple times, creating duplicative work. Other team members who might contribute to clinical notes are not adequately incorporated into current ECD workflows.

Impaired note reliability and errors: Owing to a number of the above factors, out-of-date and outright erroneous information is being included in notes. "Normal" checkboxes are checked even for items not asked or examined. When two charts are open simultaneously, information risks being written or pasted into the wrong patient's note. As a result the accuracy of note can no longer be assumed and users complain that they can no longer trust what they read in notes.

Dehumanizing the clinical encounter: To the extent that clinicians feel they need to pay more attention to the computer screen than to the patient, especially if they have weak touch-typing skills or even (ironically) have to fill in quick checkboxes which require more hand–eye focus on the computer than does free text entry, something of the special human interaction that should characterize medical visits is lost.

that better ECD can actually help improve diagnosis and prevent diagnostic errors. Just as computerized ordering of medications has decreased medication errors (while also introducing various unintended negative consequences), we maintain that better ECD has the potential to improve diagnosis and prevent diagnostic errors (Schiff and Bates, 2010; El-Kareh et al., 2013).

EIGHT GOALS FOR ECD

Here we identify eight key goals that clinical notes should aim to achieve and briefly discuss issues surrounding their status and prospects.

Accurately Record Key Information From the Encounter

Just a few years ago, any discussion of the goals of clinical notes would start and end here. "Make sure you document" xyz for ensuring a "complete" note, and document exhaustively to support billing or malpractice claim protection. These issues have not gone away and unfortunately still dominate the thinking of many physicians, consultants, and liability insurers who narrowly obsess over ill-defined note "completeness." But our understanding of the details, workflow processes, design, uses and users of notes is evolving and hopefully maturing.

First and foremost, the note should be a tool to *help clinicians to provide better care*. This means the target for note redesign should be the needs of the most important user, the caregiver who is the note author, with secondary considerations to the needs of other care team members (nursing staff, consultants, referring physician, covering clinicians, social workers, and the patient).

To the greatest extent possible, information should be automatically recorded and updated. Rather than a copy/pasted list of medicines or hemoglobin A1C values (often out of date after being carried forward multiple times), this data should be automatically updated and dynamically linked to relevant electronic data existing outside the note. Everything from the patient's age, to vital signs, weight, and laboratory values listed in a note should be dynamically linked and automatically updated. Nonetheless, there are conflicting views about what should versus should not be included in notes. For example, should the list of patient's medications be included in the note versus residing solely in the medication ordering section of the EHR? Certainly, the ability to create a time-stamped snapshot documenting the exact regimen at that moment (prior to any changes being made at that visit) could be of value, but this duplicated information can add to note-bloat. Likewise, bringing test results into the note and grappling with abnormal findings that are then reviewed and discussed with the patient during that encounter is valuable to document, but risks duplicating information residing elsewhere. We can envision creative, standardized solutions to such dilemmas, especially with more thoughtful formatting of note layout.

Produce Notes Quickly and Efficiently

An important and legitimate goal of clinicians is to quickly complete and close their note and move on to the next patient (or note!). Just-in-time production theory and methods suggest this should ideally be done in real time or close to real time (occasionally allowing delay for a pending test result, or for further reflection and research on a confusing diagnosis). Large-scale "batching," in the form of letting scores of unfinished notes stack up in one's inbox queue, is suboptimal to finishing the note while the patient is fresh in mind.

Currently there are three general formats for entering electronic notes—templated notes, free text notes, and dictated notes (Pollard et al., 2013; Edwards et al., 2014). Different specialties and physicians tend to disproportionately use/favor one over another. In one study we found that specialists more often used dictation whereas primary physicians used template and free text in equal proportions (Pollard et al., 2013). Increasingly hybrid methods are being used, and scribes and speech recognition (discussed below) are working their way into the mix. Nonetheless, in general, there is dissatisfaction with the time and resulting clinical and economic costs required for note-writing and widespread agreement that insufficient attention has been paid to better designed workflows that optimize efficiency (Ash et al., 2004; Payne and Graham, 2006; Krist et al., 2014).

Meaningfully Portray Patient's Unique Story and Clinician's Thinking

Notes need to describe and convey in narrative form what is going on with the patient—to breathe life into the record. Many templated notes give an unrecognizable generic picture of the patient, their history, the illness time course, response to treatment, and its impact on their health and quality of life (Rosenbloom et al., 2011). Clinicians need not write a novel here, but do need to skillfully and succinctly craft a narrative that captures the essence of such individualized information. And this information ought not to be buried among hundreds of checkbox defaults/responses. Some have advocated reorganizing notes to place the Assessment first (APSO; or even S-AP-O as one of us currently does), but what is needed is more than a simple rearrangement of the note order (Lin et al., 2013; Pullen, 2010). Instead, we need new ways of conceptualizing notes and better methods of recording the evolving narrative (speech recognition and scribes are rapidly growing options, see below).

Equally important is the need to capture the *clinician's thinking and rationale* for recommendations made for and with the patient. Although the

TABLE 4.2 Components of a Good Assessment 5-D's
Defining the problem(s)—describe, justify, group = > *Problem representation*
Diagnosis—DD, etiology, weighing probabilities, cause of exacerbation = > *Differential diagnosis, etiology*
Doing—how is patient doing, time course, urgency, response to Rx (treatment), interpretation of response = > *Response to treatment*
Do—what needs to be done, and why = > *Plan and its rationale*
Don't Know—what are uncertainties, need to f/up = > *Uncertainties*

Assessment in many notes is often just a skimpy single word or problem such as "CHF" (congestive heart failure), this is hardly adequate to represent a meaningful assessment of the patient or the problem from that encounter. Just adding the word "stable" after the CHF would convey an infinitely richer assessment. For assessing any major, new, or unstable problem, we created a pneumonic that emphasizes "5-D's" to craft a good assessment (Table 4.2). It begins with often implicit judgments about identifying and defining the patient's main problem(s)—so-called problem representation. This leads next to thinking about the causes of these problems, generating differential diagnoses, and weighing their various likelihoods—so-called differential diagnosis. Another key element involves assessing the patient's progress, whether they are responding to therapy, and if not, potential causes for exacerbation. Finally, it is helpful to document the rationale for next steps, as well as uncertainties and contingencies. As we stressed above, this information is first and foremost for the authoring clinician, a workspace to work out his/her thoughts, and a way to recall these assessments the next time the patient is seen as well as share with other team members involved in the patient's care.

Support Diagnostic Decision Making

Directly flowing from ECD's role in supporting and documenting the assessment is the broader role in serving as a springboard to better diagnosis. Table 4.3 summarizes a series of overlapping potential domains for redesigning notes for better diagnosis. Could better notes and interaction with the computer, for example, help avoid "premature closure" thereby preventing fixating and perpetuating a single wrong diagnosis? Redesigning clinical documentation functionality to better achieve these "stretch goals" is an important priority that has received insufficient attention according to the recent IOM Report on Improving Diagnosis in Medicine (National Academies of Sciences Engineering Medicine, 2015). Many elements are

TABLE 4.3 How Improved ECD Can Improve Diagnoses

Role for Electronic Documentation	Goals and Features of Redesigned Systems
Providing access to information	Ensure ease, speed, and selectivity of information searches; aid cognition through visual display featuring aggregation, trending, contextual relevance, and minimizing of superfluous data.
Recording and sharing assessments	Provide space for recording thoughtful, succinct assessments, differential diagnoses, contingencies, uncertainties, and unanswered questions; facilitate sharing and critical review of assessments by other clinicians as well as patients.
Maintaining dynamic patient history	Carry forward information for recall, avoiding repetitive patient querying and recording of unchanged information while highlighting new information and minimizing erroneous copying and pasting.
Maintaining problem lists	Ensure that problem lists are better organized and integrated into workflow to allow for continuous updating and incorporation into notes.
Tracking medications	Record of medications patient is actually taking, patient responses to medications, and adverse effects to ensure timely recognition of medication problems and avoid drug reactions being misdiagnosed.
Tracking tests	Integrate management of diagnostic test results into note workflow to facilitate and ensure reliable review, acknowledgment, assessment, and action in response, as well as documentation of these steps and rationale.
Ensuring coordination and continuity	Aid in aggregating, integrating, summarizing, data from all care episodes and fragmented encounters (especially "interval history") to permit thoughtful synthesis, ideally crafting of wiki-like summary.
Enabling follow-up	Facilitate patient education about plan, potential red-flag symptoms to watch for; help ensure and track any needed follow-up.
Providing feedback	Automate feedback to upstream/prior clinicians, facilitating their learning from subsequent diagnosis-related outcomes and misdiagnoses.
Providing prompts	Provide checklists to minimize reliance on memory (e.g., for ensuring key history items or differential diagnosis considerations) to direct questioning and support diagnostic thoroughness and problem solving.
Providing placeholder for resumption of work	Delineate where in diagnostic process clinician was and should resume after being interrupted to prevent lapses in data collection and diagnostic thinking.

(Continued)

TABLE 4.3 (Continued)

Role for Electronic Documentation	Goals and Features of Redesigned Systems
Calculating Bayesian probabilities	Embed calculator into notes workflow to reduce weighting errors and minimize known biases in subjective estimation of diagnostic probabilities.
Providing access to information sources	Provide instant access to knowledge resources through context-specific "infobuttons" triggered by key or highlighted words in notes that link user to textbooks and relevant guidelines.
Offering second opinion or consultation	Integrate real-time online/telemedicine access to consultants to provide just-in-time answers to questions related to referral triage, testing strategies, or expert diagnostic assessments.
Increasing efficiency	Penultimate aim that more thoughtful design, workflow integration, easing, and distribution of documentation burden would speed up charting, to free up time for enhancing communication (with patient/others) and diagnostic thinking, reflection, reading.

Source: Modified from Schiff, G.D., Bates, D.W., 2010. Can electronic clinical documentation help prevent diagnostic errors? N. Engl. J. Med. 362 (12), 1066–1069.

already in place and being used to varying extents, including the innovative integration of diagnosis decision support into electronic notes (see Fig. 4.1 screen shot of Cerner), disease-specific history taking checklists, electronic referrals that can be linked to note documentation, and infobuttons that can access online resources such as textbooks and references in real time while the note is open with the patient.

Help Ensure Problems Do Not Got Lost or Overlooked

In theory, the Problem List, which we consider an integral component of clinical documentation and electronic notes (see Problem-Based Charting discussion below), should be the essential vehicle for ensuring that patients' medical problems and abnormal findings do not get lost. In reality, electronic problem lists are problematic, plagued by multiple issues, including keeping problem lists updated, lack of clarity/agreement about what belongs on the problem list, whose responsibility it is to maintain it (should specialists be adding their diagnosis or should the primary care physician maintain ownership), confusion between simple diagnosis (diabetes) and more complex billing codes (ICD-10 coding for diabetes and presence/absence of various complications), difficulty organizing problems (being able to freely move order, group, move to inactive, etc.),

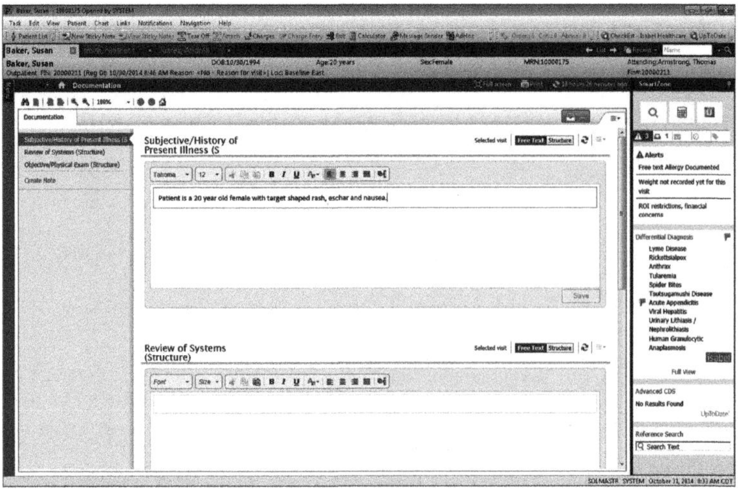

FIGURE 4.1   Illustrating the way a clinical decision diagnosis support program (Isabel) is integrated into and EMR clinical note (Cerner) and it "reads" the note as it is being written (center of screen) displaying a differential diagnosis (on right).

duplicate problems (listed with different names), etc. (Holmes, 2011a,b; Galanter et al., 2010; Wright et al., 2012, 2015).

Lost in this shuffle is the potential for action-requiring problems to be overlooked such as patient having a history of splenectomy (hence need for pneumococcal vaccination) or a pulmonary nodule (that requires reimaging in 1 year) (Gandhi et al., 2011). What a clinical note linked to the problem list can and should do well is provide highly visible and reliable closed-loop tracking abilities to ensure the clinician is reminded of key problems each visit (especially "open loops") with opportunity to review outstanding clinical issues (with the patient and subsequently when writing the note). Whereas the "Review of Systems (ROS)" was traditionally stressed as a key activity (and learning exercise for trainees), clinician encounters and notes need to be reengineered to be guided by the clinical note to go through a "Review of Problems" with ways to be interactively reminded, engaging patients and updating in the process of going through them, to ensure problems are not lost.

Succinct, Organized, Usable by Others

Now that we have potentially overloaded the clinical note with all of the above functions, how are we simultaneously going to make notes more succinct, and easier to read and use? Multiple strategies will likely have to be pursued to achieve this aim. A starting point would be better consensus around standardized organization and redesigned display of information (built off of extensive user feedback, testing, human factors studies, simulation and pilot experiments, and software testing). Notes should take advantage of the capability of electronic data to be entered one way but be displayed in another way. Key nuggets (usually free text narratives) from patient history and clinician assessment should be easily findable and formatted in a standardized way. Filters should further enable specialized looks at different slices of the note (e.g., Rheumatologist can have a view centered on the patient's rheumatology problems they are caring for) and allow tracking of temporal relationships. Clinicians' skill at crafting good notes should not be taken for granted, but instead warrants inclusion of teaching note-writing skills in medical education curriculum, continuous feedback, and tips for improvement. As Mark Twain said, "I didn't have time to write a short letter, so I wrote a long one instead"—writing concisely takes both time and skill (GoodReads, 2016).

Tool to Facilitate Coordination Across Visits, Team

We have repeatedly suggested that patient care is no longer a solo activity. Combined with the longitudinal nature of caring for patients with chronic illness (and even following up on acute illness), notes must weave together visits over time and across space (different providers, transitions of care). More than just "cc-ing" relevant team members or making notes widely accessible, serious attention needs to be directed at how notes might better coordinate care and overcome current fragmentation.

Another concept needs to be much better developed—what we (and others) term, "The Interval History"—a way to synthesize events and notes since the patient's last visit. Currently, one of us (GDS) manually looks up and pastes key parts of any specialist visits or discharge summaries (marking this content as copied from others' note) for the intervening time since last seen in primary care and then underlines key text. This primitive and labor-intensive approach needs to be supplanted by functionality whereby the EHR can help automate such summarization. And since such weaving together should not exclusively be the doctor's job, distributed systems to share responsibilities among the wider team need to be developed (Healy et al.). There is growing interest in so-called "Shared Care Plans" as key clinical document that bridges the multidisciplinary care team with the patient's goals for their health and care (Hägglund et al., 2011; Hillestad et al., 2005; AHRQ, 2016; Baker et al., 2017). The elements of such notes and how to ensure they have a genuinely beneficial impact on patients' care (rather than being a nice looking/sounding piece of additional information that just adds to the documentation burden of the staff) remain to be worked out (Ozbolt et al., 2014).

Ensure That Note Is Error/Defect-Free

Unintended negative consequences are the rule with any technology. Just as medication ordering introduced new types of errors with the implementation of computerized prescriber order entry (Schiff et al., 2015), ECD has led to the proliferation of inaccurate or out-of-date information that has in turn can cause harm. For example, clinicians can lose sight of problems such as splenectomy with potentially lethal consequences (Gandhi et al., 2011). Fortunately these are rare events, but the potential for undermining the trustworthiness and confidence is a more serious and widespread problem. Six sigma (3.4 defects per 1 million parts) levels of quality is the standard in many industries. Yet, for example, wrong patient errors (ordering or documenting on the wrong patient) currently occur in roughly 5 in 10,000 patient orders: a rate one hundred times greater than this standard from other industries (Adelman et al., 2013; Yackel and Embi, 2006; Senathirajah et al., 2014). Beyond this quantitative measurable rate, are large qualitative impacts on patient safety culture whereby clinicians become tolerant and cynical about untruthful or inaccurate information in their notes. While there are promising approaches to error reduction (e.g., for preventing wrong-patient errors by including the patient's picture on screen; manual verification steps or error checking decision support), these errors are likely to worsen before they get better unless more attention is paid to the frequency, causes, workflow, and workaround issues related to error. Thus, we can see the intimate connections between the chief complaints (Table 4.1) and these potentials for errors.

KEY EMERGING ISSUES IN ECD

Copy/Paste

Clinicians are using this functionality to both fill notes with information to satisfy billing requirements (especially in the United States) as well as speed up note-writing. In spite of the widespread deploring of copy and paste practices (and even attempts to outright prohibit or disable this functionality), we would like to suggest a more thoughtful and, frankly, balanced view of copy and paste practices (Schiff and Bates, 2010; Yackel and Embi, 2006; Senathirajah et al., 2014; Tsou et al., 2017). Yes, two cheers for the copy and paste! Positive benefits include the ability to efficiently carry forward unchanged information, avoid errors in transcribing key information, and efficiently create the structure and populate content for new notes. We view complaints about copy/paste as being more directed at the symptoms rather than the causes of poor EHR notes, usability, and charting practices. Because copy/paste is a workaround for the limitations in data entry and workflow, one can hardly blame users for taking advantage of one of most efficient tools we use every day in word processing or editing e-mails. Thus, we would urge redirecting some of the disparagement of copy/paste toward more nuanced thinking about poor EMR usability and focus on how to reengineer note production to make it radically more efficient, particularly along the lines we discuss below.

Easing Entry: Scribes (Team Documentation) and Speech Recognition

How to best enter information gleaned from the clinical encounter into the computer is the subject of much angst and controversy. Two newer approaches are spreading and likely to transform the way many clinicians document: medical scribes (team members who enter notes and may also perform other functions) (Sinsky, 2016) and speech recognition.

A growing number of practices (currently one in five in the United States) using an EHR now use medical scribes (Schiff and Zucker, 2016). Given the added time burden for clinicians to write notes, the prospect of having someone support charting thereby allowing clinicians the opportunity to better interact with patients and finish their work in a more timely way is irresistible (Sinsky, 2016; Sinsky and Beasley, 2015). While hiring additional personnel to support this function can add cost, it has been shown to pay for itself in the form of an increase in the volume (perhaps one to two additional patients/session) or enhanced documentation for capturing billing revenue. But beyond cost and efficiency, more fundamental questions about scribes' role, training, effects on interactions with patients in the exam room, and qualitative aspects of note content and production that are important to consider and remain largely unexplored. From an informatics standpoint (the orientation of this book), liberating clinicians from having stare at the computer rather than concentrating on, looking at, and communicating directly with their patients is both a plus and minus. How exactly will the clinician who is no longer at the computer interact with decision support messages or navigate to critical test or past notes if a nonclinically trained scribe serves as a "middle-man." (Schiff and Zucker, 2016). As we have argued elsewhere, many of these minuses are surmountable, but we wonder how well a scribe-authored note captures the clinical nuances for the clinician's "free-text" thinking. Even finalizing notes in real time versus later has trade-offs as doing notes later may allow time for undistracted reflection.

Speech recognition represents another transformative tool for entering information into the clinical note (Grasso, 1995; Monnich and Wetter, 2000; Hodgson and Coiera, 2015). While it can and has been used at the "back end" as part of dictated/transcribed note processing, as a "front end" version of note dictation this tool allows real-time entry of text and can even use "commands" to navigate or create structured entries. One of us (GDS) has been using speech recognition to dictate clinical notes since the late 1990s, yet despite several decades of experience and experimentation still has not settled on a satisfactory way to efficiently and effectively integrate voice recognition into the clinical encounter or note production workflow. Another informatics colleague and leader (T. Payne, personal communication 7/2016) has resorted to jerry-rigging a solution using the speech recognition capabilities of the iPhone. Given the powerful and now highly accurate ability to capture narrative text for the subjective, assessment, and plan components of the notes, we expect speech recognition to be increasingly used. And with better creative integration into the clinical workflow we look forward to overcoming many of the challenges and issues surrounding its best use.

Problem-Based Charting

As advocated by Weed more than 50 years ago, organizing notes using a dynamically updated Problem List is integral to good quality clinical documentation and vital to the care of the patient. As he stated in his classic 1968 article, these are records that "guide and teach" (Weed, 1968b). A well-curated Problem List provides a "medical portrait" of the patient and is refreshed/repainted as the patient's clinical picture evolves. It has the flexibility to highlight (or put on the back burner) critical past diagnoses, enumerate current acute and chronic problems, and remind the clinician about items (e.g., pulmonary nodule) that need periodic follow-up. But electronic notes can take paper-based SOAP notes a step further, to embrace and experiment with the concept of electronic Problem-Based Charting.

At a minimum, Problem-Based Charting involves annotation of each problem on the Problem List, providing a capsule overview of that problem. For instance, for a patient with hypertension, the clinician might document the date of onset, medications tried (and failed), relevant family history, current medication regimen and efficacy, recent cardiac testing, and whether a specialist is involved. Done once and well, this overview serves as a quick reminder to the primary clinician, as well as cross-covering providers and specialists, of the key information relevant to that problem. This approach works particularly well for chronic medical problems (such as hypertension, diabetes, hyperlipidemia, thyroid disease) where the Overview is likely to remain fairly stable once created, and where updates to the current state and future plan are likely to be minor and easily made. Ideally specialists would help maintain problems under their care (rheumatology, for the patient's systemic lupus erythematosus).

As valuable as a "stand-alone" enriched problem list can be, continuously updating such annotations obviously entails addition clinician effort. A key leap would be to have these problems dynamically updated each visit as part of charting for that encounter. Thus, the clinician is not only interacting and being reminded about the patients' problems and history, but also updating their status including the problems, current assessment, and plan.

For instance, in the patient with hypertension, the clinician might document that the patient is adherent with his/her medication regimen (or not), document any side effects from medication, whether the patient is following a low salt diet, exercising, and maintaining his/her weight, and whether they have developed any new or worrisome cardiac symptoms. While this information has typically resided in the Subjective section of traditional SOAP notes, the benefit of integrating it with the Problem List is that it becomes visually and cognitively linked with the history and overview of that particular problem, enhancing the narrative value of the note, and creating continuity across notes, as well as (in the ideal redesigned EHR workflow of the future!) creating substantial documentation efficiencies in time and duplicative effort.

In the current iterations of EHRs, we have a glimpse of ways to add an assessment and plan directly linked to that particular problem. Some clinicians have adopted this functionality to create their own workflows and variants of problem-based charting. Given that the two authors, both committed in theory to this potentially better way of charting, one has been successful (MJT) while the other has failed (GDS) in attempting to do this in using a leading US commercial EHR (Epic), writing progress notes this way, must be considered "a work in progress." Other alternatives to episodic notes, such as a wiki type of model where multiple team members are making continuous smaller edits to an evolving master document should be part of our vision (Mehta et al., 2013).

Open Notes

As we have stressed, clinical notes need to be reconceptualized and used as much more than a paper chart repository documenting isolated clinical encounters. Instead, notes are dynamic living documents that serve to coordinate the efforts of a larger team involved in the patient's care. And we now recognize that team includes the patient him/herself, who plays a central role in multiple aspects of their own care. Thus, it is only logical to permit patients to access their notes. Until recently this idea seemed impossibly scary and fraught with insurmountable concerns, at least in the United States (several European countries have a long tradition of note sharing). Pioneering "Open Notes" efforts have shown that letting patients review their own notes has not led to serious feared problems and has also resulted in multiple positive benefits including giving patients the ability to correct errors they may detect and helping build understanding (patients forget much of what they are told verbally during encounter) and relationships (reassured to see what clinician has written validating their symptoms) (Delbanco et al., 2012; Walker et al., 2011; Bell et al., 2015). More than 10 million US patients now have access to their "Open Notes" and even psychiatry notes have now been opened to patients at selected institutions (see Chapter 18: Social and Consumer Informatics). We envision a time when patient will interactively "assess" our assessments giving two-way feedback, our diagnostic assessments and their downstream diagnostic outcomes to enhance our individual and collective learning as well as learning better how to improve the quality of this important component of clinical notes. Thus, making notes both accessible and more understandable to patients is an important challenge and goal.

CONCLUSIONS—VISION FOR FUTURE

We imagine a time when ECD will move beyond simply a computerized version of the paper note, and instead leverage the advanced capabilities that transformative electronic technology offers. At the same time, producing notes will cease to be an extra time-consuming chore and instead become a joyful, efficient, interactive process that will produce much added value while taking less of the clinician's time while more productively engaging the entire health care team. Such notes will not only be searchable, but lend themselves to radically be reorganized, displayed, and aggregated across encounters and across patients for better care, research, learning, and improvement. A research agenda starting with small-scale experimentation, sharing and learning from each other's best practices, and innovations represents an important place to start today for tomorrow's notes.

REFERENCES

Adelman, J.S., Kalkut, G.E., Schechter, C.B., et al., 2013. Understanding and preventing wrong-patient electronic orders: a randomized controlled trial. J. Am. Med. Inform. Assoc. 20 (2), 305–310.

AHIMA, 2011. Fundamentals of the legal health record and designated record set. J. AHIMA 82 (2).

AHRQ, 2016. Develop a shared care plan. https://integrationacademy.ahrq.gov/playbook/develop-shared-care-plan (accessed 15.01.17).

Ash, J.S., Berg, M., Coiera, E., 2004. Some unintended consequences of information technology in health care: the nature of patient care information system-related errors. J. Am. Med. Inform. Assoc. 11 (2), 104–112.

Baker, A., Cronin, K.; Conway, P.H.; Desalvo, K.B.; Rajkumar, R.; Press, M.J., 2017. Relentless reinvention: making the comprehensive shared care plan a reality. http://catalyst.nejm.org/making-the-comprehensive-shared-care-plan-a-reality/ (accessed 19.02.17).

Bell, S.K., Folcarelli, P.H., Anselmo, M.K., Crotty, B.H., Flier, L.A., Walker, J., 2015. Connecting patients and clinicians: the anticipated effects of open notes on patient safety and quality of care. Jt Comm. J. Qual. Patient Saf. 41 (8), 378–384.

Commission TJ, 2016. Joint commission requirements. https://www.jointcommission.org/standards_information/tjc_requirements.aspx.

Cusack, C.M., Hripcsak, G., Bloomrosen, M., et al., 2013. The future state of clinical data capture and documentation: a report from AMIA's 2011, Policy Meeting. J. Am. Med. Inform. Assoc. 20 (1), 134–140.

Delbanco, T., Walker, J., Bell, S.K., et al., 2012. Inviting patients to read their doctors' notes: a quasi-experimental study and a look ahead. Ann. Intern. Med. 157 (7), 461–470.

Dick, R.S., Steen, E.B., Detmer, D.E., 1997. The Computer-Based Patient Record: An Essential Technology for Health Care. National Academies Press.

Donabedian, A., 1966. Evaluating the quality of medical care. Milbank Mem. Fund Q. 44 (3), 166–206, Suppl.

Edwards, S.T., Neri, P.M., Volk, L.A., Schiff, G.D., Bates, D.W., 2014. Association of note quality and quality of care: a cross-sectional study. BMJ Qual. Saf. 23 (5), 406–413.

El-Kareh, R., Hasan, O., Schiff, G.D., 2013. Use of health information technology to reduce diagnostic errors. BMJ Qual. Saf. 22 (Suppl. 2), ii40–ii51.

Friedberg, M.W., Chen, P.G., Aunon, F.M., et al., 2013. Factors Affecting Physician Professional Satisfaction and Their Implications for Patient Care, Health Systems, and Health Policy. Rand Corporation.

Galanter, W.L., Hier, D.B., Jao, C., Sarne, D., 2010. Computerized physician order entry of medications and clinical decision support can improve problem list documentation compliance. Int. J. Med. Inform. 79 (5), 332–338.

Gandhi, T.K., Zuccotti, G., Lee, T.H., 2011. Incomplete care—on the trail of flaws in the system. N. Engl. J. Med. 365 (6), 486–488.

GoodReads. Mark Twain Quotes. http://www.goodreads.com/author/quotes/1244.Mark_Twain (accessed 23.08.16).

Grasso, M.A., 1995. Automated speech recognition in medical applications. MD Comput. 12 (1), 16–23.

Hägglund, M., Chen, R., Koch, S., 2011. Modeling shared care plans using CONTsys and openEHR to support shared homecare of the elderly. J. Am. Med. Inform. Assoc. 18 (1), 66–69.

Hanauer, D.A., Branford, G.L., Greenberg, G., et al., 2016. Two-year longitudinal assessment of physicians' perceptions after replacement of a longstanding homegrown electronic health record: does a J-curve of satisfaction really exist? J. Am. Med. Inform. Assoc.ocw077.

Hartzband, P., Groopman, J., 2008. Off the record—avoiding the pitfalls of going electronic. N. Engl. J. Med. 358 (16), 1656–1658.

Healy T, JD–AMA VP, Pastoor SJ, MHA—Director PCC. Christine Sinsky, MD AMA, Medical Associates Clinic and Health Plans. https://www.stepsforward.org/modules/team-documentation.

Henry JW ST, Vaisali P., 2016. Adoption of Electronic Health Record Systems Among U.S. Non-Federal Acute Care Hospitals: 2008–2015.

Hillestad, R., Bigelow, J., Bower, A., et al., 2005. Can electronic medical record systems transform health care? Potential health benefits, savings, and costs. Health Aff. 24 (5), 1103–1117.

Hodgson, T., Coiera, E., 2015. Risks and benefits of speech recognition for clinical documentation: a systematic review. J. Am. Med. Inform. Assoc.ocv152.

Holmes, C., 2011a. The problem list beyond meaningful use: part I: the problems with problem lists. J. AHIMA 82 (2), 30–33.

Holmes, C., 2011b. The problem list beyond meaningful use: part 2: fixing the problem list. J. AHIMA 82 (3), 32–35.

Krist, A.H., Beasley, J.W., Crosson, J.C., et al., 2014. Electronic health record functionality needed to better support primary care. J. Am. Med. Inform. Assoc. 21 (5), 764–771.

Kuhn, T., Basch, P., Barr, M., Yackel, T., 2015. Clinical documentation in the 21st century: executive summary of a policy position paper from the American College of Physicians. Ann. Intern. Med. 162 (4), 301–303.

Lin, C.-T., McKenzie, M., Pell, J., Caplan, L., 2013. Health care provider satisfaction with a new electronic progress note format: SOAP vs APSO format. JAMA Intern. Med. 173 (2), 160–162.

Martin, S.A., Sinsky, C.A., 2016. The map is not the territory: medical records and 21st century practice. Lancet 388 (10055), 2053–2056.

Mehta, N., Vakharia, N., Wright, A., 2013. EHRs in a web 2.0 world: time to embrace a problem-list Wiki. J. Gen. Intern. Med. 29 (3), 434–436.

Monnich, G., Wetter, T., 2000. Requirements for speech recognition to support medical documentation. Methods Inf. Med. 39 (1), 63–69.

National Academies of Sciences Engineering Medicine, 2015. Improving diagnosis in health care. National Academies Press, Washington, DC. http://iom.nationalacademies.org/reports/2015/improving-diagnosis-in-healthcare (accessed 20.11.15).

Ozbolt, J., Bakken, S., Dykes, P.C., 2014. Patient-centered care systems. Biomedical Informatics. Springer, pp. 475–501.

Payne, T.H., Graham, G., 2006. Managing the life cycle of electronic clinical documents. J. Am. Med. Inform. Assoc. 13 (4), 438–445.

Payne, T.H., Corley, S., Cullen, T.A., et al., 2015. Report of the AMIA EHR 2020 task force on the status and future direction of EHRs. J. Am. Med. Inform. Assoc.ocv066.

Pollard, S.E., Neri, P.M., Wilcox, A.R., et al., 2013. How physicians document outpatient visit notes in an electronic health record. Int. J. Med. Inform. 82 (1), 39–46.

Pullen E., 2010. APSO needs to replace SOAP in EMRs.

Reiser, S.J., 1991a. The clinical record in medicine part 1: learning from cases. Ann. Intern. Med. 114 (10), 902–907.

Reiser, S.J., 1991b. The clinical record in medicine part 2: reforming content and purpose. Ann. Intern. Med. 114 (11), 980–985.

Rosenbloom, S.T., Denny, J.C., Xu, H., Lorenzi, N., Stead, W.W., Johnson, K.B., 2011. Data from clinical notes: a perspective on the tension between structure and flexible documentation. J. Am. Med. Inform. Assoc. 18 (2), 181–186.

Schiff, G.D., Bates, D.W., 2010. Can electronic clinical documentation help prevent diagnostic errors? N. Engl. J. Med. 362 (12), 1066–1069.

Schiff, G.D., Zucker, L., 2016. Medical scribes: salvation for primary care or workaround for poor EMR usability? J. Gen. Intern. Med. 31 (9), 979–981.

Schiff, G.D., Hickman, T.-T.T., Volk, L.A., Bates, D.W., Wright, A., 2015. Computerised prescribing for safer medication ordering: still a work in progress. BMJ Qual. Safe.bmjqs-2015-004677.

Senathirajah Y, Kaufman DR, Bakken S., 2014. Beyond copy and paste: clinician approaches to meeting information needs during note writing. Paper presented at MIE 2014.

Sinsky CA., 2016. Team documentation. *Steps Forward.* https://www.stepsforward.org/modules/team-documentation (accessed 19.02.17).

Sinsky, C.A., Beasley, J.W., 2015. Medical scribes and electronic health records. JAMA 314 (5), 518–519.

System CSRH, 2011. Joint Commission Medical Record Documentation Requirements 2011.

Toll, E., 2012. The cost of technology. JAMA 307 (23), 2497–2498.

Tsou, A.Y., Lehmann, C.U., Michel, J., Solomon, R., Possanza, L., Gandhi, T., 2017. Safe practices for copy and paste in the EHR. Systematic review, recommendations, and novel model for health IT collaboration. Appl. Clin. Inform. 8 (1), 12–34.

Walker, J., Leveille, S.G., Ngo, L., et al., 2011. Inviting patients to read their doctors' notes: patients and doctors look ahead: patient and physician surveys. Ann. Intern. Med. 155 (12), 811–819.

Weed, L.L., 1968a. Medical records that guide and teach. N. Engl. J. Med. 278 (11), 593–600.

Weed, L.L., 1968b. Medical records that guide and teach. N. Engl. J. Med. 278 (12), 652–657.

Weed, L.L., 1997. New connections between medical knowledge and patient care. BMJ 315 (7102), 231–235.

Wright, A., Pang, J., Feblowitz, J.C., et al., 2012. Improving completeness of electronic problem lists through clinical decision support: a randomized, controlled trial. J. Am. Med. Inform. Assoc. 19 (4), 555–561.

Wright, A., Sittig, D.F., McGowan, J., Ash, J.S., Weed, L.L., 2014. Bringing science to medicine: an interview with Larry Weed, inventor of the problem-oriented medical record. J. Am. Med. Inform. Assoc. 21 (6), 964–968.

Wright, A., McCoy, A.B., Hickman, T.-T.T., et al., 2015. Problem list completeness in electronic health records: a multi-site study and assessment of success factors. Int. J. Med. Inform. 84 (10), 784–790.

Yackel, T.R., Embi, P.J., 2006. Copy-and-paste-and-paste. JAMA 296 (19), 2315–2316.

Chapter 5

Interoperability

Mark E. Frisse

Vanderbilt University Medical Center, Nashville, TN, United States

AN OPERATIONAL DEFINITION FOR INTEROPERABILITY

Interoperability describes the extent to which two or more computer systems can act as if all data, computation, and presentation can appear to be the product of a single "interoperable" system. Ideally, truly interoperable systems possess several characteristics: they would allow any one component of a system to be substituted by a different component serving the same functions; they would enable two-directional communication; they would be based on commonly accepted data standards; they would ensure accurate data transmission; and they would maintain data security. Until recently, interoperability has largely been the concern of technologists; clinicians— and even many clinical informaticians—generally were not aware of the complexities required to make two or more systems interoperable.

THE INTEROPERABILITY IMPERATIVE

Interoperability is both a clinical and an economic imperative. Almost all organizations involved in the delivery and financing of health care must make more effective use of every available clinical and administrative data source. Each of these organizations may require the same data, but each may use these data for very different reasons: to provide more effective care, clinicians need comprehensive longitudinal data sets; to be more accountable stewards, health plans and other payers need clinical data complementing their traditional administrative claims sets; to fulfill their responsibilities, public health professionals, researchers, and policy makers also seek more comprehensive integrated data sets. Very soon, these organizations will have to find ways to collect, integrate, and incorporate into their systems a growing number of new data types emerging from an array of genetic tests, mobile devices, home sensors, and other innovative technologies. Innovative use of these newer data sources is essential for deeper understanding and more effective clinical

Key Advances in Clinical Informatics. DOI: http://dx.doi.org/10.1016/B978-0-12-809523-2.00005-4

intervention (see Chapters 8 and 11: Health Information Technology and Value; Bioinformatics and Precision Medicine).

WAYS TO EXCHANGE INFORMATION

Both collaborators and competitors recognize through mutual self-interest a need for greater data sharing. The scope of data exchange may vary. Exchange can occur within a small and tightly coupled network or it can expand to regions, states, and even national initiatives. Many of these broader initiatives create health information exchange (HIE) organizations responsible for maintaining data governance, data standards, and technologies. Maryland's CRISP and Indiana's IHIE are exemplars of this approach (CRISP, 2017; IHIE, 2017). In settings where costs are prohibitive and data exchange needs are lower, data can be transmitted and received through point-to-point methods. Examples include The Sequoia Project (formerly the Nationwide Health Information Network) (The Sequoia Project, 2016), DirectTrust (DirectTrust, 2017), the CommonWell Health Alliance (CommonWell Health Alliance, 2017), and vendor-driven initiatives like Epic Systems' Care Everywhere (Epic Systems, 2017).

AN INTEROPERABILITY FRAMEWORK

The march toward more completely interoperable systems continues. The Office of the National Coordinator for Health IT (ONC), in a 2015 Shared Nationwide Interoperability Roadmap, outlines a three-stage developmental process spanning 10 years (Table 5.1) (ONC, 2015). The first stage addresses the effective exchange of high-value data in order to improve health care quality and outcomes. This stage focuses on the efficiency and effectiveness of patient-centered clinical care. The second stage recognizes the growing variety of data elements of value to clinicians, researchers, public health, human, and community-based services. The goal of this stage is to expand both the types of data made available and the variety of individuals who can make use of these data. The third and final stage is more aspirational: the realization of an ever-evolving learning healthcare system that continuously improves care, public health, and science through real-time data access (Smith et al., 2013).

The stages proposed by ONC are not necessarily sequential; many organizations are of necessity pursuing each to some degree. Nevertheless, disciplined prioritization both in development and implementation is essential; ONC suggests that near-term work should focus on priority data domains that are most commonly used and most often represented in format standards. This will require a prioritization of data element candidates and concerted efforts to ensure standards are applied consistently across multiple standards sets and use domains (Table 5.2). Such steps will enable data to be

TABLE 5.1 Goals of the Current ONC Interoperability Roadmap

2015–17	Send, receive, find, and use priority data domains to improve health care quality and outcomes. This will enable clinical care providers to collect data elements once and use them for a variety of purposes, including sharing with individuals, sending during referrals, and contributing to quality measurements.
2018–20	Expand data sources and users in the interoperable HIT ecosystem to improve health and lower costs. A broader range of data elements will be available to researchers, public health, human, and community-based services.
2021–24	Develop learning health systems that place the person at the center of a system that can continuously improve care, public health, and science through real-time data access. Specific milestones and goals will depend on how needs and capabilities evolve.

The ONC Roadmap identifies near-term, intermediate-term, and long-term actions and roles necessary for immediate and sustained progress toward more interoperable systems. The report builds on technologies and investments made to date while encouraging new ways to support innovation so that the centrality of the EHR can be complemented or supplanted by a wider range of data sources and technologies used by individuals, providers, and researchers.
Source: Office of the National Coordinator for Health Information Technology, 2015. Connecting Health and Care for the Nation: A Shared Nationwide Interoperability Roadmap. Final Version 1.0. https://www.healthit.gov/sites/default/files/hie-interoperability/nationwide-interoperability-roadmap-final-version-1.0.pdf.

TABLE 5.2 ONC's Near-Term Priority Data Domains

- Individual Name
- Sex
- Date of Birth
- Race
- Ethnicity
- Address (Current, Historical)
- Phone Number (Current, Historical)
- Preferred Language
- Smoking Status
- Problems
- Medications
- Medication Allergies
- Laboratory Test(s)
- Laboratory Values & Result Reporting
- Vital Signs
- Procedures
- Care Team Members
- Immunizations
- Unique Device Identifier(s) for Implantable Device(s)
- Assessment and Plan of Treatment
- Goals
- Health Concerns

According to ONC, near-term work to advance semantic interoperability should focus on priority data domains that are most commonly used across many clinical and nonclinical use cases and most often represented in format standards. This will require a prioritization of data element candidates and concerted efforts to ensure standards are applied consistently across multiple standards sets and use domains.
Source: Office of the National Coordinator for Health Information Technology, 2015. Connecting Health and Care for the Nation: A Shared Nationwide Interoperability Roadmap. Final Version 1.0. https://www.healthit.gov/sites/default/files/hie-interoperability/nationwide-interoperability-roadmap-final-version-1.0.pdf.

collected once and used for many purposes and should create efficiencies in electronic health record (EHR) information sharing to a far greater extent than is the present state.

Across the globe, countries differ in their degree of interoperability and the means by which they exchange health information. Although most countries adopting EHR technologies share in certain international standards and practices, their approaches toward interoperability have differed in ways that best align with each countries' technology infrastructure and the means by which health care is financed. In general, the more centralized or uniform the technology infrastructure, the fewer interoperability challenges. Similarly, more uniform reimbursement systems employed in other countries may simplify implementations. Some efforts have been made to compare information technologies across Organization for Economic Cooperation and Development countries (Zelmer et al., 2016; Magrabi et al., 2013). Examples from other countries specifically devoted to HIE include Canada's Health Infoway, South Korea, and New Zealand's HIE initiatives (Infoway, 2017; Park et al., 2015; Park and Atalag, 2015).

TECHNICAL CONSIDERATIONS

More technically, interoperability is the result of three incremental capability levels. At the lowest level, systems transmitting and receiving data from one another must conform to standardized communication protocols. Examples of such protocols are Hypertext Transfer Protocol (the Web's widely familiar HTTP), Simple Object Access Protocol, and the increasingly popular Representational State Transfer method. These communication protocols are examples of foundational interoperability (HIMSS, 2013).

At a middle level, interoperable systems demonstrate syntactic interoperability. Systems designers must arrive at conventions for the data, message, or document structure. These conventions are defined through formal computer grammars specifying both data elements and ways in which these elements can or must be present and related to one another. Presentation of a simple Web page exemplifies this process. The Web page can be accessed through the "foundational" HTTP and represented in Hypertext Markup Language (HTML) that can be displayed through a Web browser. Perhaps the simplest expression would be a simple presentation of a text message. A slightly more complex document may have elements called "title," "author," and "content." A separate document type definition (DTD) defines the document structure with a list of legal elements and attributes. When combined, the grammar and the DTD ensure that messages are complete for the task at hand. A patient identification record, for example, may require a name, an address, and a date of birth to be present; records without these three elements would be rejected. Each of these fields in turn could be decomposed through additional rules. Syntactic rules, for example, could require a name to be composed of a family name, a given first name, and an optional middle initial or name. However defined, one needs only minimal syntax requirements to present data through a Web browser. The meaning or semantics of whatever is presented is left to the user's interpretation; the same syntactically interoperable framework could be used to present gibberish or the text of Lincoln's Gettysburg Address. Transmitting content that is only minimally structured but still meaningful to the user of a second system is the simplest form of semantic interoperability—narrative interoperability (Dolin and Alschuler, 2011).

At the highest level are the methods to ensure that two systems share a common meaning for specific terms and data elements—semantic interoperability. Retaining meaning is essential if data from disparate sources are to be combined and viewed or analyzed collectively. Many operational clinical information systems or health information exchange systems began with relatively simple syntactic and semantic rules (Stead et al., 2005; Frisse et al., 2012; Dixon, 2016).

Our appetite for more data types and our need to develop new systems to analyze and act on data will necessitate even deeper forms of syntactic and semantic expression. For example, effective electronic prescribing (e-prescribing) requires elaboration upon both expression forms. The structure of an electronic prescription must conform to a specific grammar that denotes data record elements and both necessary and allowable relationships among elements. The National Council for Prescription Drug Programs' SCRIPT standard for new prescriptions requires an explicit means of identifying the patient, the prescriber, the pharmacy, the drug, the strength, the dosage, the days of medication supplied, and a wide range of other clinical and administrative information. Some elements are essential and must be properly formed (e.g., the drug, the dose, the quantity, and the means of administration) and other elements are optional or only infrequently used. However, unless the message conforms to the standard specified through the grammar, it is not syntactically interoperable and will not be accepted.

E-prescribing systems also must take into account the numerous terms used to express similar concepts and how terms relate to one another. Prescribing clinicians and dispensing pharmacists may seek relationships between various brand and generic drugs when considering substituting one therapeutically equivalent drug with another. Every party in the prescribing—dispensing cycle relies on drug—drug and drug—allergy clinical decision support rules based on ingredients present in prescription drugs. RxNorm expresses drug semantics necessary for the wide range of activities surrounding prescription drugs and is the chosen standard for Meaningful Use certification (Nelson et al., 2011). It provides a set of names and relationships based on the drug vocabularies commonly used in pharmacy management and drug interaction software. Similar semantic standards have been employed and are often required for a range of other expressions. SNOMED

CT (Systematized Nomenclature of Medicine—Clinical Terms) can be employed for problem lists and procedures. ICD-10 (International Classification of Diseases—version 10) is often used for encounter diagnoses, and LOINC (Logical Observation Identifiers Names and Codes) is used to express laboratory test results.

NEWER STANDARDS AND APPROACHES

Over the last decade the maturation of data standards, technologies, and policies have made interoperable systems more commonplace. HL7 (Health Level Seven International) Version 3 has emerged as a comprehensive reference implementation model capable of providing clinical data in a comprehensive clinical context. For clinical activities, it declares three types of classes (Entity, Role, and Act) and a number of relationships defining these acts. HL7 Version 3 is the basis for the Clinical Document Architecture (CDA) (HL7, 2016). Returning to the spirit HTTP, the CDA is based on eXtensible Markup Language (XML). XML is a highly flexible representation form that can be parsed by machines and read by the human eye. As is the case with HTML, XML allows for both syntactic constraints and the flexibility for expressions ranging from free-form text to highly structure data. Each element in turn can refer to the specifications defining its use as data standards evolve. Increasingly, clinical communications are represented using the CDA. Complementing these efforts, organizations like Integrating the Healthcare Enterprise (IHE) bring together key stakeholders to develop a consensus on systems requirements. These organizations do not develop standards, but instead describe how standards may be used effectively. Their recommendations are expressed as "profiles" that help assure interoperable use of the CDA.

SMART AND FHIR

Fully interoperable electronic medical records systems are far from reality. Many HIE approaches achieve some degree standardized transmission and receipt of data sets, but true semantic interoperability exists only in limited settings and EHR components are certainly not substitutable. In the United States, some standards specified by Meaningful Use regulations seem excessively complex, incompletely specified, or inconsistent in their implementations. As a result, many "standard" implementations have not yet realized their intended degree of interoperability. In part as a reaction to these complexities, a growing number of vendors and developers are creating "apps" using an updated version of SMART (Substitutable Medical Apps, Reusable Technology) that exploits the data models and application programming interface specified by less complex, openly licensed HL7 standard called

Fast Health Interoperability Resources (FHIR) (see Chapter 16: An Apps-Based Information Economy in Healthcare) (Mandel et al., 2016).

OMICS

Scientifically, the success of both "omics" and population health research initiatives depends critically on the consistent expression of vast data sets derived from multiple different settings (see Chapters 11 and 15: Bioinformatics and Precision Medicine; Predictive Analytics and Population Health). Already, genetic information concerning drug metabolism has been incorporated into clinical decision support systems (Pulley et al., 2012; Peterson et al., 2016). Research use also drives interoperability. The Patient-Centered Clinical Research Network initiative exemplifies the application of interoperability standards for population health research (Collins et al., 2014). The Precision Medicine Initiative will push the boundaries of interoperability both through the evolution of standards to represent biological data and equally pressing needs to represent the notion of a patient "phenotype" in terms of clinical condition, social determinants, and new means of measurement afforded through mobile and wearable devices (Frey et al., 2016; Tenenbaum et al., 2014). The nature of this work will place even greater emphasis on a need to maintain public trust and to respect individual privacy preferences (Hudson and Collins, 2017).

Continual Evolution

Clinical needs and financial constraints necessitate sustained efforts to advance interoperability. Clinically, consumers are seeking alternative approaches to their care through retail pharmacies, stand-alone clinics, and a growing catalog of health care products relying on mobile, "wearable," and home-based technologies. These shifts provide both the opportunity and the necessity of a broader view of interoperability to ensure that every new consumer health information technology (HIT) employed will be interoperable with every other component of our health care system. Thousands of health care "apps" are available through commercial sites, and hundreds of mobile sensor systems are being designed to work with watches or other wearable technologies. New home monitoring and communication systems are being introduced by well-known security firms, telecommunication companies, and consumer technology companies seeking to integrate entertainment, communication, home management, and health into one unified platform. Although these markets are only beginning to form, interoperability must be a primary consideration as these systems strive to communicate with medical devices and with health care providers. Managerially, the growing need for secondary use of clinical and administrative data places interoperability in a new spotlight. Operationally, Medicare payment reform and other population health management efforts cannot reach their potential without secure, privacy-preserving, interoperable systems.

CONCLUSIONS

Complete interoperability is likely to remain an elusive goal. However, as advances in technology create the capability to seamlessly and affordably measure, integrate, and understand data in non-healthcare settings, the public will expect the same from our health care technology infrastructure. As demographic and financial constraints placed on our health care system reach crisis levels, new and innovative approaches to collaborative data sharing and care collaboration based on interoperable systems will be to some, literally a matter of life and death.

REFERENCES

Collins, F.S., et al., 2014. PCORnet: turning a dream into reality. J. Am. Med. Inform. Assoc. 21 (4), 576–577.

CommonWell Health Alliance, 2017 [cited 2017 January 10]. Available from: http://www.commonwellalliance.org/.

CRISP, 2017. Home Page: Chesapeake Regional Information System for our Patients [cited 2017 January 10]. Available from: https://www.crisphealth.org/.

DirectTrust, 2017. The Direct Trust Home Page [cited 2017 January 10]. Available from: https://www.directtrust.org/.

Dixon, B., 2016. Health Information Exchange: Navigating and Managing a Network of Health Information Systems. Elsevier, Boston, MA.

Dolin, R.H., Alschuler, L., 2011. Approaching semantic interoperability in Health Level Seven. J. Am. Med. Inform. Assoc. 18 (1), 99–103.

Epic Systems, 2017. The Epic Systems Care Everywhere Network [cited 2017 January 10]. Available from: https://www.epic.com/careeverywhere/.

Frey, L.J., Bernstam, E.V., Denny, J.C., 2016. Precision medicine informatics. J. Am. Med. Inform. Assoc. 23 (4), 668–670.

Frisse, M.E., et al., 2012. The financial impact of health information exchange on emergency department care. J. Am. Med. Inform. Assoc. 19 (3), 328–333.

HIMSS, 2013. What is Interoperability? [cited 2016 June 1]. Available from: http://www.himss.org/library/interoperability-standards/what-is.

HL7, 2016. CDA Release 2 [cited 2016 June 1]. Available from: http://www.hl7.org/implement/standards/product_brief.cfm?product_id=7.

Hudson, K.L., Collins, F.S., 2017. The 21st Century Cures Act—a view from the NIH. N. Engl. J. Med. 376 (2), 111–113.

IHIE, 2017. Indiana Health Information Exchange: Home Page [cited 2017 January 10]. Available from: http://www.ihie.org/.

Canada Health Infoway, 2017. Health Infoway Home Page [cited 2017 January 10]. Available from: https://www.infoway-inforoute.ca/en/component/tags/tag/1166-health-information-exchange.

Magrabi, F., et al., 2013. A comparative review of patient safety initiatives for national health information technology. Int. J. Med. Inform. 82 (5), e139–e148.

Mandel, J.C., et al., 2016. SMART on FHIR: a standards-based, interoperable apps platform for electronic health records. J. Am. Med. Inform. Assoc. 23 (5), 899–908.

Nelson, S.J., et al., 2011. Normalized names for clinical drugs: RxNorm at 6 years. J. Am. Med. Inform. Assoc. 18 (4), 441–448.

ONC, 2015. Connecting Health and Care for the Nation: A Shared Nationwide Interoperability Roadmap Final Version 1.0. Office of the National Coordinator for Health Information Technology, Washington, DC.

Park, H., et al., 2015. Can a health information exchange save healthcare costs? Evidence from a pilot program in South Korea. Int. J. Med. Inform. 84 (9), 658–666.

Park, Y.-T., Atalag, K., 2015. Current national approach to healthcare ICT standardization: focus on progress in New Zealand. Healthcare Inform. Res. 21 (3), 144–151.

Peterson, J.F., et al., 2016. Attitudes of clinicians following large-scale pharmacogenomics implementation. Pharmacogenomics J. 16 (4), 393–398.

Pulley, J.M., et al., 2012. Operational implementation of prospective genotyping for personalized medicine: the design of the Vanderbilt PREDICT project. Clin. Pharmacol. Ther. 92 (1), 87–95.

Smith, M. (Ed.), 2013. Best Care at Lower Cost: The Path to Continuously Learning Health Care in America. Committee on the Learning Health Care System in America; Institute of Medicine.

Stead, W.W., Kelly, B.J., Kolodner, R.M., 2005. Achievable steps toward building a national health information infrastructure in the United States. J. Am. Med. Inform. Assoc. 12 (2), 113–120.

Tenenbaum, J.D., Sansone, S.-A., Haendel, M., 2014. A sea of standards for omics data: sink or swim? J. Am. Med. Inform. Assoc. 21 (2), 200–203.

The Sequoia Project, 2016. The Sequoia Project [cited 2017 January 12]. Available from: http://sequoiaproject.org/.

Zelmer, J., et al., 2016. International health IT benchmarking: learning from cross-country comparisons. J. Am. Med. Inform. Assoc.

FURTHER READING

Commission on Systemic Interoperability, 2005. Ending the Document Game.

Thompson, T.G., Brailer, D.J., 2004. The Decade of Health Information Technology: Delivering Consumer-Centric and Information-Rich Health Care Framework for Strategic Action. Department of Health and Human Services, Washington, DC, p. 178.

Chapter 6

Privacy and Security

John D. Halamka

Beth Israel Deaconess Medical Center and Harvard Medical School, Boston, MA, United States

INTRODUCTION

Not a day goes by without a headline about a major privacy breach or cyber-security attack. These issues are not unique to healthcare and over the past 5 years major retail and financial institutions have been compromised. Increasingly, Boards of Directors are aware of the monetary penalties for breaches and the reputational damage they create. Healthcare organizations work hard to gain the trust of the patients. A single major security event can destroy years of good will.

Historically, healthcare organizations have spent up to 2% of their budgets on health information technology (HIT). Security spending has been less than 5% of that 2%. Hackers, seeking financial information for identity left, are increasingly finding healthcare organizations to be attractive targets because they store millions of personal identifiers such as social security numbers and their limited spending on security makes them easily penetrable.

Retail and banking organizations have bolstered their defenses and now is the time for healthcare to do the same. In this chapter, we will discuss a risk-based approach to reducing privacy and security vulnerabilities.

WHAT IS PRIVACY AND SECURITY?

Privacy is defined as the claim of individuals, groups, or institutions to determine when, how, and to what extent information about them is communicated to others (Westin, 1967).

Security is defined as the technical and policy protections put in place to help protect privacy and data integrity.

The focus of this chapter will be those technical and policy approaches that are best practices defined in widely accepted formal guidelines that mitigate the ever-increasing threats facing healthcare organizations every day.

Key Advances in Clinical Informatics. DOI: http://dx.doi.org/10.1016/B978-0-12-809523-2.00006-6

WHAT ARE THE THREATS?

Here are some case examples of the kinds of threats that have been experienced in Harvard-affiliated healthcare organizations over the past 5 years.

The Lost Laptop: Dr. Famous bought a laptop at the Boston Apple Store. He brought it back to his desk, where he downloaded 100,000 historical e-mails. He then went to a meeting and left the laptop unattended. When he returned to his office, the laptop was gone. The cost of investigation, patient notification, credit reporting, legal fees, and media management exceeded $500,000.

The Compromised Radiology Workstation: A major imaging vendor sent a technician to maintain a radiology image reading workstation. The technician connected this workstation to the Internet in order to download a software patch. He went to lunch and by the time he returned, malware had infected the device and compromised 2000 radiology images. The cost of investigation, patient notification, credit reporting, legal fees, and media management exceeded $300,000.

The Anonymous Attack: On the eve of the Boston Marathon in 2014 (1 year after the bombing), a hacking group called Anonymous decided to attack Boston area hospitals as part of a "social justice" action involving a patient. Although the attack was intended to affect one hospital, it eliminated inbound and outbound Internet communications for many hospitals and academic institutions in the Boston area. Service was restored by the time the marathon started, but US Homeland Security and the Federal Bureau of Investigation had to be engaged.

The Phishing Experience: If you received an e-mail stating "Your password is now changing" and providing the link "http://changemypassword-now.com," would you click on it? Unbelievably about 30% of people in a Harvard-related institution receiving such an e-mail did click on such a phishing attacks. Organizations are as weak as their most gullible employee (Social Engineering Red Flags).

The Boston Marathon Issues: When the marathon bombings occurred in Boston, frightened people working in hospitals wanted to keep up-to-date about what was happening. The temptation to look up a record of a Boston marathon patient was high. No one did because all staff were reminded that anyone compromising privacy would be instantly fired. However, the risks were real and the communication campaign about protecting privacy was intense.

A FRAMEWORK FOR REDUCING SECURITY RISKS IN HEALTHCARE ORGANIZATIONS?

Given the real risks described above, what can organizations do to lessen their vulnerabilities? Although there are many formal frameworks for identifying risks—NIST 800, HITRUST, COBIT, ITIL, etc., the list below includes some of the most common areas described in these widely accepted frameworks.

Risk Management

It is important to have a standard process, often using external experts or auditors, to identify the most significant security risks facing an institution each year (Baker, 2016). At Beth Israel Deaconess Medical Center (henceforth Beth Israel Deaconess) our strategy to enumerate risks used to be a mini-retreat of experts, each suggesting the emerging threats we faced and their mitigations. Today, we have a much more formal approach and we use an industry standard framework to analyze risks.

There are many types of industry standard security frameworks. NIST 800 is a US government provided rubric for ensuring compliance with the Health Insurance Portability and Accountability Act (HIPAA) security rule. HITRUST is an industry created set of best practices for building a strong security program. ISO/IEC 27001:2005 is an international standard which specifies the requirements for establishing, implementing, operating, monitoring, reviewing, maintaining, and improving security management within the context of the organization's overall business risks.

Beth Israel Deaconess has chosen to outsource this risk analysis to an expert security firm at a cost of $100,000 per year.

Additionally, we have outsourced the response to major incidents so that investigation, government reporting, media management, customer notification, and communications are handled by experts using a predefined response plan.

Finally, each year we hire a third party "white hat" hacker firm to test the security of our applications by asking them to attack us.

Identity Management

Organizations are as vulnerable as their most gullible user. If a clinician using a public computer in a hotel has their credentials stolen, many patient records could be compromised (US Department of Health and Human Services Office for Civil Rights).

Thus, we must understand who can access what data from what locations for what reasons. We must immediately deactivate accounts when individuals cease employment or clinicians are no longer credentialed. We must limit access to patient records to the minimum need to know. Although clinicians may need to access medication lists and laboratory values, there is no reason that, for example, a registration clerk needs access to clinical data. At least yearly we must ask managers to specify the level of access each of their employees needs to do their job. If employees change job roles we should immediately change their access to reflect the new role. Most companies deactivate accounts when people leave, but few companies selectively remove access when job roles change.

Beth Israel Deaconess has chosen to use an authentication platform that enables multifactor authentication. If a user is using a known computer in a known location, a username/password is enough to protect privacy. If a user is using a public computer in a remote location, multiple factors are required such as a username/password plus a one-time code sent to the user's cell phone.

Logging and Monitoring

What if someone logs in from Boston at 8:00 a.m. and Shanghai at 9:00 a.m.? Given current travel times, such an interaction is not physically possible. What if a person usually logs in 5 times per day, but logs in 50 times one day? What if a person usually looks up 50 patients per day but looks up 500 patients one day?

Detecting these kinds of changes in behavior which are likely related to stolen credentials or hacking requires specialized software called a Security Information and Event Monitoring system. The idea is simple—logs from applications, servers, networks, firewalls, and storage devices are aggregated in one place where queries can be done looking for unusual behavior. Once discovered, accounts can be reset and attacks can be blocked (Herold and Beaver, 2015).

Information Security Program Governance, Policies, and Procedures

Who can access what for what reason and from where? What websites should be blocked? What incoming and outgoing e-mails should be blocked? All of these decisions need to be made by a multi-stakeholder information security governance group (Baker, 2015).

At Beth Israel Deaconess, we have an Information Security and Privacy Committee that meets monthly comprised of clinicians, administrators, legal experts, IT experts, and patients to answer such questions and develop institutional policies.

User Awareness Training

If you found a USB drive in an airport, would you insert it in your computer and use it to transfer files? If someone "from IT," called you and asked for your password, would you provide it? Would you use a cloud sharing service (see Chapter 17: Cloud-Based Computing) such as Dropbox to exchange patient identified information?

The problem of malware/ransomware, introduced by users clicking on inappropriate links, has required intense user training as well as tools that rewrite web addresses in e-mail so that they do not directly download any threatening payloads.

All of these scenarios represent significant risks to the organization, which many employees may not immediately recognize as compliance (and common sense) violations (HHS Office of Civil Rights). At Beth Israel Deaconess, we use stickers on packaged foods in the cafeteria, posters, and digital signage to constantly educate our workforce about such threats. Also, each year we provide online security training that includes a written examination. You must pass the training in order to keep your access to IT systems. Beth Israel has nearly doubled capital and operating spending for security projects over the past 3 years.

Managed Security Services Program

If a major security threat appears on the Internet, who is going to know about it first—the two or three expert security people working in the organization or the hundreds of security professionals working for a company that focuses on identifying/eliminating threats? Sometimes it is wiser to purchase security services rather than build and staff them yourself, as they have a scale of breadth of threat intelligence that no one organization can match. Beth Israel Deaconess purchases services from experts who ensure we follow technology and policy best practices (Herzig et al., 2013).

Web Application Security and Software Development Lifecycle

Do you allow employees to create apps or bring in apps from third-party developers? How do you qualify them as secure enough?

All Beth Israel Deaconess internally developed applications or third-party apps are rigorously evaluated by our security team to validate their protection of privacy and data integrity. Our developers are using common approaches and common toolsets to minimize vulnerabilities such as formal version control and testing.

Data Ownership, Classification, and Data Protection

Do you know where your data lives? Might the electronic health record data be replicated (at least partially) in a clinical registry somewhere? Who controls access to that registry? Some data in a registry may not be controversial—such as the recording of a flu shot. Other data such as HIV status, mental health treatment, or substance abuse history might be very controversial.

Classifying data as to degrees of sensitivity, understanding where it is stored and how it is protected is a challenging task. However, documenting what data organizations have and where it is kept is critical to any security program. The crown jewels are stored in the tallest tower in a locked case, not inside the unlocked front door. We do not store high-risk data outside physically security data centers.

Configuration Management

When you buy a piece of IT hardware, there are generally many configuration options. How do you balance convenience with security? For example, allowing traffic to flow unrestricted to and from the Internet is convenient for users. It is also very likely to cause a security breach every few minutes (Cornell University Law School).

Beth Israel Deaconess uses a third-party expert to compare the configuration of every network connected device to best practices and make adjustments if necessary.

Asset Management

Many IT organizations keep a log of equipment they purchase, but do they know when it is retired? If an auditor asked for current inventory of all devices purchased by the organization, where they were and who controlled them, could you produce such a list?

Beth Israel Deaconess has partnered with a cloud-based service provider to keep all records of all devices updated on a daily basis and to ensure that devices are tracked from "birth to death."

Third-Party Risk Management

Are your vendors trustworthy? How do you know? If you send data to them, do they keep it physically and logically protected?

We have no choice but to demand third-party audits of each vendor which hosts patient identified data to ensure they are following best practices (Clarke, 2014). The HIPAA Omnibus Rule, with its concept of a chain of accountability, requires us to have business associate agreements and a deep understanding of the security protections in place with each of our vendors.

Endpoint Security

Is it possible for a nurse to download an infected copy of Angry Birds to an Android Phone and then use that phone to look up patient information? Malware on phones can intercept keystrokes and screenshots, compromising security (Baker, 2012).

At Beth Israel Deaconess, we query every phone whenever it accesses BIDMC resources to ensure it is encrypted, free of malware, and configured to allow remote deletion of all data before we allow any information to flow to it.

Enterprise Resilience

You may have invested millions in information security and ensured all staff are trained to protect data. However, what happens when a flood, fire, earthquake, civil unrest, or other event occurs involving your IT infrastructure? Can you recover without loss of data?

To address this risk, Beth Israel Deaconess has multiple data centers with replication of all data to multiple geographic locations. We can lose an entire IT building and not have significant downtime or data loss.

Physical Security

You have done everything possible in the IT department to secure data. A known fellow walks into a clinic and takes a laptop. Do you have every device locked down? Do you have cameras? Do you have security guards, locked doors, and windowless data centers?

These simple physical protections are as important as technology protections.

There you have it—a 14-point plan to improve the robustness of healthcare IT security in any institution drawing on the experiences of Beth Israel Deaconess. Organizations do not have to be the best in the country at IT security, just good enough to repel the threats they encounter every day. Our experiences illustrate that it is possible to invest limited operating and capital wisely to achieve security that meets the community standard without breaking the bank.

CONCLUSIONS

Security is not a project, it is an ongoing process (Baker, 2015). As risks are mitigated, new risks arise. The work will never be done, but each year organizations will become more resilient. The most important first step is to build awareness of threats and identify risks. Once that is done, educating the workforce and funding specific security projects will put organizations on a positive trajectory to protect patient privacy.

REFERENCES

Baker, D.B., 2012. Health information privacy and security. In: Brown, G.D., Patrick, T.B., Pasupathy, K. (Eds.), Chapter 13 in Health Informatics: A Systems Perspective. Health Administration Press, pp. 261–280. , ISBN-10: 1567934358, ISBN-13: 978-1567934359.

Baker, D.B., 2015. Trustworthy systems for safe and private healthcare. In: Saba, V., McCormick, K. (Eds.), Chapter 10 in Essentials of Nursing Informatics, sixth ed. McGraw-Hill Education, ISBN-978-0-07-182955-7.

Baker, D.B., 2016. Framework for privacy, security and confidentiality. In: McCormick, K., Guharty, B. (Eds.), Chapter 24 in Healthcare Information Technology Exam Guide for CompTIA Healthcare IT Technician and HIT Pro Certifications. https://www.amazon.com/Healthcare-Information-Technology-Certifications-Certification/dp/0071802800.

Clarke, G.E., 2014. CompTIA Security + Certification Study Guide, second ed. Mc-Graw Hill, New York.

Cornell University Law School. 45 CFR 164.312—Technical safeguards. https://www.law.cornell.edu/cfr/text/45/164.312.

Herold, R., Beaver, K., 2015. The Practical Guide to HIPAA Privacy and Security Compliance. CRC Press, Boca Raton, FL.

Herzig, T.W., Walsh, T., Gallagher, L.S. (Eds.), 2013. Implementing Information Security in Healthcare: Building a Security. HIMSS, Chicago, IL, p. 86.

HHS Office of Civil Rights. Guidance to render unsecured protected health information unusable, unreadable, or indecipherable to unauthorized individuals. https://www.hhs.gov/hipaa/for-professionals/breach-notification/guidance/index.html.

Social Engineering Red Flags. http://www.albany.edu/its/images/SocialEngineeringRedFlags.pdf.

US Department of Health and Human Services Office for Civil Rights. Breaches affecting 500 or more individuals. https://ocrportal.hhs.gov/ocr/breach/breach_report.jsf.

Westin, A.F., 1967. Privacy and Freedom. Atheneum, New York, p. 7.

RECOMMENDED FURTHER READING

Baker, D., 2006. Privacy and security in public health: maintaining the delicate balance between personal privacy and population safety. Distinguished Practitioner Paper. Proceedings of the 2006 Annual Computer Security Applications Conference. IEEE.

Baker, D.B., Kaye, J., Terry, S.F., 2016. Privacy, fairness, and respect for individuals. eGEMs (Generating Evidence & Methods to improve patient outcomes) 4 (2), Article 7. Available from http://dx.doi.org/10.13063/2327-9214.1207. Available at: http://repository.edm-forum.org/egems/vol4/iss2/7.

Masys, D., Baker, D., Butros, A., Cowles, K.E., 2002. Giving patients access to their medical records via the internet: the PCASSO experience. J. Am. Med. Inform. Assoc. 9 (2), 181–191.

McMillan M., P. Cerrato, 2014. Healthcare data breaches cost more than you think. InformationWeek Healthcare Report. http://reports.informationweek.com/abstract/105/11839/Healthcare/Healthcare-Data-Breaches-Cost-More-Than-You-Think.html.

Murphy, S.A., 2015. Healthcare Information Security and Privacy. McGraw Hill Education, New York.

Part II

Improving the Quality, Safety and Efficiency of Care

Chapter 7

Public Policy and Health Informatics

David W. Bates[1] and Aziz Sheikh[1,2]

[1]*Brigham and Women's Hospital/Harvard Medical School, Boston, MA, United States,*
[2]*The University of Edinburgh, Edinburgh, United Kingdom*

INTRODUCTION: TYPES OF POLICIES

Many policy directions can have an effect on health information technology (HIT). Among these are selections of standards, implementation of standards, financial incentives, regulating the number of electronic health records (EHRs) in a domain, certifying electronic records, requiring quality reporting, and approaches to and regulations for data sharing. More broadly, when incentives that reward better performance exist, organizations can more easily make investments.

One central role relates to selection of standards (see Chapters 5 and 12: Interoperability; Clinical Decision Support and Knowledge Management). There are now acceptable standards for most of the main types of clinical data, including images, laboratory test results, and problems, as well as for messaging. Picking one of these per data type and then requiring adherence is enormously valuable to the healthcare system, though it can be clinically complex. In most domains, there are many standards to choose from, and often several might be reasonable choices, but it is a common good if only one is selected. With some exceptions, this has not yet occurred at an international level—and rationalization internationally would be very valuable.

Another key issue is refinement of existing standards. Standards represent "works in progress" and require continuous refinement. Much of this is done by standards development organizations, which generally do a very good job, but refinement goes much faster when there is public support for the process. Some standards are relatively mature, such as LOINC (McDonald et al., 2003), the most widely used laboratory standard, but others are much newer and require more work, such as RxNorm (NIH: National Library of Medicine), a standard for representing medications. In addition, some domains change faster than others.

Key Advances in Clinical Informatics. DOI: http://dx.doi.org/10.1016/B978-0-12-809523-2.00007-8

But even having an established standard for a specific data type is insufficient to ensure that data can be transferred; many of even the best standards need improvement at the margins. One approach for improving this involves doing some testing to ensure that data in a specific field are understandable. The United Kingdom has addressed this issue by requiring what is called "conformance testing," in which a sample of data are sent to a third party to make sure they are intelligible (Scott et al., 2015). Another is to provide support for refinement of standards which are not yet mature, which include most of the current standards.

Financial incentives can also play a key role. In many situations, even small financial incentives can be sufficient to get providers over a specific threshold. In instances where there is substantial misalignment—one stakeholder, such as the payer, is getting the benefit, but providers have to make the expenditures—larger financial incentives may be needed.

Another issue has been the number of EHRs in a given setting. If too many different EHRs are implemented, especially in the outpatient setting (see Chapter 3: Outpatient Clinical Information Systems), it can be very difficult to implement clinical data exchange. Thus, many countries have elected to limit the number of EHRs in the outpatient setting in any given region to three, enabling easier exchange of data, though this does not guarantee success.

Specific Examples of National HIT Policy Initiatives

Public policy can impact on the above and a range of other related issues that can have a major impact on the success (or otherwise) of HIT programs. Below, we consider some examples of national HIT policy initiatives.

Australia

Australia had a major effort focused on increasing adoption of EHRs in primary care (Australian Department of Human Services); secondary care has not yet received as much attention and adoption levels are lower in that sector. In primary care, key approaches taken included requiring that all bills be sent electronically, providing support for practices electing to adopt, including help with both software and hardware. This resulted in increases in adoption in the outpatient setting from 15% to approximately 70% over about a 3-year period (Kidd and Mazza, 2000; McInnes et al., 2006). Providers could select any EHR they wanted, but many picked an EHR developed by an Australian primary care provider. The Australian approach, which was heavily market-based, did have some issues; for example, providers received condition-specific advertisements for medications because much of the fee for the records was paid for by the pharmaceutical industry.

United Kingdom — Primary Care

The United Kingdom implemented quite different approaches in primary and secondary care, and these will be discussed separately. The approach in primary care was much more successful. This began about 20 years with the organic adoption of home-grown EHRs that were initially developed by general practitioners (GPs) and their teams to support repeat prescribing (or refills), which was previously a very time-consuming manual role. Various waves of national policy initiatives such as GP fundholding which made GPs responsible for their own budgets and the Quality and Outcomes Framework which rewards GPs on the quality of care have greatly increased adoption and usage of EHRs. This resulted in consolidation to a handful of EHR vendors. UK primary care has as a result almost complete penetrance of EHRs with advanced decision support functionalities and extensive use of data to support clinical care, research and audit, and the planning and commissioning of healthcare.

United Kingdom — Secondary Care

The penetration of EHRs is much lower in secondary care than in primary care, this to a large part reflecting the lack of any how-grown hospital-based EHR market and the failure of aspects of a major HIT policy experiment, the National Programme for Information Technology (NPfIT; later rebadged as NHS Connecting for Health). The reasons for this very high profile and expensive failure are multifaceted, but above all underestimating the time and effort needed to move from implementation and adoption to the realization of benefits, the rushed politically driven timelines, the failure to engage adequately with healthcare professionals and hospitals, and the challenges of adopting EHRs that were poorly tailored to UK needs. Other aspects of NPfIT were however more successful—in particular, the creation of an "NHS Spine" to support the secure transfer of information between NHS sites and the successful introduction of a Picture Archiving and Communications System.

There was a major fall-out from the failure of NPfIT to implement EHRs, such that it left a policy vacuum for a number of years on account of it being too politically charged to handle. With the change in political administrations, there has more recently been renewed political interest in creating "paper-free hospitals." This led to a recent major review of England's hospital HIT strategy, which recommended the creation of a Chief Clinical Information Officer post for England, flagship sites that would serve as "global digital exemplars" and the creation of an NHS digital training academy.

United States and Meaningful Use

One of the most ambitious approaches to HIT policy has been undertaken by the United States with its meaningful use program, which targeted both primary and secondary care. The HITECH Act was passed by Congress in 2009, and it offered both hospitals and "eligible providers" financial incentives for adopting EHRs if they were meaningful users (HealthIT.gov, 2015). As described in the legislation, to be a meaningful user a required three things: (1) to use a certified EHR, (2) to exchange data with other users, and (3) to report certain quality parameters (HealthIT.gov, 2014). While this made conceptual sense, there were initially practical issues with all three of these. For example, the certification criteria needed to be defined, and an entity had to be established to perform certification. In addition, many users did not have anyone to exchange data with who was prepared to receive it. Finally, with respect to the quality parameters, providers did not have experience interacting with a reporting organization on a routine basis around quality data in the outpatient setting.

On the standards front, one especially key area was that there was coalescence around a single summary of clinical data for a patient. This summary, called a Continuity of Care Document, is especially critical for moving data from place to place because it is generally impractical to send the entire clinical record. In contrast, it is possible to send a summary to many vendors. There was a great deal of discussion regarding which standard to pick, and exactly what clinical items should be included or excluded.

One novel approach that the United States took was to require certification of EHRs. To be certified, an EHR had to meet specific criteria. In particular, many different functions were required, such as the ability to detect drug−drug interactions, or the ability to send a clinical evidence summary. Overall, there were approximately 180 areas that had to be covered. Each domain was pass-fail, and vendors had to pass for all areas to be certified. A major plus of this approach was that all vendors had to cover areas that they had not covered previously, such as creating functionality to send notifications electronically to public health authorities (Health and Human Services Department, 2016). Downsides included there being no requirement that an individual area be covered well, and perhaps more subtle, the arduous requirements led vendors to devote a large fraction of their development activity to meeting the criteria. This made it difficult for vendors to respond to other client demands, and critics have suggested that the overall effort delayed progress on domains which would have been more important and delivered higher value. Also, implementation of certification requirements mandated the development of "certifying bodies." The initial entity was called the Certification Committee for Health Information Technology. After some time, however, this group was criticized for having become too close to vendors. Subsequently, multiple certifying bodies were allowed to compete.

Overall, the program has had both successes and failures. It has achieved tremendous success in getting providers to adopt EHRs—in both inpatient and outpatient settings the adoption rates rose from approximately 20% to over 80% over a several year interval. Moreover, all the main groups improved—both hospitals that are large and well to do and also smaller and disparate share hospitals adopted, albeit it at slightly lower rates for the latter. In addition, both small and large practices also adopted (Jones and Furukawa, 2014). Another major success is that a single standard has been identified for nearly all the main types of clinical data. In contrast, the impact for clinical data exchange has been much more mixed. The levels of data exchange at this point are still relatively low, and the initial model which involved setting up of health information exchanges has not been particularly successful (Adler-Milstein et al., 2013). In addition, it is been hard to identify new quality metrics which are relevant using EHRs (Kern et al., 2009) and most payment for quality approaches still focus on quality metrics that come from claims even though it appears that quality metrics which come from EHRs will be more accurate (Tang et al., 2007). While it is early, some recent data do suggest that many quality parameters do appear to be trending in the right direction.

Hong Kong

In contrast to other regions, Hong Kong has prioritized development in secondary care. The Health Authority pays for a large proportion of the secondary care delivered and has built an internally developed system which is both high quality and relatively inexpensive. Furthermore, it enables data exchange within all hospitals in the system. Primary care has much lower levels of adoption, and approaches that let primary care providers access data from the hospital system are only now being widely implemented. In Hong Kong, the government has paid directly for clinical systems, and it allows the clinical system leadership to make the case for what information technology should be introduced next (Cheung et al., 2007).

Other Regions

Of course, many other countries and regions have taken remarkable approaches and have achieved high levels of adoption, especially in primary care. In general, in many developed countries, implementation in secondary lags behind. In primary care, many countries have limited the number of vendors that can participate in a region, typically to three or four. This has major advantages, in that some competition is retained, but the number of vendors is not so great that data exchange becomes an intractable task.

SPECIFIC DOMAINS

A number of particular areas deserve special attention. These include informatics workforce, telehealth, and personal health records (PHRs). With respect to workforce, nearly every country needs more people who are trained in the use of HIT than they have. Regarding telehealth, there are many specific issues, especially around the crossing of boundaries. PHRs are also especially important from the privacy perspective.

Regarding workforce, a number of approaches have been taken. It will be particularly important to have trained medical personnel who understand software and what can and cannot reasonably be done with it. The United States, in the meaningful use area, implemented a very large program which sponsored many without medical training to learn about software, in many instances through junior colleges (United States Department of Labor, 2014). As noted above, the United Kingdom is in the process of creating a national digital academy and there has also been substantial investment in increasing data science and analytic capabilities through the creation of the Farr Institute.

Telehealth presents issues for many reasons, notably around payment, and because it may be unclear where care is actually being delivered. Traditional medical licensing approaches are generally based on states, provinces, and countries and often do not allow providers from another country to for example read a radiograph or offer a second opinion. Some evidence from the United States does suggest that states which have more progressive regulatory approaches do have higher levels of use of telehealth (Adler-Milstein et al., 2014).

With respect to PHRs, they are now very widely used in the United States, in large part because providers had to offer a PHR to patients in order to get the meaningful use incentives. This helped many providers get through the "activation energy" issue of not wanting to offer patients a portal to their records. The main fear of physicians was that this would create a great deal of extra work for them. The evidence, however, suggests that this does not happen (Yamin et al., 2011), and that about two-thirds of contacts can be handled by someone other than the physician themselves. This does require practices to have someone who can handle messages. Another major issue with PHRs from the privacy perspective is how well issues are managed with respect to access to the data (see Chapter 18: Social and Consumer Informatics). Ideally, patients themselves should be able to choose who they want to access it.

REGULATION OF HEALTHCARE INFORMATION TECHNOLOGY

Another very important issue from the policy perspective is the extent to which HIT is regulated. It is routinely used to deliver care, and thus major HIT issues can cause harm. At the same time, regulation often slows the process of innovation, and most of the regulatory processes at the Food and Drug Administration (FDA) in the United States cannot keep up with the pace of change needed in software (Magrabi et al., 2012).

Blood banking software in the United States represents the only major clinical domain which has been regulated by the FDA for some time. This software changes much more slowly than the software in any other domain, and it appears clunky and behind the times compared to any other domain. The consequences of serious errors in blood banking are high, which was the reason for regulating this area, but it is not clear whether or not the regulation has had any impact on safety (US Department of Health and Human Services Food and Drug Administration, 2015).

A particularly controversial area is the domain of clinical decision support (CDS). This is an important area because many of the benefits of implementing HIT come from the decision support (see Chapter 12: Clinical Decision Support and Knowledge Management). It also needs to change and adapt fairly rapidly to keep up with changes in practice. To date, the FDA has not elected to regulate CDS, largely because of what is called the "learned intermediary" exception. This means that decision support is delivered to someone who has medical knowledge, and they can either accept or reject the decision support based on what they know about the patient.

One area that the FDA has elected to regulate is in the device domain, especially when software recommends a specific action to a patient. An example would be software linked with a glucometer which recommends a specific insulin dosage to a diabetic patient and might even deliver the dosage given consent from the patient. This is clearly a higher risk situation; if the software recommended too high a dosage this could be life-threatening.

Another problematic area with respect to regulation is the domain of usability. Users complain a great deal about the usability of current EHRs. However, usability is very difficult to regulate. The current approach to this problem involves requiring vendors to develop medical software based on user-centered design principles in order to be certified. Empiric evaluations, however, suggest that vendors adhere quite variably to these recommendations (Ratwani et al., 2015).

ALIGNMENT WITH BROADER PAYMENT POLICY

Alignment with how healthcare is reimbursed is pivotal in achieving high levels of system performance. Requirements that payments be requested and delivered electronically represent one powerful card that policymakers can employ. But more importantly, if payment incentives reward higher levels of quality, safety, and value, organizations will be more likely to implement changes to achieve these aims.

Measuring safety accurately is challenging and many organizations do not have a good sense of how safe the care that they provide is. Soon, however, it may be possible to use electronic approaches to improve the efficiency of measuring safety.

Quality, on the other hand, can be more readily measured at scale. Incentives that reward organizations for delivering higher quality care may be beneficial. A tricky issue here is that sicker populations tend to have worse outcomes, so outcome-oriented incentives plans may end up unintentionally penalizing organizations that care for sicker and lower income populations, such as small urban hospitals and disproportionate share hospitals.

PERVERSE CONSEQUENCES

All policies carry the potential for perverse or negative consequences. For example, if just one metric carries a financial incentive, such as mammography, but not Pap smears, providers may direct all their efforts toward the one that receives the incentive. This actually occurred in Australia. Under the US certification program, vendors had to satisfy many criteria—approximately 180—but there was no requirement that they do them well, or in ways that improved quality or safety. The unintended consequence was that many things were done to simply "tick the box," which did not necessarily have the beneficial consequences expected. It is therefore very important that public policy interventions undergo rigorous, independent evaluation.

CONCLUSIONS

Policies can have a major impact with respect to HIT, especially in situations in which incentives are misaligned. In many countries, they have resulted in large increases in adoption rates with respect to EHRs. However, realizing the desired care improvements takes longer and is more elusive, as is enabling clinical data exchange. Governments play an important role in enabling standards development and have a pivotal role in encouraging adoption of a single standard for specific data types. All policies around HIT are likely to have greater impact if they are implemented in the context of payment schemes that reward better outcomes and higher quality, safer care.

REFERENCES

Adler-Milstein, J., Bates, D.W., Jha, A.K., 2013. Operational health information exchanges show substantial growth, but long-term funding remains a concern. Health Aff.10−377.

Adler-Milstein, J., Kvedar, J., Bates, D.W., 2014. Telehealth among US hospitals: several factors, including state reimbursement and licensure policies, influence adoption. Health Aff. 33 (2), 207−215.

Australian Department of Human Services. Getting started with eHealth. https://www.humanservices.gov.au/health-professionals/subjects/getting-started-ehealth (accessed 04.01.17).

Cheung, N.T., Fung, V., Wong, W.N., Tong, A., Sek, A., Greyling, A., et al., 2007. Principles-based medical informatics for success-how Hong Kong built one of the world's largest integrated longitudinal electronic patient records. Stud. Health Technol. Inform. 129 (1), 307.

Health and Human Services Department, 2016. ONC Health IT Certification Program: Enhanced Oversight and Accountability. Federal Register: The Daily Journal of the United States Government. https://www.federalregister.gov/documents/2016/10/19/2016-24908/onc-health-it-certification-program-enhanced-oversight-and-accountability (accessed 04.01.17).

HealthIT.gov, 2014. EHR incentives and certification: what is meaningful use? https://www.healthit.gov/providers-professionals/ehr-incentives-certification (accessed 04.01.17).

HealthIT.gov, 2015. Health IT Legislation: Select Portions of the HITECH ACT and Relationship to ONC Work. https://www.healthit.gov/policy-researchers-implementers/select-portions-hitech-act-and-relationship-onc-work (accessed 04.01.17).

Jones, E.B., Furukawa, M.F., 2014. Adoption and use of electronic health records among federally qualified health centers grew substantially during 2010–12. Health Aff. 33 (7), 1254–1261.

Kern, L.M., Dhopeshwarkar, R., Barrón, Y., Wilcox, A., Pincus, H., Kaushal, R., 2009. Measuring the effects of health information technology on quality of care: a novel set of proposed metrics for electronic quality reporting. Jt Comm. J. Qual. Patient Saf. 35 (7), 359–369.

Kidd, M.R., Mazza, D., 2000. Clinical practice guidelines and the computer on your desk. Med. J. Aust. 173 (7), 373–375.

Magrabi, F., Ong, M.S., Runciman, W., Coiera, E., 2012. Using FDA reports to inform a classification for health information technology safety problems. J. Am. Med. Inform. Assoc. 19 (1), 45–53.

McDonald, C.J., Huff, S.M., Suico, J.G., Hill, G., Leavelle, D., Aller, R., et al., 2003. LOINC, a universal standard for identifying laboratory observations: a 5-year update. Clin. Chem. 49 (4), 624–633.

McInnes, D.K., Saltman, D.C., Kidd, M.R., 2006. General practitioners' use of computers for prescribing and electronic health records: results from a national survey. Med. J. Aust. 185 (2), 88.

NIH: National Library of Medicine. Unified Medical Language System: RxNorm. https://www.nlm.nih.gov/research/umls/rxnorm/ (accessed 04.01.17).

Ratwani, R.M., Benda, N.C., Hettinger, A.Z., Fairbanks, R.J., 2015. Electronic health record vendor adherence to usability certification requirements and testing standards. JAMA 314 (10), 1070–1071.

Scott, P.J., Bentley, S., Carpenter, I., Harvey, D., Hoogewerf, J., Jokhani, M., et al., 2015. Developing a conformance methodology for clinically-defined medical record headings: a preliminary report. Eur. J. Biomed. Inform. 11 (2), en23–en30.

Tang, P.C., Ralston, M., Arrigotti, M.F., Qureshi, L., Graham, J., 2007. Comparison of methodologies for calculating quality measures based on administrative data versus clinical data from an electronic health record system: implications for performance measures. J. Am. Med. Inform. Assoc. 14 (1), 10–15.

US Department of Health and Human Services Food and Drug Administration, 2015. Mobile medical applications: guidance for industry and food and drug administration staff.

United States Department of Labor, 2014. ApprenticeshipUSA Investments. https://www.dol.gov/featured/apprenticeship/grants (accessed 04.01.17).

Yamin, C.K., Emani, S., Williams, D.H., Lipsitz, S.R., Karson, A.S., Wald, J.S., et al., 2011. The digital divide in adoption and use of a personal health record. Arch. Intern. Med. 171 (6), 568–574.Slight, S.P., Bates, D.W., 2011. A risk-based regulatory framework for health IT: recommendations of the FDASIA working group. J. Am. Med. Inform. Assoc. 21 (e2), e181–e184.

RECOMMENDED FURTHER READING

Cresswell, K., Bates, D.W., Sheikh, A., 2016. Six ways for governments to get value from health IT. Lancet 387 (10033), 2074–2075.

Robertson, A., Bates, D.W., Sheikh, A., 2011. The rise and fall of England's National Programme for IT. J. R. Soc. Med. 104 (11), 434–435.

Sheikh, A., Atun, R., Bates, D.W., 2014a. The need for independent evaluations of government-led health information technology initiatives. BMJ Qual. Saf. 23 (8), 611–613.

Sheikh, A., Jha, A., Cresswell, K., Greaves, F., Bates, D.W., 2014b. Adoption of electronic health records in UK hospitals: lessons from the USA. Lancet 384 (9937), 8–9.

Chapter 8

Health Information Technology and Value

Blackford Middleton[1,2] and Ngai T. Cheung[3]

[1]*Apervita, Inc., Chicago, IL, United States,* [2]*Harvard TH Chan School of Public Health, Boston, MA, United States,* [3]*Hospital Authority of Hong Kong, Kowloon, Hong Kong*

INTRODUCTION

In this chapter, we will discuss the value proposition for healthcare information technology or simply HIT. In this context, we refer to HIT as the software and hardware used to collect, organize, analyze, and deliver clinical information to members of the care delivery team. We exclude other technologies which may also play a role in part in managing clinical information, but are not the primary technologies used by care givers. HIT includes the "electronic medical record" (EMR)—the clinical transactional application for ordering and documenting care delivery used in a care setting—and the "electronic health record" (EHR)—a more comprehensive aggregation of data from multiple care settings within a care delivery organization, data from patient interaction with a personal health record (PHR), and potentially from other care organizations as well. (In this chapter, we will use the term "electronic health record" throughout and note that it may include both EMR and PHR.) Taken together, the EHR has in many health systems and countries now replaced the paper-based record as the "system of record."

We begin with a brief review of the evidence suggesting the EHR has value in two dimensions: clinical and financial. Next, we discuss the key barriers and facilitators to achieving value with HIT and discuss certain global perspectives on the value of HIT. Finally, we conclude this chapter with a view toward the future value of HIT in the transformation of healthcare delivery.

Key Advances in Clinical Informatics. DOI: http://dx.doi.org/10.1016/B978-0-12-809523-2.00008-X

BENEFITS OF HIT

Clinical Value

To assess the clinical value of the EHR we examine the impact of information access, clinical decision support (CDS), and workflow support. A secondary clinical gain from HIT may arise when it facilitates health information exchange (HIE). As a healthcare system implements an EHR, value arises first through improved information access and documentation. The benefits of CDS typically follow and are seen in more mature implementations, and may include alerts, reminders, order sets, calculated values, drug dosing, and suggestions for follow-up. More advanced implementations may also see CDS which supports workflow, care team communications, and direct to patient reminders.

Information Access

The paper-based record has limited information management and retrieval capabilities. Physicians may not be able to find relevant information in the paper chart, and often the chart itself may not be available. The accumulated evidence suggests that simply having clinical data available in the EHR improves clinical decision making, utilization, and patient outcomes (Buntin et al., 2011; Shekelle et al., 2006). Thus, a primary determinant of value of the EHR is the extent to which it collates, organizes, and makes available all relevant patient care information for the provider.

Clinical Decision Support

The most profound impact of EHR, however, comes when CDS is provided to the clinician at the point of and time of care. This may occur in a variety of ways in an EHR; we discuss these next.

Alerts and Reminders

Review of the effectiveness and impact of various alerts and reminders shows the most predictable impact is achieved when the decision support is delivered through computer-based, patient-specific reminders that are integrated into the clinician's workflow. Large gaps persist between best evidence-based care guidelines and what is actually done in practice and evidence suggests alerts and reminders can impact disease-state specific care guidelines as well (McGlynn et al., 2003). Studies have demonstrated benefits in reminding care providers about drug selection and dosing care plan creation for chronic conditions such as diabetes, and in-hospital error prevention (Chaudhry et al., 2006; Bright et al., 2012).

Test Ordering

Using the EHR with CDS for test ordering can impact utilization of expensive tests and procedures (see Chapter 10: Medication Management, and Laboratory and Radiology Testing). Substantial research has focused on variation in rates of medical and surgical procedures, and large variations in virtually all areas have been demonstrated. Studies of test ordering have found that as much as 50% of diagnostic tests in teaching hospitals may be unnecessary, and many outpatient tests are also low yield. However, in the few studies in which high utilization has been evaluated more closely, it has not necessarily been correlated with high levels of inappropriateness; this suggests that to decrease inappropriateness, it will be necessary to do more than identify high levels of utilization: Decision support interventions must occur at the level of the individual decision, targeted at those which are likely to be inappropriate. Notable efforts pursuing this goal include diagnostic test ordering guidelines suggested by the American College of Radiology (Sistrom and American College of Radiology, 2008; Thrall, 2014), the "Choosing Wisely" guidelines suggested by the American College of Physicians (Bhatia et al., 2015; Rosenberg et al., 2015), and the "appropriate use criteria" program of the US Center for Medicare and Medicaid Services.

Using the EHR with CDS may be the best means of effecting physician behavior change with respect to inappropriate utilization. The Regenstrief Institute group demonstrated that providing information regarding charges for tests (Tierney, 1990), the time and result when they were most recently performed (McDonald, 1992), and the probability of obtaining an abnormal result (Tierney, 1988) all decreased the number of tests that were performed, by 9%−14%. Decision support can make physicians more aware of information such as costs of tests and previous results, can bring guidelines to the provider at time of ordering, and can suggest additional testing when appropriate ("corollary orders" (Overhage et al., 1997)). A careful study of the return on investment for a computer-based provider order entry system at an academic medical center demonstrated a significant savings (Kaushal et al., 2006).

Medications

Adverse drug events (ADEs) and medication errors are common in the inpatient setting (Kellaway and McCrae, 1973; Hutchinson et al., 1986) and are increasingly being recognized in the outpatient environment as well (Gandhi et al., 2003). In one study (Gandhi et al., 2000a), a cross-sectional chart review and patient survey of primary care patients found 18% of patients reported problems related to their medications. Among the ADEs detected by chart review, patients had a previously documented allergy to the medication in 11%, and 4.5% required hospitalization. A prospective study of outpatient clinics identified ADEs in 25% of patients and medication errors in 9% of prescription (Gandhi et al., 2000c). Of these, almost half had potential to harm patients.

While many strategies for preventing medication errors and ADEs have been proposed (Bates et al., 1996), the evidence supporting their efficacy is somewhat limited. Bates et al. showed that implementation of a physician order entry system resulted in a 55% decrease in the serious medication error rate (Bates et al., 1998), and others have found that delivery of CDS for antibiotics reduced costs and improved outcomes (Evans et al., 1998). Another study in the outpatient setting assessed the impact of basic computerized prescribing with only rudimentary checks (Gandhi et al., 2000c) found that computerized prescriptions contained significantly fewer medication errors and rule violations. The majority of errors could have been prevented with requiring complete prescriptions, frequency checking, and dose checking. More advanced decision support would also have prevented one-third of preventable ADEs. Other advantages of computerized prescribing systems include the ability to reduce transcription and verbal orders by automatically sending prescriptions to pharmacies and to improve the upkeep of accurate medication lists in the medical record.

Workflow

EHRs may impact clinical workflow in a variety of ways. For simplicity, we consider workflow in this context to be the actions taken to perform tasks related to patient care and communication. Many EHR features are designed to improve an individual's workflow such as data review and clinical documentation, but some may also support the workflow of care teams. We focus here on the support for tasks associated with patient care and communication among the care team.

Clinical Communication

Simple imbedded e-mail systems in an EHR allow providers to message each other conveniently and privately; these messages may form a task list for staff. Users of an EHR may also write a message to themselves or others and future date it so that it is not delivered until the future date. This forms a "tickler" or reminder system that may be used to queue future tasks, or prompt the provider to review a chart at a future date. While the value is readily apparent, however, there is little solid evidence investigating how well EHRs by themselves may support clinical communication among care teams.

Test Notification

A more advanced communication capability is the ability to notify clinicians of abnormal test results. Delays in physician response to critical laboratory results are common (Bates et al., 1996; Wyrwich, 1999). One study showed that using an inpatient computerized alerting system to notify physicians of critical laboratory results reduced the time until an appropriate treatment was ordered by 38% (Kuperman et al., 1999). Similar benefits are expected in the outpatient setting, especially since delays in obtaining test results are likely to be more common due to testing being performed off-site. Some ambulatory EHR systems message the ordering physician if an extreme abnormal laboratory result is obtained.

Referral Support

EHR systems may also support the clinical referral process—the critical communication link between primary care and subspecialty care for outpatients. Ideally, a referral communication should include a specific goal, pertinent clinical information, and an initial assessment. Feedback from the specialist should be timely so that continuity of care is not disrupted. However, evidence from physician surveys indicates that neither the initial communication nor feedback from the specialist occurs reliably. Only 76% of referrals state an explicit purpose for the referral (McPhee et al., 1984) perhaps explaining why in 14% of inpatient consultations, the referring physician and consultant disagree on the reason for the consultation (Lee et al., 1983). Sixty-three percent of primary care physicians (PCPs) in one system were dissatisfied with the referral process (Gandhi et al., 2000b). Forty percent of PCPs had not received feedback from the consulting specialist 2 weeks after the initial evaluation (Gandhi et al., 2000). Specialists are likewise dissatisfied with the referral process. In one study, 48% stated the information from the referring physician was untimely, while 43% felt the content of the referral request was inadequate.

This breakdown in physician-to-physician communication has led to delayed diagnoses, poly-pharmacy, increased litigation risk, and unnecessary testing (Epstein, 1995). Supplying the necessary information from the paper chart is difficult. Currently in the United States, every clinical encounter requires a Visit Summary to be produced and made available for the patient to view, download, or transmit to another provider, and the OpenNotes movement (see Chapter 4: Electronic Clinical Documentation) suggests that all clinical notes from the provider be made available to the patient. Such efforts will undoubtedly improve a shared understanding of care goals and objectives among members of the care team and the patient.

Health Information Exchange

The above brief review of the value of EHR in clinic and hospital settings assesses the value of these products in a myopic way: that is, the value that is obtained within a clinic, hospital, or integrated delivery network. It is analogous to considering the value of automating individual banks without linking them together through the financial networks. Work in regional healthcare information exchange and interoperability suggests that much greater value will be obtained when individual EHR systems can communicate in a secure and confidential way across local care regions and potentially across the country.

The Center for IT Leadership (CITL) performed an analysis on the Value of Healthcare Information Exchange and Interoperability (Walker et al., 2005) and found that fully standardized, interoperable systems would deliver labor savings and avoided costs related to duplicate medication and diagnostic and therapeutic test ordering of approximately $78 billion per year for the US healthcare delivery system. And we believe these potential savings will be in addition to the savings arising from EHR effects within the individual care setting. A careful analysis of HIE between emergency departments in Memphis, TN, demonstrated a reduction in costs at participating hospitals in excess of $1 million, the vast majority (97.6%) of which resulted from reduced hospitalizations (Frisse et al., 2011).

Financial Value

While most of the value of EHRs may be appreciated first in the clinical domain, important dimensions of value arise in the financial domain.

Improved Utilization of Medications and Diagnostic Tests

One of the most dramatic ways in which the EHR can produce value is through improved utilization of expensive laboratory tests, medications, and radiology procedures.

Medications

EHRs with medication decision support offered can profoundly impact utilization, reducing overuse, underuse, and misuse of medications. For example, brand to generic substitution suggestions, as well as suggestions regarding alternative cost-effective therapies, can reduce medication expenses. Complying with a patient's health plan formulary can provide savings.

The CITL modeled these effects in the ambulatory care setting (Johnston et al., 2003) and estimated that medication savings could range from $2000 annually per provider for a basic system to as much as $17,000 annually per provider with advanced decision support capabilities. If these systems were nationally adopted in the United States, they would produce savings of $3.3 billion for basic systems, $18 billion for intermediate systems, and $27 billion for advanced systems.

Laboratory

Similarly, decision support systems in laboratory order entry can improve utilization. Systems that display test costs, prior results, or the probability of future abnormal test results to physicians during order entry can each save between 5% and 15% in laboratory expenditure (Tierney et al., 1988). CITL analysis suggests that order entry systems in EHR may reduce redundant laboratory expenditures anywhere from 0.4% for basic systems all the way up to 19.4% for advanced decision support systems. CITL studies estimate reduced laboratory expenditures of as much as $32,000.00 per provider per year with advanced decision support systems. If such systems were adopted nationally, this would result in savings of as much as $5.8 billion per year.

Radiology

Decision support systems for radiology test ordering may recommend appropriate radiologic examinations, reduce inappropriate ordering, improve utilization with appropriate indications or consequent and corollary orders, and provide reminders about recent or redundant tests. CITL estimated that radiology decision support systems could avoid anywhere from 0.8% to 20% of redundant laboratory investigations with advanced systems saving $71,000 per provider per year. If such systems were adopted nationwide in the ambulatory setting alone this could result in savings of as much as $12.9 billion per year (Johnston et al., 2004).

COSTS OF HIT

This brief review prevents a detailed assessment of the costs of implementing and using HIT. However, the costs of adopting an EHR may be well over $50,000 per provider when all costs are considered and significant additional costs may arise as the users and healthcare delivery systems consider how to optimize care using the HIT. Critical challenges in this process include defining optimal care patterns, incentivizing providers toward normative care, and sharing CDS knowledge that may help to optimize care. Further, significant costs may arise as health systems consider how to increase information sharing across organizational boundaries. Finally, significant costs may arise anew if a health system seeks to switch from one EHR system to another.

BARRIERS AND FACILITATORS TO ACHIEVING HIT VALUE

There is a huge body of research demonstrating the value of HIT (Kaushal et al., 2003; Garg et al., 2005; Chaudhry et al., 2006). However even as sophisticated HIT systems became widely available commercially, it became increasingly clear that installation of such a system did not guarantee healthcare success. Articles on the unintended consequences of HIT started appearing in the literature over 10 years ago, and this theme has since been widely explored, including reports by bodies such as the Institute of Medicine (now the National Academy of Medicine) (Institute of Medicine, 2011) and the US Department of Health and Human Services (Schneider et al., 2014). Several influential studies have also called into question the earlier optimism of the benefits that HIT could achieve (Himmelstein et al., 2010; Black et al., 2011). There are many reports of dissatisfaction with current systems for doctors (Miliard, 2015), nurses (Miliard, 2014), and hospitals (Monegain, 2014). System downtime affecting care delivery and security breaches are being reported on a weekly basis. Well-funded national programs have encountered significant difficulties, including some of the notable disappointments of England's NPfIT (House of Commons Committee of Public Accounts, 2013). We will discuss key barriers and facilitators in turn.

Key Barriers and Facilitators

The complexity of healthcare

To reconcile the immense potential of HIT with the often less than stellar outcomes of its implementation, we must realize that healthcare is a complex adaptive system (Plsek and Greenhalgh, 2001), perhaps the most complex industry in the world—dealing as it does with huge variety in patients, diseases, and clinical practice and with the necessary social, economic, and ethical considerations on top of the clinical and scientific aspects. It is inevitable that the successful implementation of IT in health is even more difficult than other industries. HIT is a change agent that disrupts the complex system of healthcare in unpredictable ways and this has been widely studied as the "unintended consequences" of HIT (Koppel et al., 2005; Ash et al., 2004; Campbell et al., 2006; Kuperman and McGowan, 2013; Nijland et al., 2008).

Usability

Recent research has focused on the usability of HIT (Middleton et al., 2013). Surveys have shown low clinician satisfaction with HIT, for reasons including lost productivity, difficulty of use, and poor workflow fit (Hanover, 2013). Poor usability can lead to a wide variety of medical errors including wrong patient, wrong treatment, wrong medication, delay in treatment, unintended care, and care sequencing error (Shneiderman, 2011). Alert fatigue is a well-documented subclass of poor usability with reports of medical alert override rates ranging from 49% to 96% of cases (van der Sijs et al., 2006). Abiding by the "five rights" of decision support (the right information to the right person in the right format through the right channel at the right time) may help ameliorate the problem (Osheroff et al., 2012).

User Satisfaction

A 2013 American Medical Association report on physician professional satisfaction found that HIT could have a major impact, both positive and negative. Factors that worsened professional satisfaction including poor usability, time-consuming data entry, interference with face-to-face patient care, inefficient work content, lack of interoperability of EHR products, and degradation of clinical documentation. Factors increasing satisfaction included remote access to patient information and improvements in quality of care (Friedberg et al., 2013).

Organizational Issues

HIT is costly, disruptive, and requires constant monitoring and enhancement. Sustained top-level commitment is needed for funding, manpower, and leadership. A culture of quality and innovation, with equal emphasis on safety and reporting are required. For instance, at the Hong Kong Hospital Authority clinical IT issues are regularly discussed at the Board and top leadership levels, and the HIT governance model is explicit and constantly maintained. HIT issues are also reported on the same level as clinical incidents and receive the same systematic treatment.

Engagement and Expertise

Engagement of the clinical workforce is key to success and clinician champions are important. However, "techno-enthusiast" clinicians may not always be the best champions as they are often overly optimistic and underestimate the difficulties faced by their less enthusiastic colleagues. Another key facilitator is the presence of staff trained in health informatics such that they understand the subtleties and challenges of HIT. One good way to engage clinicians is through successful pilot projects—after all, "success breeds success" (van de Rijt et al., 2014). Another key facilitator is the presence of staff trained in health informatics and who thus understand the subtleties and challenges of HIT.

Incentives and Priorities

Change management is a big part of HIT implementation, and having the right incentives in place will make a big difference. National or organizational policies are one enabler. Financial incentives can work—the success of the EHR incentives programs of the US HITECH Act shows this—but financial incentives have their own unintended consequences. More powerful and sustainable incentives can be found where the emphasis is on clinical value including improving clinician workflow and patient outcomes. A robust system of prioritization is needed to align HIT projects with the needs of clinicians and patients and the goals of the organization.

Usability and Workflow Support

In the past, there was a lot of emphasis on the need for clinician training. However, this has always been problematic for doctors as they are often too busy or unwilling to lose their clinical time to attend training. More recently, the focus has shifted to usability and the end-user experience; with better usability and intuitive interfaces not only is the need for training reduced but also the safety of the systems improves. Good support for workflow is also crucial, and a system should improve communications and task efficiency while minimizing interruptions and unnecessary steps (Lee and Shartzer, 2005).

Interoperability and Standards

Interoperability of data and HIE are considered to be key facilitators of HIT and crucial to reaping the full benefits of HIT (Brailer, 2005), and perhaps essential to the sustainability of national healthcare systems. In 2005, Walker et al. estimated that fully interoperable EHR systems could save 5% of the US healthcare costs, but that partial interoperability would accrue much lower benefits (Walker et al., 2005). Achieving interoperability is a difficult challenge and the adoption of national standards for the representation and transfer of information is one of the prerequisites (see Chapter 5: Interoperability). However, interoperability is not just a technical challenge and there are leadership, political, organizational, legal, and market issues (Hovenga and Garde, 2010).

The Benefits of Scale

Many of the benefits of HIT can only be seen at a healthcare system level, and there exists a significant "societal good" that does not accrue to any one stakeholder, but may be shared by all—analogies may be drawn with an interstate highway system (Middleton, 2005). HIT can improve care coordination, decrease variation in practice, and provide data to inform planning and management. At the national level, HIT can be a tool to help implement policy. Having scale also mean that the necessary infrastructure, expertise, and technology can be developed and deployed far more economically than with a myriad of separate siloed efforts.

GLOBAL PERSPECTIVES ON HIT VALUE

The importance of HIT is recognized globally. The World Health Organization's (WHO) Framework for Action published in 2007 highlights a well-functioning health information system as one of the six building blocks of a health system (World Health Organization, 2007). This Framework applies to low-, medium-, and high-resource nations, and today the WHO Global Observatory for eHealth lists 133 countries which either have a national HIT or eHealth strategy or which have a national EHR in place (World Health Organization, 2016). Although there have been significant successes in the use of HIT in low-resource settings, for reasons of space the remainder of this chapter will focus on high-resource settings.

North America

In North America, the United States has long been a leader in biomedical informatics research (see Chapter 11: Bioinformatics and Precision Medicine), but adoption of HIT has been surprisingly slow (Berner et al., 2005). It has only been in the last few years with the stimulus of the HITECH Act that adoption of EHRs in the hospitals has reached a high level (Blumenthal, 2010; Henry et al., 2016). Canada's Health Infoway was formed in 2001 to act as a strategic investor in HIT with $2.1 billion invested to date (Canada Health Infoway, 2016). Although progress has been slower than originally envisaged, it seems to have now passed the tipping point, and Canadian iEHRs are well on their way (Gheorghiu and Hagens, 2016).

Europe

Medical informatics has a long history in Europe. Papers discussing the potential of HIT can be found starting from the 1960s (Healy, 1968), and there has been copious European medical informatics research since then. A European eHealth action plan was developed in 2004 calling for the majority of European health organizations and health regions to provide a variety of online services including e-prescription, tele-consultation, e-referral, and telecare by 2008 (Commission of the European Communities, 2004). Although actual progress fell well short of this, Europe has still been at the forefront of national HIT initiatives with Denmark and Estonia widely regarded as the most successful. Overall the uptake of HIT is increasing, but lack of interoperability remains an issue (European Union, 2014).

Asia

Asia also has seen significant HIT successes. Hong Kong was the first to have large-scale EMR implementation with complete interoperability in the public sector (Cheung et al., 2001), followed by the territory-wide EHR for all healthcare providers. Singapore has also deployed EMRs in all the public hospitals and has launched the national EHR in 2011 (Rowlands, 2012). Taiwan and South Korea are also developing interoperable networks among the hospitals which have widely implemented independent EMRs, and both are leveraging data from the national insurance claims systems to create national patient portals. Australia and New Zealand have had greater HIT success in the general practice clinic setting; Australia launched the $1 billion national Personally Controlled EHR in 2012 but struggled with adoption and is now relaunching the system as "My Health Record" with significant changes (Gartrell, 2015).

There is no question that we have turned the corner on HIT and global adoption is near-inevitable. We have enough experience and technology to ensure that we can embrace the facilitators and avoid the barriers to HIT success. As Bitton et al. (2013) noted, the question now is not whether to adopt HIT, but how best to do it.

FUTURE DIRECTIONS: ENHANCING THE VALUE OF EHR

Beyond the clinical and financial value of the HIT discussed above, several areas will undoubtedly contribute to the value of the HIT in the future. HIT value will be increased by increased information exchange, improved usability, and through the implementation of more sophisticated methods of CDS. The last is essential for the notion of personalized or precision medicine. However, these advanced CDS methods are not likely to arise within the context of the EHR itself—rather, they are likely to arise and be available to patients and clinicians alike when the EHR becomes more open to externalized knowledge-based, or data-driven, tools and services. Such as vision for "substitutable apps" has already been realized in the consumer technology space with app stores for Apple and Google products, but is not yet a reality in HIT (see Chapter 16: An Apps-Based Information Economy in Healthcare) (Mandl and Kohane, 2012; Mandl et al., 2015).

CONCLUSIONS

HIT has become an essential component of the provider's armamentarium to deliver care. The value of HIT is hard to measure, however, and will mean different things to different stakeholders in healthcare, whether patient, clinician, administrator, or researcher. Like IT in many other sectors of modern life—communications, travel, finance—it will undoubtedly continue to evolve and deliver increasing value to the patient, the provider and health system, and society overall.

REFERENCES

Ash, J.S., Berg, M., Coiera, E., 2004. Some unintended consequences of information technology in health care: the nature of patient care information system-related errors. J. Am. Med. Inform. Assoc. 11 (2), 104–112.

Bates, D.W., Burrows, A.M., Grossman, D.G., Schneider, P.J., Strom, B.L., 1996. Top-priority actions for preventing adverse drug events in hospitals. Recommendations of an expert panel. Am. J. Health. Syst. Pharm. 53, 747–751.

Bates, D.W., Leape, L.L., Cullen, D.J., Laird, N., Petersen, L.A., Teich, J.M., et al., 1998. Effect of computerized physician order entry and a team intervention on prevention of serious medication errors. JAMA 280 (15), 1311–1316.

Berner, E.S., Detmer, D.E., Simborg, D., 2005. Will the wave finally break? A brief view of the adoption of electronic medical records in the United States. J. Am. Med. Inform. Assoc. 12 (1), 3–7.

Bhatia, R.S., Levinson, W., Shortt, S., et al., 2015. Measuring the effect of Choosing Wisely: an integrated framework to assess campaign impact on low-value care. BMJ Qual. Saf. 24 (8), 523–531.

Bitton, A., Flier, L.A., Jha, A.K., 2013. Health information technology in the era of care delivery reform: to what end? JAMA 307 (24), 2593–2594.

Black, A.D., Car, J., Pagliari, C., Anandan, C., Cresswell, K., Bokun, T., et al., 2011. The impact of eHealth on the quality and safety of health care: a systematic overview. PLoS Med. 8 (1), e1000387.

Blumenthal, D., 2010. Launching HITECH. N. Engl. J. Med. 362 (5), 382–385.

Brailer, D.J., 2005. Interoperability: the key to the future health care system. Health Aff.Suppl Web Exclusives, W5–19–W5–21.

Bright, T.J., Wong, A., Dhurjati, R., et al., 2012. Effect of clinical decision-support systems: a systematic review. Ann. Intern. Med. 157 (1), 29–43.

Buntin, M.B., Burke, M.F., Hoaglin, M.C., Blumenthal, D., 2011. The benefits of health information technology: a review of the recent literature shows predominantly positive results. Health Aff. 30 (3), 464–471.

Campbell, E.M., Sittig, D.F., Ash, J.S., Guappone, K.P., Dykstra, R.H., 2006. Types of unintended consequences related to computerized provider order entry. J. Am. Med. Inform. Assoc. 13 (5), 547–556.

Canada Health Infoway, 2016. Investment Programs. https://www.infoway-inforoute.ca/en/what-we-do/progress-in-canada/investment-programs (accessed 22.03.16).

Chaudhry, B., Wang, J., Wu, S., et al., 2006. Systematic review: impact of health information technology on quality, efficiency, and costs of medical care. Ann. Intern. Med. 144 (10), 742–752.

Cheung, N.T., Fung, K.W., Wong, K.C., Cheung, A., Cheung, J., Ho, W., et al., 2001. Medical informatics—the state of the art in the Hospital Authority. Int. J. Med. Inform. 62 (2–3), 113–119.

Commission of the European Communities, 2004. COM 356: e-Health—making health care better for European citizens: an action plan for a European e-Health area. Brussels.

Epstein, R.M., 1995. Communication between primary care physicians and consultants. Arch. Fam. Med. 4 (5), 403–409.

European Union, 2014. European hospital survey: benchmarking deployment of eHealth services (2012–2013). Final report. Publications Office of the European Union: European Commission Joint Research Centre Institute for Prospective Technological Studies, Luxembourg.

Evans, R.S., Pestotnik, S.L., Classen, D.C., Clemmer, T.P., Weaver, L.K., Orme Jr., J.F., et al., 1998. A computer-assisted management program for antibiotics and other antiinfective agents. N. Engl. J. Med. 338 (4), 232–238.

Friedberg, M.W., Chen, P.G., Van Busum, K.R., et al., 2013. Factors Affecting Physician Professional Satisfaction and Their Implications for Patient Care, Health Systems, and Health Policy. RAND Corporation, Santa Monica, CA, <http://www.rand.org/pubs/research_reports/RR439.html>.

Frisse, M.E., Johnson, K.B., Nian, H., et al., 2011. The financial impact of health information exchange on emergency department care. J. Am. Med. Inform. Assoc. 19 (3), 328–333.

Gandhi, T.K., Burstin, H.R., Cook, E.F., Puopolo, A.L., Haas, J.S., Brennan, T.A., et al., 2000a. Drug complications in outpatients. J. Gen. Intern. Med. 15 (3), 149−154.

Gandhi, T.K., Sittig, D.F., Franklin, M., Sussman, A.J., Fairchild, D.G., Bates, D.W., 2000b. Communication breakdown in the outpatient referral process. J. Gen. Intern. Med. 15 (9), 626−631.

Gandhi, T.K., Weingart, S.N., Seger, A., Seger, D.S., Borus, J.S., Burdick, E., et al., 2000c. Impact of basic computerized prescribing on outpatient medication errors and adverse drug events. J. Gen. Intern. Med. 16 (Suppl. 1), 195.

Gandhi, T.K., Weingart, S.N., Borus, J., Seger, A.C., Peterson, J., Burdick, E., et al., 2003. Adverse drug events in ambulatory care. N. Engl. J. Med. 348 (16), 1556−1564.

Garg, A.X., Adhikari, N.K., McDonald, H., Rosas-Arellano, M.P., Devereaux, P.J., Beyene, J., et al., 2005. Effects of computerized clinical decision support systems on practitioner performance and patient outcomes: a systematic review. JAMA 293 (10), 1223−1238.

Gartrell, A., 2015. Australians to benefit from Sussan Ley's ehealth health records revamp. *Sydney Morning Herald*, May 10, 2015. http://www.smh.com.au/federal-politics/political-news/australians-to-benefit-from-sussan-leys-ehealth-health-records-revamp-20150508-ggxkew (accessed 21.03.16).

Gheorghiu, B., Hagens, S., 2016. Measuring interoperable EHR adoption and maturity: a Canadian example. BMC Med. Inform. Decis. Mak1−7.

Hanover, J., 2013. Business strategy: the current state of ambulatory EHR buyer satisfaction. *IDC Health Insights* (Doc #HI244027), November 2013.

Healy, M.J., 1968. Mathematics, computers, and the doctor. Br. Med. J. 1 (5586), 243−245.

Henry, J., Pylypchuk, Y., Searcy, T., Patel, V., 2016. Adoption of Electronic Health Record Systems among U.S. Non-Federal Acute Care Hospitals: 2008−2015. Office of the National Coordinator for Health Information Technology, Washington, DC, ONC Data Brief, no.35.

Himmelstein, D.U., Wright, A., Woolhandler, S., 2010. Hospital computing and the costs and quality of care: a national study. Am. J. Med. 123 (1), 40−46.

House of Commons Committee of Public Accounts, 2013. The dismantled National Programme for IT in the NHS. http://www.publications.parliament.uk/pa/cm201314/cmselect/cmpubacc/294/294.pdf.

Hovenga, E.J., Garde, S., 2010. Electronic health records, semantic interoperability and politics. Electron. J. Health Inform. 5 (1), e2.

Hutchinson, T.A., Flegel, K.M., Kramer, M.S., Leduc, D.G., Kong, H.H., 1986. Frequency, severity and risk factors for adverse drug reactions in adult out-patients: a prospective study. J. Chronic Dis. 39, 533−542.

Institute of Medicine, 2011. Health IT and Patient Safety: Building Safer Systems for Better Care. National Academies Press, Washington, DC.

Johnston, D., Pan, E., Walker, J.D., Bates, D.W., Middleton, B., 2003. The Value of Computerized Provider Order Entry in Ambulatory Settings. Center for Information Technology Leadership, Boston, MA.

Johnston, D., Pan, E., Walker, J., 2004. The value of CPOE in ambulatory settings. J. Healthc. Inf. Manag. 18 (1), 4−8.

Kaushal, R., Shojania, K.G., Bates, D.W., 2003. Effects of computerized physician order entry and clinical decision support systems on medication safety: a systematic review. Arch. Intern. Med. 163 (12), 1409−1416.

Kaushal, R., Jha, A.K., Franz, C., et al., 2006. Return on investment for a computerized physician order entry system. J. Am. Med. Inform. Assoc. 13 (3), 261−266.

Kellaway, G.S., McCrae, E., 1973. Intensive monitoring for adverse drug effects in patients discharged from acute medical wards. N. Z. Med. J. 78, 525–528.

Koppel, R., Metlay, J.P., Cohen, A., Abaluck, B., Localio, A.R., Kimmel, S.E., et al., 2005. Role of computerized physician order entry systems in facilitating medication errors. JAMA 293 (10), 1197–1203.

Kuperman, G.J., McGowan, J.J., 2013. Potential unintended consequences of health information exchange. J. Gen. Intern. Med. 28 (12), 1663–1666.

Kuperman, G.J., Teich, J.M., Tanasijevic, M.J., Ma'luf, N., Rittenberg, E., Fiskio, J., et al., 1999. Improving response to critical laboratory results with automation: results of a randomized controlled trial. J. Am. Med. Inform. Assoc. 6, 512–522.

Lee J., Shartzer A., 2005. "Health IT and Workflow in Small Physicians' Practices," NICHM Foundation Questions and Answers Brief April 2005. www.nihcm.org/AHRQ-QandA.pdf (02.08.05).

Lee, T., Pappius, E.M., Goldman, L., 1983. Impact of inter-physician communication on the effectiveness of medical consultations. Am. J. Med. 74 (1), 106–112.

Mandl, K.D., Kohane, I.S., 2012. Escaping the EHR trap—the future of health IT. N. Engl. J. Med. 366 (24), 2240–2242.

Mandl, K.D., Mandel, J.C., Kohane, I.S., 2015. Driving innovation in health systems through an apps-based information economy. Cell Syst. 1 (1), 8–13.

McDonald, C.J., 1992. Effects of computer reminders for influenza vaccination on morbidity during influenza epidemics. MD Comput. 9, 304–312.

McGlynn, E.A., Asch, S.M., Adams, J., et al., 2003. The quality of health care delivered to adults in the United States. N. Engl. J. Med. 348 (26), 2635–2645.

McPhee, S.J., Lo, B., Saika, G.Y., Meltzer, R., 1984. How good is communication between primary care physicians and subspecialty consultants? Arch. Intern. Med. 144 (6), 1265–1268.

Middleton, B., 2005. Achieving U.S. Health information technology adoption: the need for a third hand. Health Aff. 24 (5), 1269–1272.

Middleton, B., Bloomrosen, M., Dente, M.A., Hashmat, B., Koppel, R., Overhage, J.M., et al., 2013. Enhancing patient safety and quality of care by improving the usability of electronic health record systems: recommendations from AMIA. J. Am. Med. Inform. Assoc. 20 (1), 2–8.

Miliard, M., 2014. Nurses not happy with hospital EHRs. *Healthcare IT News*, Oct 20, 2014. http://www.healthcareitnews.com/news/nurses-not-happy-hospital-ehrs.

Miliard, M., 2015. 'Dissatisfaction' leading to EHR replacement trend. *Healthcare IT News*, June 17, 2015. http://www.healthcareitnews.com/news/dissatisfaction-leading-ehr-replacement-trend.

Monegain, B., 2014. Premier survey shows EHR buyers' remorse. *Healthcare IT News*, June 2, 2014. http://www.healthcareitnews.com/news/premier-survey-shows-ehr-buyers-remorse.

Nijland, N., van Gemert-Pijnen, J., Boer, H., Steehouder, M.F., Seydel, E.R., 2008. Evaluation of internet-based technology for supporting self-care: problems encountered by patients and caregivers when using self-care applications. J. Med. Internet Res. 10 (2), e13.

Osheroff, J.A., Levick, D.L., Saldana, L., Velasco, F.T., Sittig, D.F., Rogers, K.M., et al., 2012. Improving Outcomes with Clinical Decision Support: An Implementer's Guide, second ed. Healthcare Information and Management Systems Society, Chicago, IL.

Overhage, J.M., Tierney, W.M., Zhou, X.H., McDonald, C.J., 1997. A randomized trial of "corollary orders" to prevent errors of omission. J. Am. Med. Inform. Assoc. 4 (5), 364–375.

Plsek, P.E., Greenhalgh, T., 2001. Complexity science: the challenge of complexity in health care. BMJ 323 (7313), 625–628.

Rosenberg, A., Agiro, A., Gottlieb, M., et al., 2015. Early trends among seven recommendations from the choosing wisely campaign. JAMA Intern. Med. 175 (12), 1913–1920.

Rowlands, D., 2012. National eHealth record systems—the Singapore experience. *Pulse + IT Magazine.* http://www.pulseitmagazine.com.au/index.php?option = com_content&view = article&id = 1139 (accessed 21.03.16).

Schneider, E.C., Ridgely, M.S., Meeker, D., Hunter, L.E., Khodyakov, D., Rudin, R., 2014. Promoting Patient Safety Through Effective Health Information Technology Risk Management. RAND Corporation.

Shekelle, P.G., Morton, S.C., Keeler, E.B., 2006. Costs and benefits of health information technology. Evid. Rep. Technol. Assess. (Full Rep) 132, 1–71.

Shneiderman, B., 2011. Tragic errors: usability and electronic health records. Interactions 18 (6), 60–63.

Sistrom, C.L., American College of Radiology, 2008. In support of the ACR Appropriateness Criteria. J. Am. Coll. Radiol. 5 (5), 630–635.

Thrall, J.H., 2014. Appropriateness and imaging utilization: "computerized provider order entry and decision support". Acad. Radiol. 21 (9), 1083–1087.

Tierney, W.M., 1988. Computer predictions of abnormal test results. Effects on outpatient testing. JAMA 259, 1194–1198.

Tierney, W.M., 1990. The effect on test ordering of informing physicians of the charges for outpatient diagnostic tests. N. Engl. J. Med. 322, 1499–1504.

Tierney, W.M., McDonald, C.J., Hui, S.L., Martin, D.K., 1988. Computerized display of abnormal test results: effects on outpatient testing. JAMA 259, 1194–1198.

van de Rijt, A., Kang, S.M., Restivo, M., Patil, A., 2014. Field experiments of success-breeds-success dynamics. Proc. Natl Acad. Sci. 111 (19), 6934–6939.

van der Sijs, H., Aarts, J., Vulto, A., Berg, M., 2006. Overriding of drug safety alerts in computerized physician order entry. J. Am. Med. Inform. Assoc. 13 (2), 138–147.

Walker, J., Pan, E., Johnston, D., Adler-Milstein, J., Bates, D.W., Middleton, B., 2005. The value of health care information exchange and interoperability. Health Aff. (Millwood)Suppl Web Exclusives:W5–10–W5–18. Available from http://dx.doi.org/10.1377/hlthaff.w5.10.

World Health Organization, 2007. Everybody's Business: Strengthening Health Systems to Improve Health Outcomes: WHO's Framework for Action. World Health Organization, Geneva.

World Health Organization, 2016. Directory of eHealth policies. Global Observatory for eHealth. http://www.who.int/goe/policies/countries/en/ (accessed 20.03.16).

Wyrwich, K.W., 1999. Linking clinical relevance and statistical significance in evaluating intra-individual changes in health-related quality of life. Med. Care 37, 469–478.

Chapter 9

Organizational and Behavioral Issues

Joan S. Ash[1] and Nancy M. Lorenzi[2]

[1]Oregon Health & Science University, Portland, OR, United States, [2]Vanderbilt University Medical Center, Nashville, TN, United States

INTRODUCTION

Organizational and behavioral issues impact everything that is done in health informatics. Informatics is interdisciplinary, with professionals coming from different educational and disciplinary backgrounds. Whether those backgrounds are primarily healthcare, information technology (IT), or informatics, each profession is inculcated with different professional values during training. The disciplines also have different vocabularies and views of the world. Informaticians are professionals who manage health information, so a major responsibility of informaticians is to bridge the gap between worlds and to understand different cultures and professional vocabularies (Ash et al., 2003; Wright et al., 2014). Therefore, behavioral and organizational issues are especially critical for everyone working in this interdisciplinary area. We believe that these issues deserve more systematic study, recognition, and attention than ever before.

Organizational behavior is a discipline in itself, with a solid research base and history (Robbins and Judge, 2015; Borkowski, 2016). Usually it is described at three levels: individual behavior is at the first level and influences group and team behavior, which is the second level, which influences behavior at the third organizational level. This chapter focuses on organizational behavior at this third and highest level. At this level we are including building blocks of culture, structure, human resources management, leadership, power and politics, communication, and motivation. These obviously depend on individual and group behaviors. Also, organizational behavior includes "extra-organizational" behavior, meaning organizations interacting with one another, an especially important concept when we consider sharing resources and data. While informaticians generally acknowledge the importance of organizational issues, they often do not deal with these issues in a

Key Advances in Clinical Informatics. DOI: http://dx.doi.org/10.1016/B978-0-12-809523-2.00009-1

systematic or strategic way. The result can be a painful and expensive implementation failure. In a classic example, the University of Virginia had to discontinue computerized provider order entry (CPOE) for a year after users protested that it impeded workflow (Massaro, 1993a,b).

We propose a model, the Golden Circle of Health Informatics Behavior, for thinking about these issues. Our model uses organizational behavior building blocks, but recommends assessing the issues at hand and then choosing appropriate targeted strategies for addressing these. The basis for our global model is that proposed by Simon Sinek as a leadership model (Sinek, 2009a,b). However, we have adapted it as a framework for thinking about complex organizational issues in health informatics.

THE GOLDEN CIRCLE OF HEALTH INFORMATICS BEHAVIOR

The visualization of the model is as a series of concentric circles. As shown in Fig. 9.1, which is our adaptation of the model, the core of this concentric model is what Sinek calls "The Why," which is the significant chief purpose of the organization or the product or the implementation. It is where the passion rests, e.g., for quality, for patient care, etc. In our model, "The Why" also includes chief motivators for change that come from outside the organization, such as payment reform, new government regulations, and financial incentives.

The next circle includes "The How," which is how an organization or a group within the organization does what it does. We elaborate on "The How" in this chapter, since we believe there are building blocks for "The How" processes and there are ways to select from and balance those building

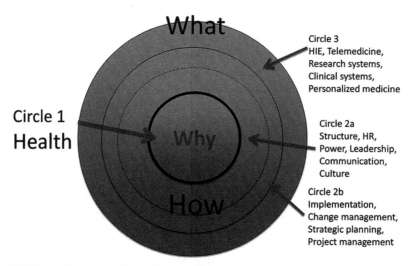

FIGURE 9.1   The Golden Circle of Health Informatics Behavior.

blocks to fit individual contexts. In Fig. 9.1 "The How" concentric circle outlines the topic areas that form the building blocks.

Finally, the outermost circle is "The What," which usually receives all the attention. This is what the organization does or produces. "The What" can be products (an information system) or efforts/services (telemedicine, health information exchange (HIE)). "The What" is compelling because it is what sells, but it cannot be produced efficiently and effectively if "The How" processes are deficient. Sinek contends that people do things for "The Why," not "The What." Unfortunately, however, most of us focus on "The How" and "The What" without ever considering "The Why."

Why Are People Motivated?

The core of what is done in health informatics is to improve the quality, safety, and efficiency of care. Those attracted to the field believe in this and are motivated to achieve these goals. While Sinek notes that in business, "The Why" is usually unclear and neglected, this is generally not the case in either healthcare or informatics. However, those in the field need reminders from time to time; managers can consider restating these goals as motivators for staff. For example, it is possible to reduce the "pain" caused by a difficult implementation process that could last for years by reminding those involved that they make a difference in patients' lives everyday. For them, leaders need to remove obstacles so that people can do what they are trained to do—help people. Sinek claims that when employees believe in "The Why," they are more satisfied, and the end result is good for the organization—efficiencies such as the ability to see more patients, increase the throughput of the patients, and improve the workflow will benefit patients as well as providers.

How Can People Succeed?

Individuals who focus on the people–process components in health informatics possess capabilities that are unique to our field: They bridge the gap between the clinical and IT worlds. Professionals in the field need to exercise their unique "bridger" capabilities in informatics (Ash et al., 2003). First, they can draw from the field of organizational behavior for some basic building blocks related to human behavior in organizations (shown in circle 2a of Fig. 9.1) and, capitalizing on their bridger knowledge, they can balance and mix those building blocks to suit the situation at hand, such as an implementation or change management process (shown in circle 2b of Fig. 9.1). For example, when planning an implementation of a tool for medication reconciliation, the implementation team needs to make decisions about a structure for managing the project (organizational structure), manpower needs (human resources), whom to involve and how (power), the extent of buy-in by executives (leadership), development of a communication and training effort (communication), and how the project can be framed to fit the organizational culture.

In a study by Novak and others, researchers discovered that the implementation team members not only completed the items listed above, but they also assumed the responsibility of being mediators who intervened in the use of technology and impacted the trajectory of use patterns. Their additional efforts included resolving challenges related to coordination, integrating the physical aspects of bar code medication administration into everyday practice, and providing strong advocacy (Novak et al., 2012).

Basic Organizational Ingredients for Success (Circle 2a)

As noted earlier, the basic building blocks exist at three levels: the individual, the group, and the organizational levels. Fig. 9.2 summarizes the building blocks. Individuals make up the organization and they bring to their work a set of professional skills and backgrounds, but also a set of personal attributes and experiences. Each has his or her own personality. For users, this diversity includes different learning styles so that training needs to target more than one style. For informatics staff, this means that people need to be selected, trained, and rewarded according to their preferences and styles. Each individual also brings a diversity of cultures to the organization, whether they are professional, national, or ethnic cultures.

At the group level, individual differences blend and behavior changes. We refer here to formal groups defined by the organization chart and, especially in informatics, formal groups can also take the form of committees such as clinical decision support (CDS) oversight committees. Since such a committee is focused on clinical content and experience, it is quite different from, for example, an established department within the organization. It therefore needs to be managed differently, with a more data and

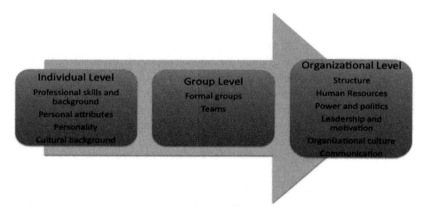

FIGURE 9.2 Basic organizational ingredients and building blocks for success.

consensus-driven approach. What is important is that after years of research in the organizational behavior field, a great deal is already known about how to manage committees, departments, and informatics projects. With a trend toward collaboration (Elliott Rose and Lackie, 2014; Johnson, 2013; Peterson et al., 2013; Pickering et al., 2012; Ruscitto, 2015), the patient-centered medical home (Finkelstein et al., 2011; O'Kane and Kinggard, 2014) and growing use of non-provider clinical staff (Brown, 2015; Yan et al., 2016), in addition to increasing IT use, some providers are feeling somewhat disenfranchised, and involvement in committee work can generate good will as well as better care (Ash et al., 2006). Teams are a special kind of group, one that has synergy, and team members need to be selected, trained, and rewarded for teamwork.

At the organizational level, the building blocks are organizational structure, human resources, power and politics, leadership and motivation, culture, and communication. Organizational structure is the formal way the organization is broken down into groups (Kinston, 1983). Informatics is interdisciplinary and relatively new as a discipline, so there are many different models for where it sits in an organization. In some hospitals, for example, informatics is a separate department, but in others it is part of the IT or quality department, depending on the local context. Human resources management involves hiring, training, tracking, and paying individuals. Any manager in informatics must be familiar with these processes. Important aspects are fitting the right person to the job, career development, and performance evaluation. The existence of unions in many healthcare organizations creates certain unique issues as well. For example, in one hospital system with which we are familiar, when nurses went on strike, all of the temporary replacement nurses had to be trained to use the electronic health record (EHR) at once, straining already stretched informatics and IT resources.

While discussion about power, politics, conflict, and negotiation is often avoided, these topics can be handled skillfully and for good purpose by leaders and managers who take the time to learn about them. They are, after all, facts of life in any organization. Leadership and motivation are topics that have received a good deal of attention in the research literature. Leaders are not all good managers and vice versa. The leader is generally "The Why" person, the one who articulates the vision and reason for existence of the organization, while the manager is "The How" person, who can get things done. Increasing attention is being paid to followership as well, since without followers there would be no organization. Individuals can play the follower role well (or not). What is important about motivation in organizations is that it must meet individual as well as organizational needs.

Organizational culture is critical in organizations and it is essential that informatics staff understand its importance. Culture is a "system of shared meaning held by members that distinguishes that organization from other organizations" (Robbins and Judge, 2015, p. 465). Organizations also have subcultures, meaning that different units or professions have their own cultures. As "bridgers," informatics professionals try to understand and bring cultures together. Culture is critical for implementation and change efforts as leaders must identify stakeholders' cultures and norms to involve and motivate them. In addition, a culture of safety needs to exist and should actively be promoted if we are to succeed in our work (Pickering et al., 2012).

Finally, communication is a basic building block of organizational behavior, which, for those in informatics, is important because individual backgrounds are interdisciplinary and therefore leaders must go the extra mile in translating across cultures. Excellent communication is also important for providing information and mechanisms for feedback from users. Most important is that leaders (project or senior) need to select the most appropriate channel of communication for the situation at hand and that they overcommunicate rather than undercommunicate (Dykstra, 2002).

The Contingency Theory Approach—Tailoring to the Situation

We mentioned the importance of matching the organizational strategy to the situation. While this concept is found in today's Contingency Theory, which says that there is no one right best way to lead or manage, it actually started in the 1920s with Mary Parker Follett and her transformational or situational leadership strategies (Phelps and Olson, 2007). Much of the skill of handling organizational issues is to be able to size up the situation, to diagnose the problem, and then pick from the building blocks for treatment or prevention purposes.

Using the Basic Ingredients to Develop Successful Strategies

The building blocks can be used in various ways to assist those in informatics in doing strategic planning, project management, change management, and implementation. Our first example concerns implementation of CPOE in a hospital.

The hypothetical Mercer Medical Center case described in 2000 (Ash et al., 2000) challenges us to deal with a situation in which outside pressures have caused the hospital administrator to demand that CPOE be implemented in a 9-month period rather than the planned 2-year period, although staff members are not ready. The Chief Medical Information Officer (CMIO) must decide whether to dig in her heels and refuse to consider it, accept the new timeline and alienate staff, or quickly develop a compromise plan. Presumably, a strategic plan is already in place, but by using project management principles and including change management strategies, all based on organizational behavior building blocks, within the project plan, the CMIO can provide the administration with an alternative. For example, she can show that only with a generous influx of additional resources and a careful plan that includes detailed pilot testing can the 9-month timeline be met.

The CMIO should first consider organizational structure. She presumably has no informatics department working with her, so she needs to work closely with the Chief Information Officer (CIO), who holds the resources under his control. Because the impetus for the timeline change is rooted in a healthcare quality issue, she must convince the quality improvement department to work with her as well. The project management structure and capability of the organization will be important to a successful CPOE implementation, and involvement of knowledgeable project management personnel will be critical for assessing alternative timelines.

The CMIO must consider human resources when thinking about the skills and numbers of staff needed for the implementation on the part of those doing the implementation but also from the point of view of users. For example, the workload of users may need to be lightened during the implementation period and the organization will need to expect lower productivity at that time, so manpower needs must be assessed and negotiated.

The CMIO's political skills with her colleagues will be essential. If she has no resources within her domain, all she has to work with are her negotiating skills. She knows her power lies in her relations with the medical staff, who have been supportive in the past, and she owes past success to her ability to bring different stakeholders to the table. This seems like an ideal situation for bringing administration, quality, clinicians, IT, and even patient and community representatives together.

Leadership comes into play in this situation on the part of not only the CMIO, but also the Chief Executive Officer (CEO). It is not a good sign that the CEO is demanding a diminished timeline without having involved the CMIO before this point, but at least the CEO had made a prior commitment to support CPOE implementation and to hire a talented CMIO. At this point, the CMIO needs to provide strong leadership on behalf of the medical staff, with the support of the CIO, to negotiate a compromise solution.

Organizational culture is a positive driving force in this organization. There is a sense of pride in being on the cutting edge of health information technology (HIT), the medical staff is on board, and there seems to be a culture of safety since the prime reason for the implementation is patient safety. The CMIO can use these values to her advantage as she negotiates a compromise that can preserve organizational values and fend off a risky shortening of the timeline.

Finally, the CMIO needs to consider communication and, as a consummate "bridger," she can assure that the outside entities that are pushing for a fast implementation, plus all of the internal stakeholders, and patient representatives, are identified and targeted with a communication strategy for each step in the selection, implementation, and optimization life cycle.

A Three-Phase Change Management Process to Support "The How"

We believe that knowledge and use of change theory is necessary for successful accomplishment of "The How." As noted, the building blocks in circle 2a of Fig. 9.1 can be used for change management purposes to accomplish "The How." Kurt Lewin, known as the father of social psychology, developed a well-known three-step model for organizational change (Lewin, 1958). His theory includes three phases. The cycle begins with sensing a need to change and beginning to assess the current state and plan for the future state. This is the "Unfreezing" stage. The second phase is the transformation phase. This phase is the actual implementation period that includes both organizational structure and behavioral components. He calls this phase "Changing." The third of Lewin's phases is "Refreezing." However, in our constantly changing world of informatics, refreezing rarely really happens because an organization needs to constantly be ready for the next change. In the spirit of modifying models so that they fit the discipline of health informatics, we are offering, in Table 9.1, an adaptation of the Lewin model specifically for informatics.

A key difference in our model in Table 9.1 is the transitional component between each phase. This means that neither people nor organizations "jump" into the next phase, because it is more of a gradual, carefully planned process. It is our experience that many of the failures or ideas not being successfully implemented occur because the planners and organizers are enthusiastic, but their excitement is not carried over to others in the organization who must change the processes that have been routine for a long time. Clinical champions can be instrumental in helping providers to adjust to the changes (Shea and Belden, 2016). The transitional component is similar to the translational phase in the "bench-to-bedside" model which outlines the way results from research (the bench) may influence health practice (bedside) and eventually population health. Drolet and Lorenzi (2011) suggest that the translational period is a black box during which active discussion and decision making for the next phase remain vague, but these activities provide a necessary bridge to the next phase.

Box 9.1 offers a second informatics example of "The How" process, based on the adapted Lewin model and targeted at precision medicine. This example uses the model to guide implementation of a project to use precision medicine to better prescribe the most appropriate medications for patients. For example, sometimes a patient is given a widely used drug, but the person reacts poorly to the drug with risky consequences. Or a patient may have a significant number of procedures leading to administration of a drug that works well for others but does nothing for this particular patient. Both situations could have been avoided if the patient's genetics were known so that the medications could be targeted with more precision.

TABLE 9.1 Adaptation of Kurt Lewin's Systems Model for Change

Desire to Implement Something Different	Critical Transition Phase	Implementation of the Desired Change	Critical Transition Phase	Outcome/Impact of the Desired Change
Creating the "why"	This is the preparing the people, the process, and the organization for the pending change	Starting the project as a pilot	This is the preparing the people, the process, and the organization for the pending change	Modifying or expanding the project
Understanding the options available for implementation		Expanding the pilot and reinforcing the change		
Gathering data—quantitative and/or qualitative		Gathering data-feedback from all participants—quantitative and/or qualitative		Gathering data-feedback from all participants—quantitative and/or qualitative
Analyzing the results		Analyzing the results		Analyzing the results
Planning for the implementation		Planning for expansion		Modifying this change or planning for the next change
Phase 1: Unfreezing		Phase 2: Starting the New		Phase 3: Living in the "Slush" of Constant Change

BOX 9.1 Case Study Example Illustrating Selection of "The How" for Precision Medicine

In the world of precision medicine, organizations and their informatics units will need to effectively negotiate the connection of people–process technology. The following example to give an overview of the combination of organizational skills and technology needed for success when implementing a precision medicine concept. The example is linked to the Golden Circle of Health Informatics Behavior. It starts with "The Why" statement. When the "The Why" is clearly presented to a patient as an option, the patient will likely be receptive to the new option.

What if the organization has a goal for prescribing a medication based an individual's DNA profile?

The Why

Healthcare providers may inadvertently hurt people by prescribing a drug that is "traditionally" used as the treatment of choice for a specific disease without knowing if it is appropriate for this individual. The "Why" is the desire to use a person's DNA to determine whether a drug is the best choice for the person with the disease. With use of the targeted and more effective drug, the patient's health would benefit.

We adapted the following based on Lewin's model to guide the implementation effort for this project.

The How and The What Combined

Phase 1

- The organization articulates the goal and selects one or two drugs or disease types for the initial implementation.
- The pilot area is selected and readiness to proceed is assessed.
- Plans are made to ensure (1) the clinicians involved know what is happening and are willing to personally be in the test group, (2) that the physician's patients are willing to be in the test system pilot, (3) that the people in the selected group have their DNA profiled, and (4) correct data are listed in the EHR.
- Plans for the four components will require clarity of purpose and actions for each one.

Phase 2

- There is a continuous need to monitor the process.
- The pilot will require a change of practice behavior. For example, some physicians who said they will be part of the system will likely not routinely participate.
- If patients need to pay extra for the DNA test, they might not be prepared to participate unless they truly understand "The Why."
- There is continuous examination of how information is recorded in the EHR.
- This phase includes sharing information about the process and ensuring that all the people–process and technology components are working as they should.

(Continued)

BOX 9.1 (Continued)

Phase 3

- An evaluation is conducted to determine success of the pilot.
- The evaluation will have a qualitative component, namely, interviews with participants—physicians, patients, others connected.
- The evaluation will have a quantitative component, regarding data quality, availability, etc.
- It will also have a clinical component—measuring outcomes, quality, etc.
- And the technology component, the link between this effort and the informatics systems, is assessed.

In this age of precision medicine (see Chapter 11: Bioinformatics and Precision Medicine), many organizations are at the "Unfreezing" phase (Tekola-Ayele and Rotimi, 2015; Vicini et al., 2016). In other words, they see the potential changes that will occur because of the "new" precision medicine, but they have not yet actually changed their organizations. To be in the "Changing" phase, organizations need to plan to unfreeze, transition to implementing the change, and transition to partially solidifying the change until further change is needed.

What Is It That We Are Trying to Accomplish?

Research on organizational issues in health informatics has largely focused on the implementation of clinical systems (Ash, 1997; Aydin et al., 1998; Cresswell and Sheikh, 2013; Fritz et al., 2015; Goldstein et al., 2004; Kaplan, 1997; Lorenzi and Riley, 2003; Lorenzi, 2005; Lorenzi et al., 2008; Mair et al., 2012; McAlearney et al., 2014). It is now widely recognized that attention to such issues is critical for successful implementation. Despite decades of research, however, there is no detailed recipe for success. We have learned what implementers need to consider, but only they can determine what applies to their own sites. For example, if nurses are unionized in an organization, the implementation plan must include working with union representatives, but that need not be a consideration for a hospital with no nurses' union.

More recent research has also been conducted on other informatics products or services such as telemedicine (Bangert and Doktor, 2000; Johnson, 2010; Whitten et al., 2010), HIE (Vest et al., 2013; Overhage, et al, 2005), research data systems (Ash et al., 2008; Embi et al., 2013; Peute et al., 2010), and systems related to personalized medicine (Vicini et al., 2016). Each one of these "What" areas has unique aspects that pose new and different challenges. All involve organizational (or multiorganizational) structure, human resources, power and politics, leadership, and communication issues and challenges. Telemedicine and personalized medicine are similar to EHR applications in that they are generally implemented within individual organizations and therefore follow general principles of implementation throughout an implementation and optimization life cycle. Both are extremely dependent on the engagement of the clinician user population. HIE and research data systems, though dependent on the EHRs of individual organizations, involve multiorganizational structures for data sharing. Of the organizational building blocks described above, leadership skills are probably the most important for these two "Whats." In both cases, the wise leader will pay attention to "The Why" and then figure out human resources, the power and politics within and between organizations, leadership strengths and weaknesses among the many players, possible clashes of organizational cultures, and cross-organizational communication strategies.

Of "The What" areas outlined, the least studied from the organizational point of view is personalized medicine. The report on the American Medical Informatics Association Health Policy Invitational Meeting held in 2014 describes numerous policy and technical issues related to informatics and personalized medicine (Wiley et al., 2016). Policy issues regarding data protection, standardization of terminology and data models, and data integrity involve organizational issues, but they are not described in detail nor have they been researched extensively. One particularly fruitful and necessary focus of future research is the end user perspective. The report notes that clinical expertise in interpreting reports resulting from genetic testing is needed and could be assisted by CDS embedded in EHRs, but readiness on the part of practicing clinicians and the impact on their workflows is as yet unknown.

As healthcare and informatics professionals become more excited about the possibilities of health and healthcare through precision/personalized medicine, it is critical to develop organizational behavior processes within the implementation of the new initiatives for both the clinicians who will "prescribe" or "order" the new and the patient's who will "use" the new.

LOOKING AHEAD—FUTURE TRENDS AND DEVELOPMENTS

As has been emphasized throughout this volume, the sociotechnical and constantly evolving nature of HIT means that its complexity will grow rather than diminish over time. As adoption continues to increase and implementation of the many "Whats" that make up the ever-growing catalogue of HIT products and services becomes even more widespread, organizations will move into the optimization phase of an expanded "implementation" life cycle or begin a new life cycle if "The What" is significantly new. For example, a number of organizations with established locally developed EHRs are in the process of abandoning them in favor of commercial products. Whether this is considered by an organization to be a new implementation, a re-implementation, or an optimization, it will be vastly different from the first implementation and there will be many lessons learned as a result.

Adding to the complexity of EHRs will be how the new precision medicine, DNA, tests, data, etc. will be stored, accessed, and from a culture standpoint—used. Careful evaluation research about this next largely unstudied future phase will hopefully yield knowledge about best practices for organizations that need to do this later.

The mixing and matching of the building blocks for organizational success cannot be done using a detailed prescriptive recipe—it takes knowledge of organizational behavior, of healthcare, and of HIT. However, there are theory-based methods for doing an organizational assessment. The Golden Circle of Health Informatics Behavior and the building blocks described here are essential for increasing understanding and successfully implementing "The New."

Informatics professionals entering the field through graduate and training programs are learning the health and IT pieces, but many are not exposed to any formal training in organizational and management issues. There is widespread recognition that knowledge and skills in this third organizational area are needed. For example, for medical subspecialty certification in informatics for physicians, competence in leadership and management knowledge are required and assessed in an examination. The more training and experience they and all other leaders in informatics can absorb in the field of organizational behavior and management, and the more research and evaluation that can be conducted in those areas, the better our implementations, our "What," will be in the future.

CONCLUSIONS

Organizational issues in health informatics are complex and there are no detailed recipes for guaranteed success because every organization is different. However, leaders can use the building blocks (the ingredients) in the Golden Circle of Health Informatics Behavior to formulate their own recipes for success. Informatics leaders bridge the gap between the clinical and IT worlds. We urge them to seek an appreciation and understanding of organizational behavior as well so that they can become even more effective bridgers.

REFERENCES

Ash, J.S., 1997. Organizational factors that influence information technology diffusion in academic health sciences centers. J. Am. Med. Inform. Assoc. 4 (2), 102−111.

Ash, J.S., Anderson, J.G., et al., 2000. Managing change: analysis of a hypothetical case. J. Am. Med. Inform. Assoc. 7 (2), 125−134.

Ash, J.S., Stavri, P.Z., Dykstra, R., Fournier, L., 2003. Implementing computerized physician order entry: the importance of special people. Int. J. Med. Inform. 69, 235–250.

Ash, J.S., Sittig, D.F., Campbell, E., Guappone, K., Dykstra, R., 2006. An unintended consequence of CPOE implementation: shifts in power, control, and autonomy. Proc. Am. Med. Inform. Assoc. 2006, 11–15.

Ash, J.S., Anderson, N.R., Tarczy-Hornoch, P., 2008. People and organizational issues in research systems implementation. J. Am. Med. Inform. 15 (3), 283–289.

Aydin, C.E., Anderson, J.G., Rosen, P.N., Felitti, V.J., Weng, H.C., 1998. Computers in the consulting room: a case study of clinician and patient perspectives. Health Care Manag. Sci. 1 (1), 61–74.

Bangert, D., Doktor, R., 2000. Implementing store-and-forward telemedicine: organizational issues. Telemed. J. E-Health 6 (3), 355–360.

Borkowski, N., 2016. Organizational Behavior in Health Care, 3rd ed. Jones & Bartlett, Burlington, MA.

Brown, L., 2015. Expanding opportunities through patient care: pharmacists provide care...let me tell you about it. J. Am. Pharm. Assoc. 55 (4), 348–353.

Cresswell, K., Sheikh, A., 2013. Organizational issues in the implementation and adoption of health information technology innovations: an interpretative review. Int. J. Med. Inform. 82 (5), e73–e86.

Drolet, B.C., Lorenzi, N.M., 2011. Translational research: understanding the continuum from bench to bedside. Transl. Res. 157 (1), 1–5.

Dykstra, R., 2002. Computerized physician order entry and communication: reciprocal impacts. Proc. Am. Med. Inform. Assoc. 2002, 230–234.

Elliott Rose, A., Lackie, K., 2014. From me to we and back again: creating health system transformation through authentic collaboration within and beyond nursing. Nurs. Leadersh. 27 (1), 21–25.

Embi, P.J., Hebert, C., Gordillo, G., Kelleher, K., Payne, P.R.O., 2013. Knowledge management and informatics considerations for comparative effectiveness research: a case-driven exploration. Med. Care 51 (8 Suppl. 3), S38–S44.

Finkelstein, J., Barr, M.S., Kothari, P.P., Nace, D.K., Quinn, M., 2011. Patient-centered medical home cyberinfrastructure current and future landscape. Am. J. Prev. Med. 40 (5 Suppl. 2), S225–S233.

Fritz, F., Tilahun, B., Dugas, M., 2015. Success criteria for electronic medical record implementations in low-resource settings: a systematic review. J. Am. Med. Inform. Assoc. 22 (2), 479–488.

Goldstein, M.K., Coleman, R.W., Tu, S.W., Shankar, R.D., O'Connor, M.J., Musen, M.A., et al., 2004. Translating research into practice: organizational issues in implementing automated decision support for hypertension in three medical centers. J. Am. Med. Inform. Assoc. 11 (5), 368–376.

Johnson, J.E., 2013. Working together in the best interest of patients. J. Am. Board Fam. Med. 26 (3), 241–243.

Johnson, N.D., 2010. Teleradiology 2010: technical and organizational issues. Ped. Rad. 40 (6), 1052–1055.

Kaplan, B., 1997. Addressing organizational issues into the evaluation of medical systems. J. Am. Med. Inform. Assoc. 4 (2), 94–101.

Kinston, W., 1983. Hospital organization and structure and its effect on inter-professional behavior and the delivery of care. Soc. Sci. Med. 17 (16), 1159–1170.

Lewin, K., 1958. Group Decision and Social Change. Holt, Rinehart and Winston, New York.

Lorenzi, N.M., 2005. Clinical adoption, Aspects of Electronic Health Record Systems, second edition Springer, New York, pp. 378–397., Health Informatics Series.

Lorenzi, N.M., Riley, R.T., 2003. Managing Technological Change: Organizational Aspects of Health Informatics. Springer, New York.

Lorenzi, N.M., Novak, L.L., Weiss, J.B., Gadd, C.S., Unertl, K.M., 2008. Crossing the implementation chasm: a proposal for bold action. J. Am. Med. Inform. Assoc. 15 (3), 290–296.

Mair, F.S., May, C., O'Donnell, C., Finch, T., Sullivan, F., Murray, E., 2012. Factors that promote or inhibit the implementation of e-health systems: an explanatory systematic review. Bull. World Health Organ. 90 (5), 357–364.

Massaro, T.A., 1993a. Introducing physician order entry at a major academic medical center: I. Impact on organizational culture and behavior. Acad. Med. 68, 20–25.

Massaro, T.A., 1993b. Introducing physician order entry at a major academic medical center: II. Impact on medical education. Acad. Med. 68, 25–30.

McAlearney, A.S., Hefner, J.L., Sieck, C., Rizer, M., Huerta, T.R., 2014. Evidence-based management of ambulatory electronic health record system implementation: an assessment of conceptual support and qualitative evidence. Int. J. Med. Inform. 83 (7), 484–494.

Novak, L.L., Anders, S., Gadd, C.S., Lorenzi, N.M., 2012. Mediation of adoption and use: a key strategy for mitigating unintended consequences of health IT implementation. J. Am. Med. Inform. Assoc. 19 (6), 1043–1049.

O'Kane, C.C., Kinggard, T.E., 2014. Patient centered medical home update. J. Ark. Med. Soc. 110 (12), 252–253.

Overhage, J.M., Evans, L., Marchibroda, J., 2005. Communities' readiness for health information exchange: the National Landscape in 2004. J. Am. Med. Inform. Assoc. 12 (2), 107–112.

Peterson, L.E., Phillips, R.L., Puffer, J.C., Bazemore, A., Petterson, S., 2013. Most family physicians work routinely with nurse practitioners, physician assistants, or certified nurse midwives. J. Am. Board Fam. Med. 26 (3), 244–245.

Peute, L.W., Aarts, J., Bakker, P.J., Jaspers, M.W., 2010. Anatomy of a failure: a sociotechnical evaluation of a laboratory physician order entry system implementation. Int. J. Med. Inform. 79 (4), e58–e70.

Phelps L. D., Olson B.J., 2007. Edwards Deming, Mary P. Follett and Frederick W. Taylor: reconciliation of differences in organizational and strategic leadership. Acad. Strat. Manag. J. 6, 1–14.

Pickering, B.W., Litell, J.M., Herasevich, V., Gajic, O., 2012. Clinical review: the hospital of the future—building intelligent environments to facilitate safe and effective acute care delivery. Crit. Care 16 (2), 220.

Robbins, S.P., Judge, T.A., 2015. Organizational Behavior, 16th ed. Pearson, Upper Saddle River, NJ.

Ruscitto, K.H., 2015. The future has arrived: shared, collaborative and sustainable. Health Prog. 96 (5), 18–21.

Shea, C.M., Belden, C.M., 2016. What is the extent of research on the characteristics, behaviors, and impacts of health information technology champions? A scoping review. BMC Med. Inform. Decis. Mak. 16, 2.

Sinek, S., 2009a. Start With Why: How Great Leaders Inspire Everyone to Take Action. Penguin Group, New York.

Sinek S., 2009b (date posted September 28, 2009). Start with why—how great leaders inspire action. TEDx Talks YouTube video file. Retrieved January 20, 2017 from https://www.youtube.com/watch?v = u4ZoJKF_VuA&t = 75s.

Tekola-Ayele, F., Rotimi, C.N., 2015. Translational genomics in low- and middle-income countries: opportunities and challenges. Public Health Genomics 18 (4), 242–247.

Vest, J.R., Campion, R.R., Kaushal, R., 2013. Challenges, alternatives, and paths to sustainability for health information exchange efforts. J. Med. Syst. 37 (6), 9987.

Vicini, P., Fields, O., Lai, E., Litwack, E.D., Martin, A.M., Morgan, T.M., et al., 2016. Precision medicine in the age of big data: the present and future role of large-scale unbiased sequencing in drug discovery and development. Clin. Pharmacol. Ther. 99 (2), 198−207.

Whitten, P., Holtz, B., Nguyen, L., 2010. Keys to a successful and sustainable telemedicine program. Int. J. Technol. Assess. Health Care 26 (2), 211−216.

Wiley, L.K., Tarczy-Hornoch, P., Denny, J.C., Freimuth, R.R., Overby, C.L., Shah, N., et al., 2016. Harnessing next-generation informatics for personalizing medicine: a report from AMIA's 2014 Health Policy Invitational Meeting. J. Am. Med. Inform. Assoc. 23 (2), 413−419.

Wright, A., Ash, J.S., Erickson, J.L., Wasserman, J., Bunce, A., Stanescu, A., et al., 2014. A qualitative study of the activities performed by people involved in clinical decision support: recommended practices for success. J. Am. Med. Inform. Assoc. 21 (3), 464−472.

Yan, C., Rose, S., Rothberg, M.B., Mercer, M.B., Goodman, K., Misra-Hebert, A.D., 2016. Physician, scribe, and patient perspective on clinical scribes in primary care. J. Gen. Intern. Med. 31 (9), 990−995.

RECOMMENDED FURTHER READING

Strauss, A.T., Martinez, D.A., Garcia-Arce, A., Taylor, S., Mateja, C., Fabri, P.J., et al., 2015. A user needs assessment to inform health information exchange design and implementation. BMC Med. Inform. Decis. Mak. 15, 81.

Chapter 10

Medication Management, and Laboratory and Radiology Testing

Sarah P. Slight[1] and David W. Bates[2]
[1]Highlander House, Ovington, Northumberland, United Kingdom, [2]Brigham and Women's Hospital/Harvard Medical School, Boston, MA, United States

INTRODUCTION

In this chapter, we review three key clinical areas—imaging, laboratory testing, and prescribing—and, in so doing, highlight problems with their use, the frequency of errors, why these occur, and how health information technology (HIT) is being used to improve the safety and quality of care in these areas. We discuss how natural language processing (NLP) can assist with the extraction of free-text medication-related information from clinical notes and a novel, web-based system—*Renal Patient View* (RPV)—can provide patients with convenient access to their results. Finally, we consider the very positive findings from the use of a new secure web-based communication tool—*Alert Notification of Critical Results (ANCR)*—to facilitate the transfer of critical and clinically significant imaging results between care providers.

MEDICATIONS

Rate and Types of Medication Errors

Most of what health care providers do is governed by the health care system in which they work (Errors et al., 2007). Numerous examples have been reported in the literature showing how underlying system failures can shape individual behavior and create conditions under which medication errors occur. One such case was the death of a day-old infant in Denver in 1996, which was caused by the injection of a 10-fold overdose of penicillin G benzathine intravenously rather than intramuscularly (Errors et al., 2007). Later, upon autopsy, it was confirmed that this infant never had the condition for which penicillin G benzathine was prescribed, and thus never needed

Key Advances in Clinical Informatics. DOI: http://dx.doi.org/10.1016/B978-0-12-809523-2.00010-8
131

treatment. Another catastrophic case was that of 18-year-old Wayne Jowett in England in 2001, who was injected with a dose of the cytotoxic drug vincristine intrathecally rather than intravenously, leading to his death (Dyer, 2001; Bates and Slight, 2014). In these two cases, medication errors were never the fault of a single individual or caused by the failure of a single element; they were multifactorial.

Medication errors are common and have been found to occur at an unacceptably high rate. The incidence of medication errors also varies greatly across health care settings, with hospital care being the most researched. The rates of prescribing errors for adults in US hospitals were found to vary considerably from 0.6 to 53.0 per 1000 orders (Errors et al., 2007; Bates et al., 1995; Lesar, 2002; Lesar et al., 1990, 1997). Much of this variation can be explained by the use of different definitions of error and detection methods. Bates et al. used comprehensive identification methods and found a rate of 1400 prescribing errors per 1000 patient admissions or 0.3 prescribing errors per patient per day (Bates et al., 1995). Kaushal et al. used similar detection methods in pediatric units and found a rate of 405 prescribing errors per 1000 patient admissions or 0.1 prescribing errors per patient per day (Kaushal et al., 2001). In the primary care setting, Avery et al. conducted a retrospective case note review of unique prescription items in 15 general practices in England and detected prescribing or monitoring errors in 1 in 20 of these items (Avery et al., 2013). The vast majority of errors were judged to be of mild to moderate severity, and a number of factors associated with their occurrence, including age less than 15 years or more than 64 years, and higher numbers of unique medication items prescribed. Several staff from these general practices also participated in interviews and focus groups and felt that: (1) the working environment with its extensive workload, time pressures, and interruptions and (2) problems with the timeliness, legibility, content, and layout of secondary care correspondence, possibly contributed to the occurrence of these errors (Slight et al., 2013).

The heterogeneity of clinical data can also present challenges for health care providers in all types of settings. Electronic health records (EHRs) often contain large amounts of medication-related information in free-text fields (see Chapter 4: Electronic Clinical Documentation). This information may be necessary to create an accurate medication profile of the patient and provide safe personalized care, but is often inaccessible to computerized applications that rely on structured data.

The Use of HIT to Overcome Key Medication-Related Challenges

Computerized provider order entry (CPOE) and computerized decision support (CDS) systems can play a key role in intercepting errors and overcome key medication-related challenges. CDS can provide physicians with relevant, timely, treatment-related information and decrease the costs of medications (see Chapter 12: Clinical Decision Support and Knowledge Management). CDS can keep track of complex formularies and help providers suggest the appropriate drug for an individual patient. They can also be used to implement guidelines for expensive medications, which can be used to improve the likelihood that the appropriate patients will get these medications. Another influential, but more subtle approach to reducing medication costs is to suggest the appropriate dosage for a specific patient, which is often lower that providers may have ordered left to their own devices (see Chapter 8: Health Information Technology and Value).

Natural Language Processing

One example of a key technological advance with implications for many areas including medications is NLP. NLP is the ability of a computer program to understand human speech as it is spoken. While NLP has been available for some time, it has recently become much more sophisticated and accurate and it is, as a result, now increasingly being used in clinical care. Many studies have worked on extracting free-text medication-related information from clinical notes and discharge summaries using NLP (Chhieng et al., 2007; Levin et al., 2007; Gold et al., 2008). Uzuner et al. found that different NLP systems performed well at extracting data on medications, dosages, routes, frequencies, but not so well at detecting how long the medicine was to be administered for and the medical reasons for which it was given (Uzuner et al., 2010). Vanderbilt University Medical Center developed a new NLP system (MedEx) that was superior to systems for the task of extracting medication information reported in previous studies, but which also generated a number of false positives (Xu et al., 2010). Xu et al. explained how these false positives could have been caused by a number of different factors, including ambiguous drug names, e.g., "potassium is normal," where "potassium" refers to a laboratory test instead of a drug (supplement); or "Vital" which could be a drug name or an adjective in the English language (Xu et al., 2010). Some regular terms, like "06" (meaning 6:00 a.m.), were not recognized by the system. The authors also evaluated MedEx's performance using two different data sets: discharge summaries and outpatient clinic visit notes. Small drops on recall were observed when MedEx was applied to outpatient clinic visit notes, which may have been due to more spelling errors and abbreviations of drug names in clinic visit notes (Xu et al., 2010). NLP systems may also assist with the mapping of data elements present in unstructured text to structured fields in an EHR in order to improve data quality. These systems may also "learn" over time, monitoring the results of users' previous interactions (as feedback) and noting where expectations were not met. Clearly, NLP systems can play a valuable role in capturing textual information and making it accessible to users. However, there is still quite a lot of work to do to overcome the challenges of processing complex medication-related textual information in diverse settings.

LABORATORY TESTING

Laboratory medicine has a prominent role to play in patients' safety. As in other medical fields, errors can occur in diagnostics; the vast majority of all medical diagnoses rely on laboratory data. Over the past 20 years, the number of laboratory tests available to US clinicians has more than doubled to at least 3500 tests (Hickner et al., 2014). This increased volume and complexity can present clinicians with new challenges in accurately ordering and interpreting diagnostic tests.

Rate and Types of Laboratory Errors

The laboratory testing process can be divided into three stages: *preanalytical*, *analytical*, and *postanalytical* (Fig. 10.1). Some authors have also included a *pre-preanalytical* and *post-postanalytical* stage to incorporate activities associated with the initial selection and interpretation of tests, respectively (Laposata and Dighe, 2007). The definition of these terms can often vary between institutions (Hawkins, 2012), and evidence suggests that the frequency of laboratory errors in these different stages can vary greatly, with the majority of errors occurring in the *pre-* and *postanalytical* stages (Carraro and Plebani, 2007; Plebani, 2010; Astion et al., 2003; Kalra, 2004).

Preanalytical Errors

In the preanalytical stage, hemolysis (the pathological breakdown of red blood cells) was found to be responsible for the vast majority of samples being rejected in a clinical chemistry laboratory (Chawla et al., 2010). In most cases, hemolysis is due to a failure to follow the correct procedures for sample collection, handling, and storage (Lippi et al., 2011). Samples may not be accompanied with appropriate requisition slips, or the slips not filled out correctly (wrong names or identifiers (ID)), and insufficient amount of blood supplied (Chawla et al., 2010; Upreti et al., 2013). Other common

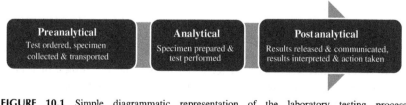

Preanalytical
Test ordered, specimen collected & transported

Analytical
Specimen prepared & test performed

Postanalytical
Results released & communicated, results interpreted & action taken

FIGURE 10.1 Simple diagrammatic representation of the laboratory testing process (Hammerling, 2012).

preanalytical errors included ordering the wrong test, the correct test but for the wrong patient, or supplying samples in an incorrect container (Carraro and Plebani, 2007; Hammerling, 2012; Plebani and Carraro, 1997). A recent review of the literature reported how failures to order appropriate diagnostic tests, including laboratory tests, accounted for 55% of observed incidents of missed and delayed diagnoses in the ambulatory setting and 58% of errors in the emergency department (ED) (Plebani, 2010).

Analytical Errors

Common *analytical* errors included the mix-up of samples, undetected failure in quality control, and procedures not followed correctly (Carraro and Plebani, 2007; Abdollahi et al., 2014). Sakyi et al. retrospectively analyzed data obtained from a clinical biochemistry laboratory over a 3-year period and uncovered 108 (0.1%) *analytical* errors, the main cause of these being equipment malfunction (Sakyi et al., 2015). Abdollahi et al. found a much higher error rate of 23.2% ($n = 35{,}531$) over a similar time period, the main causes relating to the reagent used and interference of other substances (Abdollahi et al., 2014).

Postanalytical Errors

Postanalytical errors can include sending the result(s) to the wrong patient and interpreting the result incorrectly (Laposata and Dighe, 2007; Plebani, 2010). Incorrect interpretation of laboratory tests in particular was found to be responsible for a high percentage of errors in primary care (37%), internal medicine (38%), and the ED (37%) (Plebani, 2010). General practitioners were found to want more interpretative comments from UK laboratories and believed this was an invaluable part of the "laboratory service" (Kilpatrick and Freedman, 2011). The overall error rate of the laboratory testing process was found to decrease significantly over time, with changes occurring in the types and frequencies of errors, particularly in the *preanalytical* phase (Carraro and Plebani, 2007). The authors linked this to the introduction of a CPOE system (the benefits of which are discussed in the next section).

The Use of HIT to Overcome Key Laboratory Challenges

Clinicians are primarily responsible for test ordering, the first key step of the laboratory testing process. However, clinicians may face several challenges, including an expanding menu of available tests that has been linked to the expected rise in volume and complexity of tests over time (Hickner et al., 2014). Most CPOE systems provide search functionality, which can return a narrowed-down list of corresponding tests from a specific search term. To provide patient-specific testing advice, the CPOE system must have real-time electronic access to patient data that includes other test results and medications. CPOE systems therefore need to interface with other EHR systems, such as the Laboratory Information Management System (LIMS) and clinical data repository. This allows laboratory orders to be directly transmitted and received electronically, thus reducing transcription errors (Baron and Dighe, 2011). The LIMS can also generate bar-coded specimen labels for samples at the point-of-care by scanning a patient's identification wristband, thus reducing the risk of mislabeled specimens (Baron and Dighe, 2011).

CPOE systems have also demonstrated improvements in the overall appropriateness of laboratory orders and utilization (Nightingale et al., 1994). Neilson et al. analyzed the ordering habits of clinical staff at Vanderbilt University Hospital and used these findings to make two modifications to their CPOE system (Neilson et al., 2004). These included introducing a daily prompt in the system that asked providers whether they wanted to discontinue tests scheduled beyond 72 h and, second, unbundling panel tests into single components (e.g., sodium, potassium) so that they could be ordered separately. The authors reported a 24% reduction in test ordering beyond 72 h (first modification), with an accumulative reduction of up to 51% for panel component tests (following the second modification) (Neilson et al., 2004). This study showed how it was possible to curtail overutilization of inpatient laboratory tests with a simple modification to the ordering screen of a CPOE system.

RENAL PATIENT VIEW

The recently published *Pathology Modernisation Programme* illustrated some key ways in which digital creativity and innovation can make a real difference to patient care (NHS England). There is an increasing desire for patients to have access to their own pathology reports; the third generation of internet-based services, which emphasize machine-facilitated understanding of information, can provide patients with access to their data from anywhere, including cloud and smartphones, and a more intuitive user experience (NHS England). The RPV is a novel, web-based system in the United Kingdom where kidney patients are currently able to access their test results along with information and advice on their condition. The service is now used by 80% of renal units and appears to be relatively unique to both the United Kingdom and nephrology as a specialty (NHS England; Woywodt et al., 2014). Woywodt et al. reported that most patients accessed RPV on average one to five times per month (78%) and found it easy to use (92%) (Woywodt et al., 2014). They used RPV to monitor their kidney function (73%), check creatinine levels (81%), check potassium results (57%), and felt better informed and more in control of their condition (93%). By harnessing the potential of digital channels, it is clear that online patient portals like RPV can provide patients with access that was more appropriate and convenient to their lifestyles and needs. This, is turn, has the potential to reduce face-to-face contact between clinicians and patients at times where it may not be considered necessary.

RADIOLOGY TESTING

Rate and Types of Radiological Errors

A literature review published on radiological errors found a "real-time" error rate among radiologists in their day-to-day practices of 3%−5% (Berlin, 2007). However, by concentrating only on radiology studies that harbored significant abnormalities, the error or "miss" rate was found to be much higher at 30% (Berlin, 2007). The error rate can vary depending on the specific radiological investigation performed (Goddard et al., 2001). Harvey et al. screened 152 previous "normal" mammograms in which breast carcinomas were later detected in subsequent mammograms for these patients and found that as many as 75% of the previous "normal" mammograms showed evidence of carcinoma on subsequent review (Harvey et al., 1993). However, the authors were careful to conclude that these may not necessarily have been true radiological errors as the majority of mammographic abnormalities manifest only as asymmetric densities, which occur very frequently and could be easily overlooked. Radiological imaging results require expert interpretation often under conditions of uncertainty. According to Brady et al., errors are considered inevitable and the concept of necessary fallibility must be accepted (Brady et al., 2012).

Radiological errors can be grouped into four main types: observer errors, errors in interpretation, failure to communicate the result in a timely and clinically appropriate manner, and failure to suggest the next appropriate procedure (Pinto and Brunese, 2010). Observer errors included failure to fixate in the area of the lesion (scanning error) or a failure to detect it (recognition error) (Kundel et al., 1978). Failures to detect a lesion in a film reading could be related to excessive visual and mental fatigue of radiologists, as a consequence of continuous and prolonged decision making (Krupinski et al., 2010; Gaba and Howard, 2002). This has become an increasingly important issue as the volume and complexity of images has grown over the past number of years (Vertinsky and Forster, 2005). Lesion size and shape were also considered important factors influencing whether a cancer was detected or not.

After detecting the lesion, it is possible for the radiologist to incorrectly interpret a malignant lesion as a normal structure (Kundel et al., 1978). There were a number of different reasons why radiologists may make this type of error, including the level of vigilance of the interpreter, the reading room environment, patient's clinical history, and the presence or absence of previous studies (Pinto and Brunese, 2010). Lee et al. suggested that radiologists should seek more clinical information from the ordering physician when the interpretation of an image is tightly coupled with clinical context (Lee et al., 2013). Whang et al. reviewed the malpractice history of 8401 radiologists across 47 US states to determine the most frequent causes of malpractice suits against radiologists (Whang et al., 2013). This study found that interpretive errors were, by far, the most common cause with the most frequently missed findings relating to breast cancer. Although less common, errors of communication with either the patient or ordering physician were also reported as reasons for initiating a lawsuit (Whang et al., 2013).

HIT Interventions to Improve the Efficiency of Radiology Test Utilization

The costs of radiology testing continue to rise, and over time newer imaging technologies tend to replace older ones, although the marginal increase in value of newer technologies is not always clear (Matin et al., 2006). A variety of interventions have been implemented to attempt to improve the use of radiology testing, including utilization management (Otero et al., 2006), CDS, requiring radiology consultation (Ingraham et al., 2016), and image-sharing technologies (Vest et al., 2015). These have been met with mixed success; simply displaying the charges for tests did not have any effect in one study (Bates et al., 1997). However, required consultation for certain higher cost tests appears to have some impact; in one study Ingraham found that 4% of high-cost imaging requests were rejected, and 9% received recommendation for an alternate investigation (Ingraham et al., 2016). Decision support for high-cost tests appears likely to be beneficial. With data exchange, image sharing technologies do appear to reduce the rates of repeat and unnecessary imaging (Vest et al., 2015).

Alert Notification of Critical Results

The Joint Commission recognized the importance of timely communication of critical test results and urged hospitals to develop, implement, and evaluate procedures for managing and reporting these clinically significant results (The Joint Commission, 2016). The Brigham and Women's Hospital (Boston, MA) designed a new secure web-based communication tool—ANCR—to facilitate the transfer of critical and clinically significant imaging results between care providers (Lacson et al., 2014). Three levels of alerts (red, orange, or yellow) were designed and dependent on the urgency with which the critical result needed to be communicated and addressed. The timeline for notification also varied between alert types, as too did the preferred manner by which the results should be communicated. For example, a tension pneumothorax on chest radiography was considered urgently life-threatening (red alert) and should be communicated within 60 min through interruptive mechanisms (e.g., in-person, paging, or by phone). It was decided that less urgent findings should be noninterruptive, thus helping to reduce inefficiency and unnecessary disruptions in the workflow of the ordering physician. An example of a yellow alert was a new incidental pulmonary nodule, which was less time-sensitive and could be asynchronously communicated (e.g., secure e-mail) within 15 days. The ANCR system also recorded when the alert was created and when it was acknowledged by ordering providers. This was an important consideration, as the system continued to generate subsequent pages and/or e-mails until the communication loop was closed.

Lacson et al. evaluated the impact of this ANCR system at the Brigham and Women's Hospital, a 753-bed urban adult tertiary referral academic medical center in Boston, MA (Lacson et al., 2014). The system was implemented in January 2010 and integrated with the institution's multiple clinical information systems, including the Picture Archiving and Communication System, physician directory service, paging and e-mail systems, and patient electronic medical records. This integration helped to reduce the amount of data input required by the radiologists and ordering providers, and improved workflow. A random sample of radiology reports ($n = 37,604$) were manually reviewed after the ANCR implementation and compared with baseline outcomes 1 year before ($n = 9430$). The authors observed a ninefold increase in the communication of critical imaging results via the new ANCR system within the first 4 years after implementation (chi-square trend test, $P < 0.0001$) (Lacson et al., 2014). The tool was also shown to reduce workflow interruptions with the utilization of 41,445 noninterruptive yellow-level alerts, which was approximately 41% of all ANCR alerts over the 4 years (Lacson et al., 2014). Where ANCR was used to communicate critical results to referring providers, these providers used the new tool to acknowledge result receipt in more than 98% of cases.

Although other institutions may have different practices and policies, the authors reported how a number of factors positively influenced ANCR adoption at the Brigham and Women's Hospital. Firstly, there was substantial institutional and executive support for the timely communication of critical test results, thus specifically addressing the potential breakdowns in the patient care process that can lead to malpractice claims. Second, ANCR was integrated within the existing workflow of radiologists and ordering providers, and the radiologists could track whether the ordering provider or indeed another covering clinician followed up the alert. Third, noninterruptive ANCR communication was used for less urgent findings, thus replacing the more intrusive and time-consuming person-to-person or telephone communication that had been previously used. If we reflect on the fact that this institution performs more than 400,000 radiological examinations annually, the ANCR implementation brought about the timely closed-loop communication of an additional 28,800 reports (7.2% improvement). However, this study did not evaluate whether the required follow-up actions were taken in response to the critical imaging results, nor the direct role of patients receiving these alerts.

CONCLUSIONS

In this chapter, we have discussed the frequency and types of medication, laboratory and radiology errors, as well as the evidence that HIT can improve efficiency in these areas. Clinicians typically face continually increasing workloads and pressures, which can severely limit the time available to comprehend the contents of medication-related correspondence, and pathology or radiology reports. We focused in particular on key technological advances that can assist and guide both clinicians with the accessibility of information (NLP) and patients with their test results (PRV), the selection of pathology tests (CPOE), and facilitate the transfer of critical and clinically significant imaging results (ANCR). Digital technology seems a natural fit for helping both clinicians and patients not only to respectively treat but also self-manage their diseases, thus leading to a greater sense of control and potentially better patient outcomes.

REFERENCES

Abdollahi, A., Saffar, H., Saffar, H., 2014. Types and frequency of errors during different phases of testing at a clinical medical laboratory of a teaching hospital in Tehran, Iran. N. Am. J. Med. Sci. 6 (5), 224−228.

Astion, M.L., Shojania, K.G., Hamill, T.R., Kim, S., Ng, V.L., 2003. Classifying laboratory incident reports to identify problems that jeopardize patient safety. Am. J. Clin. Pathol. 120 (1), 18−26.

Avery, A.J., Ghaleb, M., Barber, N., Dean Franklin, B., Armstrong, S.J., Serumaga, B., et al., 2013. The prevalence and nature of prescribing and monitoring errors in English general practice: a retrospective case note review. Br. J. Gen. Pract. 63 (613), e543−e553.

Baron, J.M., Dighe, A.S., 2011. Computerized provider order entry in the clinical laboratory. J. Pathol. Inform. 2, 35.

Bates, D.W., Slight, S.P., 2014. Medication errors: what is their impact? Mayo Clin. Proc. 89 (8), 1027−1029.

Bates, D.W., Boyle, D.L., Vander Vliet, M.B., Schneider, J., Leape, L., 1995. Relationship between medication errors and adverse drug events. J. Gen. Intern. Med. 10 (4), 199−205.

Bates, D.W., Kuperman, G.J., Jha, A., Teich, J.M., Orav, E.J., Ma'luf, N., et al., 1997. Does the computerized display of charges affect inpatient ancillary test utilization? Arch. Intern. Med. 157 (21), 2501−2508.

Berlin, L., 2007. Accuracy of diagnostic procedures: has it improved over the past five decades? AJR Am. J. Roentgenol. 188 (5), 1173−1178.

Brady, A., Laoide, R.O., McCarthy, P., McDermott, R., 2012. Discrepancy and error in radiology: concepts, causes and consequences. Ulster Med. J. 81 (1), 3−9.

Carraro, P., Plebani, M., 2007. Errors in a stat laboratory: types and frequencies 10 years later. Clin. Chem. 53 (7), 1338−1342.

Chawla, R., Goswami, B., Tayal, D., Mallika, V., 2010. Identification of the types of preanalytical errors in the clinical chemistry laboratory: 1-year study at G.B. Pant Hospital. Lab Med. 41, 89−92.

Chhieng, D., Day, T., Gordon, G., Hicks, J., 2007. Use of natural language programming to extract medication from unstructured electronic medical records. AMIA Annu. Symp. Proc.908.

Dyer, C., 2001. Government to introduce safer administration of cancer drugs after fatal error. BMJ 322 (7293), 1013.

Committee on identifying and preventing medication. In: Errors, A.P., Wolcott, J., Bootman, J. L., Cronenwett, L.R. (Eds.), Preventing Medication Errors: Quality Chasm Series. The National Academies Press, Washington, DC.

Gaba, D.M., Howard, S.K., 2002. Patient safety: fatigue among clinicians and the safety of patients. N. Engl. J. Med. 347 (16), 1249−1255.

Goddard, P., Leslie, A., Jones, A., Wakeley, C., Kabala, J., 2001. Error in radiology. Br. J. Radiol. 74 (886), 949−951.

Gold, S., Elhadad, N., Zhu, X., Cimino, J.J., Hripcsak, G., 2008. Extracting structured medication event information from discharge summaries. AMIA Annu. Symp. Proc.237−241.

Hammerling, J.A., 2012. A review of medical errors in laboratory diagnostics and where we are today. Lab Med. 43 (2).

Harvey, J.A., Fajardo, L.L., Innis, C.A., 1993. Previous mammograms in patients with impalpable breast carcinoma: retrospective vs blinded interpretation. 1993 ARRS President's Award. AJR Am. J. Roentgenol. 161 (6), 1167−1172.

Hawkins, R., 2012. Managing the pre- and post-analytical phases of the total testing process. Ann. Lab. Med. 32 (1), 5−16.

Hickner, J., Thompson, P.J., Wilkinson, T., Epner, P., Sheehan, M., Pollock, A.M., et al., 2014. Primary care physicians' challenges in ordering clinical laboratory tests and interpreting results. J. Am. Board Fam. Med. 27 (2), 268−274.

Ingraham, B., Miller, K., Iaia, A., Sneider, M.B., Naqvi, S., Evans, K., et al., 2016. Reductions in high-end imaging utilization with radiology review and consultation. J. Am. Coll. Radiol. 13 (9), 1079−1082.

Kalra, J., 2004. Medical errors: impact on clinical laboratories and other critical areas. Clin. Biochem. 37 (12), 1052−1062.

Kaushal, R., Bates, D.W., Landrigan, C., McKenna, K.J., Clapp, M.D., Federico, F., et al., 2001. Medication errors and adverse drug events in pediatric inpatients. JAMA 285 (16), 2114−2120.

Kilpatrick, E.S., Freedman, D.B., National Clinical Biochemistry Audit Group, 2011. A national survey of interpretative reporting in the UK. Ann. Clin. Biochem. 48 (Pt 4), 317−320.

Krupinski, E.A., Berbaum, K.S., Caldwell, R.T., Schartz, K.M., Kim, J., 2010. Long radiology workdays reduce detection and accommodation accuracy. J. Am. Coll. Radiol. 7 (9), 698−704.

Kundel, H.L., Nodine, C.F., Carmody, D., 1978. Visual scanning, pattern recognition and decision-making in pulmonary nodule detection. Invest. Radiol. 13 (3), 175−181.

Lacson, R., Prevedello, L.M., Andriole, K.P., O'Connor, S.D., Roy, C., Gandhi, T., et al., 2014. Four-year impact of an alert notification system on closed-loop communication of critical test results. AJR Am. J. Roentgenol. 203 (5), 933−938.

Laposata, M., Dighe, A., 2007. "Pre-pre" and "post-post" analytical error: high-incidence patient safety hazards involving the clinical laboratory. Clin. Chem. Lab. Med. 45 (6), 712−719.

Lee, C.S., Nagy, P.G., Weaver, S.J., Newman-Toker, D.E., 2013. Cognitive and system factors contributing to diagnostic errors in radiology. AJR Am. J. Roentgenol. 201 (3), 611–617.

Lesar, T.S., 2002. Prescribing errors involving medication dosage forms. J. Gen. Intern. Med. 17 (8), 579–587.

Lesar, T.S., Briceland, L.L., Delcoure, K., Parmalee, J.C., Masta-Gornic, V., Pohl, H., 1990. Medication prescribing errors in a teaching hospital. JAMA 263 (17), 2329–2334.

Lesar, T.S., Lomaestro, B.M., Pohl, H., 1997. Medication-prescribing errors in a teaching hospital. A 9-year experience. Arch. Intern. Med. 157 (14), 1569–1576.

Levin, M.A., Krol, M., Doshi, A.M., Reich, D.L., 2007. Extraction and mapping of drug names from free text to a standardized nomenclature. AMIA Annu. Symp. Proc.438–442.

Lippi, G., Plebani, M., Di Somma, S., Cervellin, G., 2011. Hemolyzed specimens: a major challenge for emergency departments and clinical laboratories. Crit. Rev. Clin. Lab. Sci. 48 (3), 143–153.

Matin, A., Bates, D.W., Sussman, A., Ros, P., Hanson, R., Khorasani, R., 2006. Inpatient radiology utilization: trends over the past decade. AJR Am. J. Roentgenol. 186 (1), 7–11.

Neilson, E.G., Johnson, K.B., Rosenbloom, S.T., Dupont, W.D., Talbert, D., Giuse, D.A., et al., 2004. The impact of peer management on test-ordering behavior. Ann. Intern. Med. 141 (3), 196–204.

NHS England, National Pathology Programme Digital First: Clinical Transformation through Pathology Innovation.

Nightingale, P.G., Peters, M., Mutimer, D., Neuberger, J.M., 1994. Effects of a computerised protocol management system on ordering of clinical tests. Qual. Health Care 3 (1), 23–28.

Otero, H.J., Ondategui-Parra, S., Nathanson, E.M., Erturk, S.M., Ros, P.R., 2006. Utilization management in radiology: basic concepts and applications. J. Am. Coll. Radiol. 3 (5), 351–357.

Pinto, A., Brunese, L., 2010. Spectrum of diagnostic errors in radiology. World J. Radiol. 2 (10), 377–383.

Plebani, M., 2010. The detection and prevention of errors in laboratory medicine. Ann. Clin. Biochem. 47 (Pt 2), 101–110.

Plebani, M., Carraro, P., 1997. Mistakes in a stat laboratory: types and frequency. Clin. Chem. 43 (8 Pt 1), 1348–1351.

Sakyi, A., Laing, E., Ephraim, R., Asibey, O., Sadique, O., 2015. Evaluation of analytical errors in a clinical chemistry laboratory: a 3 year experience. Ann. Med. Health Sci. Res. 5 (1), 8–12.

Slight, S.P., Howard, R., Ghaleb, M., Barber, N., Franklin, B.D., Avery, A.J., 2013. The causes of prescribing errors in English general practices: a qualitative study. Br. J. Gen. Pract. 63 (615), e713–e720.

The Joint Commission, 2016. Hospital National Patient Safety Goals.

Upreti, S., Upreti, S., Bansal, R., Jeelani, N., Bharat, V., 2013. Types and frequency of preanalytical errors in haematology lab. J. Clin. Diagn. Res. 7 (11), 2491–2493.

Uzuner, O., Solti, I., Cadag, E., 2010. Extracting medication information from clinical text. J. Am. Med. Inform. Assoc. 17 (5), 514–518.

Vertinsky, T., Forster, B., 2005. Prevalence of eye strain among radiologists: influence of viewing variables on symptoms. AJR Am. J. Roentgenol. 184 (2), 681–686.

Vest, J.R., Jung, H.Y., Ostrovsky, A., Das, L.T., McGinty, G.B., 2015. Image sharing technologies and reduction of imaging utilization: a systematic review and meta-analysis. J. Am. Coll. Radiol. 12 (12 Pt B), 1371–1379 e3.

Whang, J.S., Baker, S.R., Patel, R., Luk, L., Castro III, A., 2013. The causes of medical malpractice suits against radiologists in the United States. Radiology 266 (2), 548–554.

Woywodt, A., Vythelingum, K., Rayner, S., Anderton, J., Ahmed, A., 2014. Single-centre experience with Renal PatientView, a web-based system that provides patients with access to their laboratory results. J. Nephrol. 27 (5), 521–527.

Xu, H., Stenner, S.P., Doan, S., Johnson, K.B., Waitman, L.R., Denny, J.C., 2010. MedEx: a medication information extraction system for clinical narratives. J. Am. Med. Inform. Assoc. 17 (1), 19–24.

Chapter 11

Bioinformatics and Precision Medicine

Arjun K. Manrai and Isaac S. Kohane
Harvard Medical School, Boston, MA, United States

INTRODUCTION

President Obama unveiled the Precision Medicine Initiative in early 2015, announcing a large-scale commitment to research that aims to individualize diagnosis and therapy. The announcement draws on a report commissioned by the National Academy of Sciences in 2011 (NRC, 2011) that argues for a patient-centric view of clinical and molecular data stored in a large, shared database—an "Information Commons." Similar efforts to measure and disseminate multiscale variation across populations are taking place worldwide, including for example the Precision Medicine Catapult in the United Kingdom. A major goal of precision medicine is to use this variation across diverse population segments in order to derive therapeutically and diagnostically useful molecular subtypes of traditionally monolithic diseases.

The promise of precision medicine is illustrated by three examples. First, the parents of a young boy with a severe undiagnosed disorder used social media and meticulous observation of their son's symptoms to help scientists end a long diagnostic odyssey and discover causal mutations in the *NGLY1* gene (Might and Wilsey, 2014). Second, non-small lung carcinoma, historically treated as a single disease, has seen a paradigm shift in targeted therapy based upon genetic driver mutations (Pao and Girard, 2011). Third, a wife–husband team recently leveraged large-scale sequence data to compute disease risk estimates for prion disease, a fatal genetic disorder affecting the wife (Minikel et al., 2016; Lebo et al., 2016). Two themes emerge from these cases: (1) informatics methods dictate the scale and scope of questions that investigators are able to ask today and (2) these investigators increasingly include patients and families—citizen science is already here.

In this chapter, we describe why these exemplars are not merely one-offs, but the early signs of medicine's information age. We first review the intellectual foundations of precision medicine and the central role that

Key Advances in Clinical Informatics. DOI: http://dx.doi.org/10.1016/B978-0-12-809523-2.00011-X

bioinformatics has played in achievements such as the Human Genome Project. While the idea of individualized medicine is not new (Collins and Varmus, 2015), at present there is a confluence of data and talent to accelerate progress toward the decades-old vision of personalizing the definition and management of human disease. We believe that understanding the intellectual roots of precision medicine will help prioritize investment and efficiently scale up current efforts. Throughout this chapter, we describe elements of clinical practice and research that have been transformed by bioinformatics.

NOSOLOGY

Many discussions of precision medicine and bioinformatics begin with advances in DNA sequencing (Schuster, 2008) and human genetics (Visscher et al., 2012; Stessman et al., 2014). We adopt a different entry point here by starting with selected historical developments in disease definition and classification—nosology.

Ontological Commitments

What is a disease? It is tempting to answer that disease is the absence of health, but equally tempting to state that health is the absence of disease. Our question is prone to circular reasoning, and quickly becomes philosophically and practically thorny (Smith, 2002). A useful view of a disease is that it is a *knowledge representation* (Davis et al., 1993), and therefore an *ontological commitment*. Such "commitments are in effect a strong pair of glasses that determine what we can see, bringing some part of the world into sharp focus, at the expense of blurring other parts" (Davis et al., 1993). As such a lens, the clinical criteria used to define a disease, by virtue of delineating which individuals are brought into focus, fundamentally shape investigations of correlated clinical, demographic, environmental, lifestyle, laboratory, and molecular measurements, including of course genetic sequencing. While it is tempting to think that the study of one of these axes of variation, for example, heritability, is a specialized pursuit divorced from the other data types and indeed the disease definition itself, this perspective may be limited in statistical power at best (Kohane, 2014) and harmful at worst. For example, we recently showed how African Americans were disproportionately misdiagnosed using genetic variants once thought to cause the inherited heart disease hypertrophic cardiomyopathy (Manrai et al., 2016a,b). Inclusion of ancestrally diverse populations in early control cohorts would likely have prevented these misclassifications, as would a more complete understanding of the phenotypic spectrum of individuals carrying these variants.

Early Days of Disease Taxonomy

The ontological commitments inherent to defining disease are illustrated by the works of the 17th century English physician Thomas Sydenham and the 18th century Swedish taxonomist Carolus Linnaeus. Sydenham argued that diseases had essential characters in nature, independent of observation, and thus could be grouped similarly to how plants and animals are grouped (Sydenham and Latham, 1850). Linnaeus is renowned for this type of classification, but less well known is that Linnaeus was also a physician. In 1763, he published a classification system of disease in *Genera Morborum* (Linnaeus, 1763). Most diseases in Linnaeus' system are grouped symptomatically, as opposed to etiologically or anatomically. This classification of disease can be viewed as choosing (and committing to) a high-dimensional feature space used to represent disease entities (i.e., the space of all possible symptoms) as well as a distance metric to define how similar diseases are to one another. Symptoms have historically been, and often remain, the only available lamppost under which to look.

The 20th century Canadian physician Sir William Osler advocated for an approach to disease diagnosis and therapy that extends the observation and grouping of clusters of symptoms to their correlation with pathological findings (Osler, 1892). Diseases tend to be grouped by the main organ system affected, and disease classification follows from inductive generalization and an appeal to parsimony (Loscalzo and Barabasi, 2011). This approach is an important advance over previous approaches and still underlies much of current clinical practice. It has been argued that although this view is useful, it has fundamental limits including, but not limited to, that classical Oslerian correlations do not account for preclinical disease and explain why individuals can manifest the same disease in very different ways, failing to incorporate molecular, cellular, physiological, and environmental context to individualize disease presentation (Loscalzo and Barabasi, 2011).

Disease Definition in the Era of Precision Medicine

Advances in molecular profiling, epitomized by next-generation sequencing (Schuster, 2008), have made it routine to acquire a detailed assessment of an individual's molecular state. The measurable array of molecular information includes genomic, transcriptomic, epigenetic, metabolomic, microbiomic, and other panomic data types. These data give rise to a daunting hypothesis space that is often winnowed either by targeted (Tabor et al., 2002) or agnostic (Visscher et al., 2012) statistical approaches that correlate a single marker or group of markers with a disease or trait. Coupled with functional assays that may help transmute these correlations into causal mechanisms, large-scale molecular approaches have been successful in uncovering the etiology of many common and rare diseases. A major goal of precision medicine is to scale up this dissection of traditionally monolithic diseases into their constituent entities (NRC, 2011). The reclassification of several human cancers has been used to motivate this approach generally. For example, rapid progress has been achieved in the molecular dissection of non-small lung carcinoma (Pao and Girard, 2011), the identification of *ARID1A* mutations in gastric cancer (Wang et al., 2011), and *SPOP* mutations in prostate cancer (Barbieri et al., 2012).

Mechanistic causality should be distinguished from the correlation of molecular markers with disease. The latter often takes the form of an odds ratio or "penetrance," the probability of disease in an individual with a given mutation. Hard won lessons from decades of genetic testing have shown that a causal association of a "pathogenic" variant does not equate (and often overstates) the type of simpler correlations that physicians and patients often need in using genetic data (Beutler et al., 2002; Manrai et al., 2016b). Moreover, sources of bias, both technical and clinical, as well as the abundance of variation in a typical genome (MacArthur et al., 2012), may generate the appearance of association or causality for variation that is in reality benign—there is ample "narrative potential" in the human genome (Goldstein et al., 2013).

Bioinformatics and Nosology

Bioinformatics has been previously defined as "conceptualizing biology in terms of macromolecules (in the sense of physical-chemistry) and then applying "informatics" techniques (derived from disciplines such as applied maths, computer science, and statistics) to understand and organize the information associated with these molecules, on a large-scale" (Luscombe et al., 2001). Biomedical informatics leverages many of the same informatics techniques and tools for problems in medicine. Nosology, bioinformatics, and biomedical informatics are intertwined fields today. Clinical vocabularies and hierarchies encode thousands of years of knowledge about the definition, interconnections, and granularity of human disease. These knowledge representations include the International Statistical Classification of Diseases and Related Health Problems (ICD) (World Health Organization, 2009), Medical Subject Headings (Lipscomb, 2000), Systematized Nomenclature of Medicine—Clinical Terms (Donnelly, 2006), and the compendium mapping between these vocabularies known as the Unified Medical Language System (Lindberg et al., 1993; Bodenreider, 2004). These controlled vocabularies are expressions of collective knowledge and ontological commitments with profound consequences.

Disease vocabularies such as the ICD coupled with procedural vocabularies such as the Current Procedural Terminology constrain provider—payer dialogue and reimbursement, drive electronic health records (EHRs), and dictate the scope of epidemiological and population health studies that exploit these data (Kohane, 2011). For example, a healthcare provider can justify a procedure performed in her clinic to her patient's insurance company only within the expressivity limits of a disease diagnostic vocabulary. Similarly, a researcher repurposing insurance claims data to study temporal trends in genetic testing in the United States can isolate a single gene test only when a specific code exists for that test, as opposed to nonspecific concepts for generic molecular pathologies. As these vocabularies expand, as they did in the recent federal mandate from ICD-9 to ICD-10, myriad bioinformatics challenges emerge in concept mapping, database architecture, information retrieval, and reproducing and extending previous research analyses (Boyd et al., 2013). In summary, bioinformatics helps to preserve and advance our knowledge representation of nosology and supports the healthcare enterprise at multiple scales (Fig. 11.1).

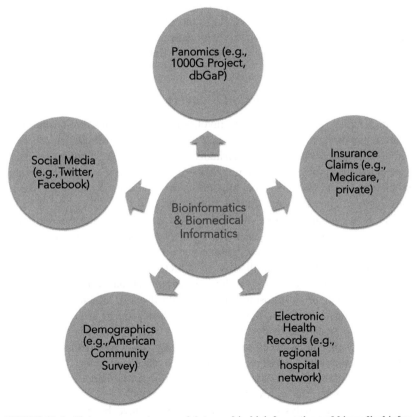

FIGURE 11.1   **Heterogeneous streams of data used in bioinformatics and biomedical informatics.** Statistical methods and computational tools from bioinformatics and biomedical informatics are used to manage, harmonize, and analyze a diverse set of data types, including data from social media, panomics (e.g., genomics, proteomics), insurance claims, EHRs, and demographics. 1000G, 1000 Genomes Project (http://www.internationalgenome.org/); dbGaP, Database of Genotypes and Phenotypes (https://www.ncbi.nlm.nih.gov/gap).

HUMAN GENETICS LEADING TO PRECISION MEDICINE

Having started with nosology, we turn to human genetics. As noted above, the notion of using genetic data to personalize medicine is not new (Collins and Varmus, 2015), but the current availability of representative, large-scale sequencing data has reached an inflection point, switching the discovery bottleneck from data to researcher time. After decades of progress, genetics research today continues to rapidly uncover biological insights into hitherto poorly characterized diseases (Visscher et al., 2012; Bamshad et al., 2011).

We note that genetics is a single axis of information about an individual, and that it is necessary to study other molecular, lifestyle, and environmental exposures (Patel and Ioannidis, 2014) that an individual encounters from conception to death to attain a comprehensive perspective on health and disease (Manrai et al., 2017). Nonetheless, recent progress in human genetics, particularly that of large consortia with standardized methods and shared data, serves as an excellent template that other fields may choose to emulate.

Useful Dichotomies and Early Discoveries

Human genetics research frequently invokes two dichotomies. First, inherited diseases are often thought of as having either a simple, Mendelian (monogenic) or complex, polygenic inheritance pattern. Second, human genetics research is often viewed as being either "agnostic" (e.g., a genome-wide association study (GWAS)) or hypothesis-driven (e.g., a candidate gene study). Both dichotomies are limited, but useful.

The twin human genetics dichotomies help frame the successes of the techniques of genetic linkage analysis and positional cloning. Since the 1980s, linkage analysis has used genome-wide markers to agnostically search for loci that segregate with disease in affected families (Botstein et al., 1980). Linkage analysis leverages the few recombination events per meiosis in order to associate chromosomal regions with disease, although usually at coarse mapping resolution. These regions are then mapped at higher resolution using the technique of positional cloning. These techniques have led to the discovery of causal genes and mutations for many Mendelian diseases, including cystic fibrosis (Kerem et al., 1989; Riordan et al., 1989), nail patella syndrome (Dreyer et al., 1998), hemochromatosis (Feder et al., 1996), and Duchenne muscular dystrophy (Koenig et al., 1987).

While linkage analysis and positional cloning have been successful in mapping Mendelian traits (single-gene disorders), they have been met with limited success for common, complex diseases such as type 2 diabetes. In response, many researchers have posited a priori hypotheses in the form of "candidate genes" from the approximately 20K human protein-coding genes. These biological bets have seen some successes (Tabor et al., 2002), but the collection of findings as a whole has disappointed, with one estimate of the replication rate being just 1.2% (Ioannidis et al., 2011). There are many plausible explanations (Ioannidis, 2005) for the irreproducibility of these biological bets, including unaddressed multiple testing, ascertainment bias, and confounding due to mismatched ancestry between cases and controls.

The Human Genome Project

On June 26, 2000, Francis Collins, Craig Venter, and then President Bill Clinton announced the completion of the Human Genome Project. The venture was big data before "Big Data"—a sociotechnical feat that is now an archetype for large-scale consortium efforts in genomics. The project led to many gains beyond the value of the sequence itself. For example, the cost of a typical positional cloning study, once estimated to cost tens of millions of dollars, dropped to around $100,000 in the years following the project (Botstein and Risch, 2003). Of the many lessons gleaned, few are more valuable than the prescient insistence on a precompetitive data release policy that required data and materials be made available within 6 months of generation (Collins et al., 2003). The Human Genome Project also highlighted the importance of formal investment into the ethical and legal consequences of large-scale biology, allocating 3%−5% of the annual budget to the Ethical, Legal, and Social Implications component of the National Human Genome Research Institute (NHGRI) (Thomson et al., 1997).

The origins of bioinformatics are often linked to genomics and the Human Genome Project in particular, but computers had been used in molecular biology for decades prior, exemplified by lasting advances in protein biochemistry during the 1960s (Hagen, 2000). But bioinformatics has arguably been nowhere more essential to success than in the Human Genome Project. Informatics methods were developed and used extensively in processing, storing, analyzing, and providing rapid widespread access to a trove of accumulating sequencing data. In recognition of this role, "Bioinformatics and Computational Biology" was listed as a major goal in the 1998−2003 published goals for the Human Genome Project (Collins et al., 1998). In 1998, Collins et al. noted:

> *Especially critical is the shortage of individuals trained in bioinformatics. Also needed are scientists trained in the management skills required to lead large data-production efforts. Another urgent need is for scholars who are trained to undertake studies on the societal impact of genetic discoveries. Such scholars should be knowledgeable in both genome-related sciences and in the social sciences.*
>
> Collins et al., 1998.

In response to the shortage noted by Collins et al. and many others, a generation of scientists has been trained in genomics and bioinformatics.

Interestingly, many of the claims in the paragraph above might still be justifiably made in 2017, particularly with regard to the societal impact of genetics research.

Accelerating the Discovery of Disease Genetics: GWASs and Next-Generation Sequencing

A GWAS adopts an "agnostic" approach to discovering genomic loci associated with disease. The approach is (mostly) agnostic to chromosomal position, but it is not agnostic to the frequency or effect size of disease-associated genetic variation (Altshuler et al., 2008). In a typical GWAS, allele frequencies across genome-wide loci (e.g., 1M single nucleotide polymorphism markers) are compared between unrelated cases and controls while controlling for multiple testing (Pe'er et al., 2008) and confounding by ancestry (Price et al., 2006). This study design rests on the haplotype structure of the human genome, which was in large part catalogued by the International HapMap Project (Gibbs et al., 2003). In 1996, a paper by Risch and Merikangas demonstrated the theoretical possibility and utility of a GWAS (Risch and Merikangas, 1996). Merely a decade later, several successful GWAS had been performed (Dewan et al., 2006; Klein et al., 2005; Wellcome Trust Case Control Consortium, 2007). Today, there are thousands of genetic associations with disease stored in the NHGRI GWAS Catalog (Welter et al., 2014), providing biological leads for follow-up studies and explaining a portion of the heritability of many complex diseases, though still incomplete (Eichler et al., 2010). GWASs have been applied across the spectrum of human disease, including diseases such as Crohn's disease (Franke et al., 2010) and type 2 diabetes (Frayling, 2007), and traits such as height (Lango Allen et al., 2010) and fasting glucose levels in nondiabetic individuals (Saxena et al., 2010).

In recent years, massive parallel sequencing has allowed investigators to move from chip-based genotyping to direct sequencing for disease gene discovery. Cost-effective, accurate sequencing of the exome, the protein-coding portion of the genome, as well as the whole genome, are now mainstays in both clinical practice (Yang et al., 2013) and biomedical research (Bamshad et al., 2011). These techniques have led to the discovery of the molecular basis of many rare and hitherto poorly understood diseases, including Miller syndrome (Ng et al., 2010b), Kabuki syndrome (Ng et al., 2010a), human FADD deficiency (Bolze et al., 2010), Hajdu−Cheney syndrome (Simpson et al., 2011), and many other Mendelian diseases. Further, exome sequencing has been used to identify molecular subtypes of diseases that have often been treated homogeneously. Molecular subtyping has been fruitful in cancer, for example, in the identification of *ARID1A* mutations in gastric cancer (Wang et al., 2011), *SPOP* mutations in prostate cancer (Barbieri et al., 2012), and driver mutations that stratify treatment for non-small lung carcinoma (Pao and Girard, 2011). Beyond producing insights into the diagnosis and treatment of Mendelian diseases, the identification of rare molecular lesions can provide a window into distinct diseases with shared etiology and even common diseases, where differing molecular insults to a single pathway may manifest in profoundly different clinical presentations.

Bioinformatics That Enables Human Genetics Research Today

As human genetics has evolved from early low-resolution linkage analyses to the rapid sequencing of whole genomes, bioinformatics methods have become increasingly central. Methods and tools to facilitate data storage, retrieval, and computation have needed to keep pace with burgeoning genotyping and sequencing data, which can be combined with gene expression measurements (Gusev et al., 2016), epigenetic data (Dawson and Kouzarides, 2012), three-dimensional folding maps (Rao et al., 2014), and many other types of data. Resources such as the UCSC Genome Browser (Kent et al., 2002; Karolchik et al., 2003) have been essential in making data available to experimental and computational scientists alike. Open-source software powers many contemporary large-scale genetic association studies with popular tools such as PLINK (Purcell et al., 2007). Software such as GATK (McKenna et al., 2010) is critical for variant discovery in sequencing data. A recurring theme in these pursuits is a concerted effort to continuously improve the standardization of data types, methods, and even file formats (Marcketta and Auton, 2011).

REPRODUCIBLE PRECISION MEDICINE AND CITIZEN SCIENCE

Reproducible discoveries in human genetics reflect in large part the benefits of consortia, data sharing, and protocol standardization. While there are obvious gains in statistical power from pooling samples, consortia come with additional mechanisms to enhance reproducible discovery. As an example, consortia enable unique approaches to quality assurance and control—the Human Genome Project conducted "round robin" exercises where random sequence files were validated by multiple other centers (Collins et al., 2003). To ensure reproducible discoveries in precision medicine going forward, a sustained commitment is required to community efforts for data sharing, funding streams, and training.

Responsible Data Sharing

The Human Genome Project prioritized data sharing from the start. The NHGRI and US Department of Energy initially adopted a policy for data to be released within 6 months. In 1996, the Bermuda Principles led to the automatic release of data within 24 h for sequence assemblies that were greater than 1−2 kb in length (Collins et al., 2003). These principles are not limited to genomic sequencing data, but have become norms across several areas of high-throughput biology. For gene expression data, standards such as the Minimum Information About a Microarray Experiment (Brazma et al., 2001) coupled with centralized databases such as the Gene Expression Omnibus (Edgar et al., 2002) enable both independent verification and repurposing of original data (Schmid et al., 2012). Data sharing can be encouraged both by bottom−up and top−down approaches; individual researchers as well as funding agencies and journals are stakeholders in promoting reproducible research practices. Importantly, even when data sharing policies exist, adherence will not be perfect, impeding both corroboration and extension of the original results by external investigators (Alsheikh-Ali et al., 2011).

In recent years, large catalogues of exome and genomic sequence variation have become widely accessible (Abecasis et al., 2012; Lek et al., 2016). These databases provide much-needed validation of sequence variants believed to be pathogenic and have been used to systematically reevaluate prior assertions regarding pathogenicity (Andreasen et al., 2013; Piton et al., 2013; Jabbari et al., 2013), as well as to compute quantitative measures of disease risk (Minikel et al., 2016).

Unique Challenges of Precision Medicine

Some perennial threats to reproducibility may become pronounced in the era of precision medicine. First, data analyzed by investigators are often dense and highly correlated (Ioannidis et al., 2009), heterogeneous (Madigan et al., 2013), and increasingly divorced from the data producers. Second, while frictionless access to centralized, privacy protecting, and freely accessible large integrated datasets is a vision shared by many researchers, in practice, linking and harmonizing datasets from different investigators or institutions, each often with its own proprietary schema, seems to be becoming more onerous and expensive. Furthermore, sociotechnical challenges of data sharing are likely here to stay insofar as the academic reward system does not explicitly value these efforts (Ioannidis, 2014). Third, subgroup analyses that make claims about precise population segments are very common, but often lack contextualization to determine the scope of the implicit file drawer problem (Rosenthal, 1979). It is often argued that sharing data and code will help ameliorate these concerns. Other possible solutions include postpublication review, increasing large-scale consortia efforts, and modifying the rewards for science (Ioannidis, 2014). Practices may also be borrowed from fields outside of biomedicine. For example, in the particle physics and cosmology communities, there is extensive use of blind analysis to minimize bias (MacCoun and Perlmutter, 2015).

Citizen Science

Investigators today increasingly include patients and their family members (Might and Wilsey, 2014; Minikel et al., 2016). Many other citizen scientists have contributed to other scientific projects within and outside of

An interested citizen scientist can acquire an accurate and inexpensive readout of (parts of) his or her genome from multiple commercial vendors today. In other circumstances, a patient or family member might learn about a particular genetic variant that is believed to be the cause of an inherited disease that runs in the family, such as hypertrophic cardiomyopathy (HCM). Widespread access to personal genomic data has been enabled by decades of progress punctuated by paradigm shifts in genotyping and sequencing technology (Schuster 2008).

Below, we outline steps that a citizen scientist could take to analyze his or her own genome. Often, the genome will be assessed at a large number of single nucleotide polymorphism (SNP) sites. A home testing kit with a simple cheek swab is usually sufficient to obtain the results. The dataset typically looks like:

SNP id (e.g. rsID) | chromosome | position | genotype

Four steps that can be taken to analyze the genome and interrogate any claims made with the data, now or in the future, include:

1. Data wrangling
Many vendors bundle or sell web-accessible interpretation services along with the actual genotyping or sequencing data. There are also third party applications that let users directly upload the data. However, the most powerful and flexible approach in using one's genomic data requires a few skills that are worth the time investment: (a) command line use (e.g. a bash terminal); (b) a scripting language (e.g. Python or Perl); and (c) a basic statistical framework (e.g. R). Fortunately, all of these are freely accessible and have free online courses (e.g. Coursera, Codecademy).

2. Annotating
It is routine for a genotype chip to measure hundreds of thousands of markers across the genome. An array of informatics tools exists to annotate the potential biological and clinical significance of these markers, for example ANNOVAR or SNPedia.

3. Querying variant frequency across ancestrally diverse populations
A particularly useful set of resources now exists which enables users to query the frequency of particular genomic site or set of sites across diverse populations, for example the 1000 Genomes Project (1000G) and gnomAD. These databases can often be used via annotation software such as ANNOVAR.

4. Encapsulating an analysis
Analyses performed using one's personal genome can be described in detail and saved for future use or even shared with family members or others as a "digital notebook" using e.g. the Jupyter platform or tools like R Markdown; code can be shared widely via GitHub.

RESOURCES

Python: https://www.python.org/
Perl: https://www.perl.org/
R: https://cran.r-project.org/
GitHub: https://github.com/
Coursera: https://www.coursera.org/
Codecademy: https://www.codecademy.com/

ANNOVAR: http://annovar.openbioinformatics.org/
1000G: http://www.internationalgenome.org/
SNPedia: https://www.snpedia.com/
gnomAD http://gnomad.broadinstitute.org/
Jupyter: http://jupyter.org/
R Markdown: http://rmarkdown.rstudio.com/

FIGURE 11.2 A reproducible personal genome analysis in the era of citizen science.

biomedicine (Savage, 2012). Individuals interested in genomics and precision medicine have access to an unprecedented set of publicly accessible resources to conduct their own investigations. In Fig. 11.2, we outline the steps that an interested citizen scientist could take to conduct their own reproducible genomics analysis.

CONCLUSIONS

Early successes in precision medicine (Might and Wilsey, 2014; Pao and Girard, 2011; Minikel et al., 2016) foreshadow the benefits of scaling up current progress. Sustained success will require continued investment into informatics research, training of biomedical data scientists, data sharing, and reproducible science practices exemplified by landmark efforts such as the Human Genome Project. To realize precision medicine, and to guard against the false and even harmful claims that may be made with these same data, we anticipate that quantitative reasoning and computational fluency will eventually become as important to physicians and citizens as literacy, and that research efforts will increasingly be led by citizen scientists. Indeed the early signs are already here, where invested patients and families have made lasting contributions.

REFERENCES

Abecasis, G.R., Auton, A., Brooks, L.D., DePristo, M.A., Durbin, R.M., Handsaker, R.E., et al., 2012. An integrated map of genetic variation from 1,092 human genomes. Nature 491 (7422), 56−65.

Alsheikh-Ali, A.A., Qureshi, W., Al-Mallah, M.H., Ioannidis, J.P., 2011. Public availability of published research data in high-impact journals. PLoS One 6 (9), e24357.

Altshuler, D., Daly, M.J., Lander, E.S., 2008. Genetic mapping in human disease. Science 322 (5903), 881−888.

Andreasen, C., Nielsen, J.B., Refsgaard, L., Holst, A.G., Christensen, A.H., Andreasen, L., et al., 2013. New population-based exome data are questioning the pathogenicity of previously cardiomyopathy-associated genetic variants. Eur. J. Hum. Genet. 21 (9), 918−928.

Bamshad, M.J., Ng, S.B., Bigham, A.W., Tabor, H.K., Emond, M.J., Nickerson, D.A., et al., 2011. Exome sequencing as a tool for Mendelian disease gene discovery. Nat. Rev. Genet. 12 (11), 745−755.

Barbieri, C.E., Baca, S.C., Lawrence,, M.S., Demichelis, F., Blattner, M., Theurillat, J.P., et al., 2012. Exome sequencing identifies recurrent SPOP, FOXA1 and MED12 mutations in prostate cancer. Nat. Genet. 44 (6), 685−689.

Beutler, E., Felitti, V.J., Koziol, J.A., Ho, N.J., Gelbart, T., 2002. Penetrance of 845G--> A (C282Y) HFE hereditary haemochromatosis mutation in the USA. Lancet 359 (9302), 211−218.

Bodenreider, O., 2004. The Unified Medical Language System (UMLS): integrating biomedical terminology. Nucleic Acids Res. 32 (Database issue), D267−D270.

Bolze, A., Byun, M., McDonald, D., Morgan, N.V., 2010. Whole-exome-sequencing-based discovery of human FADD deficiency. Am. J. Hum. Genet. 87 (6), 873−881.

Botstein, D., Risch, N., 2003. Discovering genotypes underlying human phenotypes: past successes for Mendelian disease, future approaches for complex disease. Nat. Genet. 33, 228−237.

Botstein, D., White, R.L., Skolnick, M., 1980. Construction of a genetic linkage map in man using restriction fragment length polymorphisms. Am. J. Hum. Genet. 32 (3), 314−331.

Boyd, A.D., Li, J.J., Burton, M.D., Jonen, M., Gardeux, V., Achour, I., et al., 2013. The discriminatory cost of ICD-10-CM transition between clinical specialties: metrics, case study, and mitigating tools. J. Am. Med. Inform. Assoc. 20 (4), 708−717.

Brazma, A., Hingamp, P., Quackenbush, J., Sherlock, G., Spellman, P., Stoeckert, C., et al., 2001. Minimum information about a microarray experiment (MIAME)-toward standards for microarray data. Nat. Genet. 29 (4), 365−371.

Collins, F.S., Varmus, H., 2015. A new initiative on precision medicine. N. Engl. J. Med. 372 (9), 793−795.

Collins, F.S., Patrinos, A., Jordan, E., Chakravarti, A., Gesteland, R., Walters, L., 1998. New goals for the U.S. Human Genome Project: 1998−2003. Science 282 (5389), 682−689.

Collins, F.S., Morgan, M., Patrinos, A., Watson, J.D., Olson, M.V., Collins, F.S., et al., 2003. The Human Genome Project: lessons from large-scale biology. Science 300 (5617), 286−290.

Davis, R., Shrobe, H., Szolovits, P., 1993. What is a knowledge representation? AI Magazine.

Dawson, M.A., Kouzarides, T., 2012. Cancer epigenetics: from mechanism to therapy. Cell 150 (1), 12−27.

Dewan, A., Liu, M., Hartman, S., Zhang, S.S., Liu, D.T., Zhao, C., et al., 2006. HTRA1 promoter polymorphism in wet age-related macular degeneration. Science 314 (5801), 989−992.

Donnelly, K., 2006. SNOMED-CT: the advanced terminology and coding system for eHealth. Stud. Health Technol. Inform. 121, 279−290.

Dreyer, S.D., Zhou, G., Baldini, A., Winterpacht, A., Zabel, B., Cole, W., et al., 1998. Mutations in LMX1B cause abnormal skeletal patterning and renal dysplasia in nail patella syndrome. Nat. Genet. 19 (1), 47−50.

Edgar, R., Domrachev, M., Lash, A.E., 2002. Gene Expression Omnibus: NCBI gene expression and hybridization array data repository. Nucleic Acids Res. 30 (1), 207−210.

Eichler, E.E., Flint, J., Gibson, G., Kong, A., Leal, S.M., Moore, J.H., et al., 2010. Missing heritability and strategies for finding the underlying causes of complex disease. Nat. Rev. Genet. 11 (6), 446−450.

Feder, J.N., Gnirke, A., Thomas, W., Tsuchihashi, Z., Ruddy, D.A., Basava, A., et al., 1996. A novel MHC class I-like gene is mutated in patients with hereditary haemochromatosis. Nat. Genet. 13 (4), 399−408.

Franke, A., McGovern, D.P., Barrett, J.C., Wang, K., Radford-Smith, G.L., Ahmad, T., et al., 2010. Genome-wide meta-analysis increases to 71 the number of confirmed Crohn's disease susceptibility loci. Nat. Genet. 42 (12), 1118−1125.

Frayling, T.M., 2007. Genome-wide association studies provide new insights into type 2 diabetes aetiology. Nat. Rev. Genet. 8 (9), 657−662.

Gibbs, R.A., Belmont, J.W., Hardenbol, P., Willis, T.D., Yu, F., Yang, H., et al., 2003. The International HapMap Project. Nature 426 (6968), 789−796.

Goldstein, D.B., Allen, A., Keebler, J., Margulies, E.H., Petrou, S., Petrovski, S., et al., 2013. Sequencing studies in human genetics: design and interpretation. Nat. Rev. Genet. 14 (7), 460−470.

Gusev, A., Ko, A., Shi, H., Bhatia, G., Chung, W., 2016. Integrative approaches for large-scale transcriptome-wide association studies. Nature 48 (3), 245−252.

Hagen, J.B., 2000. The origins of bioinformatics. Nat. Rev. Genet. 1 (3), 231−236.

Ioannidis, J.P., 2005. Why most published research findings are false. PLoS Med. 2 (8), e124.

Ioannidis, J.P., 2014. How to make more published research true. PLoS Med. 11 (10), e1001747.

Ioannidis, J.P., Loy, E.Y., Poulton, R., Chia, K.S., 2009. Researching genetic versus nongenetic determinants of disease: a comparison and proposed unification. Sci. Transl. Med. 1 (7), 7ps8.

Ioannidis, J.P., Tarone, R., McLaughlin, J.K., 2011. The false-positive to false-negative ratio in epidemiologic studies. Epidemiology 22 (4), 450−456.

Jabbari, J., Jabbari, R., Nielsen, M.W., Holst, A.G., Nielsen, J.B., Haunsø, S., et al., 2013. New exome data question the pathogenicity of genetic variants previously associated with catecholaminergic polymorphic ventricular tachycardia. Circ. Cardiovasc. Genet. 6 (5), 481−489.

Karolchik, D., Baertsch, R., Diekhans, M., Furey, T.S., Hinrichs, A., Lu, Y.T., et al., 2003. The UCSC Genome Browser Database. Nucleic Acids Res. 31 (1), 51−54.

Kent, W.J., Sugnet, C.W., Furey, T.S., Roskin, K.M., Pringle, T.H., Zahler, A.M., et al., 2002. The human genome browser at UCSC. Genome Res. 12 (6), 996−1006.

Kerem, B., Rommens, J.M., Buchanan, J.A., Markiewicz, D., Cox, T.K., Chakravarti, A., et al., 1989. Identification of the cystic fibrosis gene: genetic analysis. Science 245 (4922), 1073−1080.

Klein, R.J., Zeiss, C., Chew, E.Y., Tsai, J.-Y., Sackler, R.S., Haynes, C., et al., 2005. Complement factor H polymorphism in age-related macular degeneration. Science 308 (5720), 385−389.

Koenig, M., Hoffman, E.P., Bertelson, C.J., Monaco, A.P., Feener, C., Kunkel, L.M., 1987. Complete cloning of the Duchenne muscular dystrophy (DMD) cDNA and preliminary genomic organization of the DMD gene in normal and affected individuals. Cell 50 (3), 509−517.

Kohane, I.S., 2011. Using electronic health records to drive discovery in disease genomics. Nat. Rev. Genet. 12 (6), 417−428.

Kohane, I.S., 2014. Deeper, longer phenotyping to accelerate the discovery of the genetic architectures of diseases. Genome Biol. 15 (5), 115.

Lango Allen, H., Estrada, K., Lettre, G., Berndt, S.I., Weedon, M.N., Rivadeneira, F., et al., 2010. Hundreds of variants clustered in genomic loci and biological pathways affect human height. Nature 467 (7317), 832−838.

Lebo, M.S., Sutti, S., Green, R.C., 2016. 'Big Data' Gets Personal. Sci. Transl. Med. 8 (322), 322fs3-3fs3.

Lek, M., Karczewski, K.J., Minikel, E.V., Samocha, K.E., Banks, E., Fennell, T., et al., 2016. Analysis of protein-coding genetic variation in 60,706 humans. Nature 536 (7616), 285−291.

Lindberg, D.A., Humphreys, B.L., McCray, A.T., 1993. The Unified Medical Language System. Methods Inf. Med. 32 (4), 281−291.

Linnaeus, C. Genera morborum: in auditorum usum, Upsaliae : Apud Christ. Erh. Steinert, 1763.

Lipscomb, C.E., 2000. Medical subject headings (MeSH). Bull. Med. Library Assoc.

Loscalzo, J., Barabasi, A.-L., 2011. Systems biology and the future of medicine. Wiley Interdiscip. Rev. Syst. Biol. Med. 3 (6), 619−627.

Luscombe, N.M., Greenbaum, D., Gerstein, M., 2001. What is bioinformatics? A proposed definition and overview of the field. Methods Inf. Med. 40 (4), 346−358.

MacArthur, D.G., Balasubramanian, S., Frankish, A., Huang, N., Morris, J., Walter, K., et al., 2012. A systematic survey of loss-of-function variants in human protein-coding genes. Science 335 (6070), 823−828.

MacCoun, R., Perlmutter, S., 2015. Hide results to seek the truth. Nature 526 (7572), 187–189.

Madigan, D., Ryan, P.B., Schuemie, M., Stang, P.E., Overhage, J.M., Hartzema, A.G., et al., 2013. Evaluating the impact of database heterogeneity on observational study results. Am. J. Epidemiol. 178 (4), 645–651.

Manrai, A.K., Funke, B.H., Rehm, H.L., Olesen, M.S., Maron, B.A., Szolovits, P., et al., 2016a. Genetic misdiagnoses and the potential for health disparities. N. Engl. J. Med. 375 (7), 655–665.

Manrai, A.K., Ioannidis, J.P., Kohane, I.S., 2016b. Clinical genomics: from pathogenicity claims to quantitative risk estimates. JAMA 315 (12), 1233–1234.

Manrai, A.K., Cui, Y., Bushel, P.R., Hall, M., Karakitsios, S., Mattingly, C.J., et al., 2017. Informatics and data analytics to support exposome-based discovery for public health. Annu. Rev. Public Health 38 (1), http://dx.doi.org/10.1146/annurev-publhealth-082516-012737.

Marcketta, A., Auton, A., 2011. Variant call format, binary variant call format and VCFtools. 2156 (August), 10461.

McKenna, A., Hanna, M., Banks, E., Sivachenko, A., et al., 2010. The Genome Analysis Toolkit: a MapReduce framework for analyzing next-generation DNA sequencing data. Genome Res. 20 (9), 1297–1303.

Might, M., Wilsey, M., 2014. The shifting model in clinical diagnostics: how next-generation sequencing and families are altering the way rare diseases are discovered, studied, and treated. Genet. Med. 16 (10), 736–737.

Minikel, E.V., Vallabh, S.M., Lek, M., Estrada, K., Samocha, K.E., Sathirapongsasuti, J.F., et al., 2016. Quantifying prion disease penetrance using large population control cohorts. Sci. Transl. Med. 8 (322), 322ra9.

Ng, S.B., Bigham, A.W., Buckingham, K.J., Hannibal, M.C., McMillin, M.J., Gildersleeve, H.I., et al., 2010a. Exome sequencing identifies MLL2 mutations as a cause of Kabuki syndrome. Nat. Genet. 42 (9), 790–793.

Ng, S.B., Buckingham, K.J., Lee, C., Bigham, A.W., Tabor, H.K., Dent, K.M., et al., 2010b. Exome sequencing identifies the cause of a Mendelian disorder. Nat. Genet. 42 (1), 30–35.

NRC, 2011. Toward Precision Medicine: Building a Knowledge Network for Biomedical Research and a New Taxonomy of Disease. National Academies Press, Washington, DC.

Osler W. The Principles and Practice of Medicine. New York: Appleton; 1892.

Pao, W., Girard, N., 2011. New driver mutations in non-small-cell lung cancer. Lancet Oncol. 12 (2), 175–180.

Patel, C.J., Ioannidis, J.P., 2014. Studying the elusive environment in large scale. JAMA 311 (21), 2173–2174.

Pe'er, I., Yelensky, R., Altshuler, D., Daly, M.J., 2008. Estimation of the multiple testing burden for genomewide association studies of nearly all common variants. Genet. Epidemiol. 32 (4), 381–385.

Piton, A., Redin, C., Mandel, J.-L., 2013. XLID-causing mutations and associated genes challenged in light of data from large-scale human exome sequencing. Am. J. Hum. Genet. 93 (2), 368–383.

Price, A.L., Patterson, N.J., Plenge, R.M., Weinblatt, M.E., Shadick, N.A., Reich, D., 2006. Principal components analysis corrects for stratification in genome-wide association studies. Nat. Genet. 38 (8), 904–909.

Purcell, S., Neale, B., Todd-Brown, K., Thomas, L., Ferreira, M.A., Bender, D., et al., 2007. PLINK: a tool set for whole-genome association and population-based linkage analyses. Am. J. Hum. Genet. 81 (3), 559–575.

Rao, S.S., Huntley, M.H., Durand, N.C., Stamenova, E.K., et al., 2014. A 3D map of the human genome at kilobase resolution reveals principles of chromatin looping. Cell 159 (7), 1665−1680.

Riordan, J.R., Rommens, J.M., Kerem, B., Alon, N., Rozmahel, R., Grzelczak, Z., et al., 1989. Identification of the cystic fibrosis gene: cloning and characterization of complementary DNA. Science 245 (4922), 1066−1073.

Risch, N., Merikangas, K., 1996. The future of genetic studies of complex human diseases. Science 273 (5281), 1516−1517.

Rosenthal, R., 1979. The file drawer problem and tolerance for null results. Psychol. Bull. 86 (3), 638−641.

Savage, N., 2012. Gaining wisdom from crowds. Commun. ACM 55, 13. Available from: http://dx.doi.org/10.1145/2093548.2093553.

Saxena, R., Hivert, M.-F., Langenberg, C., Tanaka, T., Pankow, J.S., Vollenweider, P., et al., 2010. Genetic variation in GIPR influences the glucose and insulin responses to an oral glucose challenge. Nat. Genet. 42 (2), 142−148.

Schmid, P.R., Palmer, N.P., Kohane, I.S., Berger, B., 2012. Making sense out of massive data by going beyond differential expression. Proc. Natl. Acad. Sci. 109 (15), 5594−5599.

Schuster, S.C., 2008. Next-generation sequencing transforms today's biology. Nat. Methods 5 (1), 16−18.

Simpson, M.A., Irving, M.D., Asilmaz, E., Gray, M.J., Dafou, D., et al., 2011. Mutations in NOTCH2 cause Hajdu-Cheney syndrome, a disorder of severe and progressive bone loss. Nat. Genet. 43 (4), 303−305.

Smith, R., 2002. In search of 'non-disease'. BMJ 324 (7342), 883−885.

Stessman, H.A., Bernier, R., Eichler, E.E., 2014. A genotype-first approach to defining the subtypes of a complex disease. Cell. http://www.sciencedirect.com/science/article/pii/S0092867414001573.

Sydenham, T., R.G. Latham, 1850. The Works of Thomas Sydenham, MD.

Tabor, H.K., Risch, N.J., Myers, R.M., 2002. Candidate-gene approaches for studying complex genetic traits: practical considerations. Nat. Rev. Genet. 3 (5), 391−397.

Thomson, E.J., Boyer, J.T., Meslin, E.M., 1997. The ethical, legal, and social implications research program at the National Human Genome Research Institute. Kennedy Inst. Ethics J. 7 (3), 291−298.

Visscher, P.M., Brown, M.A., McCarthy, M.I., Yang, J., 2012. Five years of GWAS discovery. Am. J. Hum. Genet. 90 (1), 7−24.

Wang, K., Kan, J., Yuen, S.T., Shi, S.T., Chu, K.M., Law, S., et al., 2011. Exome sequencing identifies frequent mutation of ARID1A in molecular subtypes of gastric cancer. Nat. Genet. 43 (12), 1219−1223.

Wellcome Trust Case Control Consortium, 2007. Genome-wide association study of 14,000 cases of seven common diseases and 3,000 shared controls. Nature 447 (7145), 661−678.

Welter, D., MacArthur, J., Morales, J., Burdett, T., Hall, P., Junkins, H., et al., 2014. The NHGRI GWAS Catalog, a curated resource of SNP-trait associations. Nucleic Acids Res. 42 (Database issue), D1001−D1006.

World Health Organization, 2009. International Statistical Classification of Diseases and Related Health Problems.

Yang, Y., Muzny, D.M., Reid, J.G., Bainbridge, M.N., Willis, A., Ward, P.A., et al., 2013. Clinical whole-exome sequencing for the diagnosis of Mendelian disorders. N. Engl. J. Med. 369 (16), 1502−1511.

Chapter 12

Clinical Decision Support and Knowledge Management

Robert A. Greenes
Arizona State University and Mayo Clinic, Scottsdale, AZ, United States

INTRODUCTION

Scope of This Chapter

The desire to provide computer-based clinical decision support (CDS) has been an important motivator for health system automation over the last 50+ years in which health information technology (HIT)—particularly, electronic health record (EHR) systems—has evolved. Our goal in this chapter is to provide an appreciation of the range of possibilities for delivering CDS, an understanding of which approaches have been most successful, the limitations of current approaches, and recognition of how the needs for CDS are changing as a result of transformations in the health system itself.

The motivations for CDS have shifted over the decades and have never been greater. In the midst of a knowledge explosion from precision medicine, a plethora of new data not only from genomics but from imaging, natural language processing of narrative records, and the availability of personal devices and sensors, and the expanding capabilities for data analytics to provide ever increasing specificity of targeted CDS, the opportunities and needs for CDS are growing (Greenes, 2015). This is particularly relevant given the significant transformation to a "learning health system" that, although still in its early stages, is gaining momentum (LHS, 2015). Key aspects of this transformation are a shift from largely reactive disease-oriented healthcare to more proactive care oriented toward promoting wellness, prevention of disease, and early intervention for disease, and in the United States in particular, a transition from fee-for-service to pay for value, and from care provided by disparate entities to a more articulated health system (Greenes, 2016).

Thus it is important to understand not only what capabilities for CDS are currently available but the extent to which they either address or fall short of current needs. Where capabilities are lacking, it is important for us as health

Key Advances in Clinical Informatics. DOI: http://dx.doi.org/10.1016/B978-0-12-809523-2.00012-1

professionals to recognize that and help promote advances that will help overcome these limitations and expand the usefulness of CDS.

We will use the concept of a "clinical guideline" as a framework for this discussion. A guideline can be generally defined as a recommended set of processes for addressing a clinical problem, as it unfolds over time. An example would be the management of a diabetic patient as he or she progresses through stages of development and treatment of the disease and its complications. Most CDS capabilities address one of more aspects of a guideline, either at particular junctures in time or as the problem is addressed and evolves (days to weeks or years).

Long before the recent transformations of our health system had begun (which mainly got going toward the middle of the previous decade—the 2000s), the history of CDS had not been an entirely satisfying story—with many limitations and obstacles among the noteworthy successes that have occurred. Various institutions have had success with introducing computer-based provider order entry (CPOE), order sets, alerts and reminders, and infobutton-based access to context-specific information resources. CPOE, order sets, and infobutton management have had the highest penetration. But successes have been greatest at the institutions that first developed particular CDS interventions, largely academic medical centers, and the ability to widely disseminate and adopt these approaches has been limited. Biggest limitations have been in wide adoption of alerts and reminders, and ability to share these and other knowledge artifacts broadly.

We will focus mainly on the key motivations embodied in meeting the quadruple aim (Bodenheimer and Sinsky, 2014), which consists of the long-recognized triple aim of improving population health, enhancing patient experience, and reducing per capita cost, as well as the more recently recognized aim of improving the provider experience. These are all essential aspects embodied in the goal of continuity and coordination of the health and healthcare process.

We will also address the knowledge needed for CDS—where it comes from, how it should be represented to support CDS, and how it should be managed and updated. This topic is particularly challenging, because the "should" in the above statement is far from the reality of how knowledge "is" represented, managed, and updated.

Knowledge management requires an understanding of where a CDS artifact (e.g., a rule, order set, or other deliverable component of advice) has come from, its evidence base or derivation, and its lifecycle as it gets implemented or needs to get updated, to appreciate opportunities and challenges going forward. The transformation of the health system is again particularly relevant in considering this topic, as we consider new sources of knowledge arising from the science and technical advances of precision medicine, the increasing capabilities for big data analytics, and the role of the patient's preferences and constraints in the decision-making process.

We will focus on four main topics in this chapter:

1. The kinds of knowledge needed in today's systems for CDS
2. The current state and how we got here—motivations and drivers
3. Approaches to managing the knowledge
4. Where we need to go—the path forward

THE KINDS OF CDS AND THE KNOWLEDGE NEEDED FOR IT

A Brief History and Inventory of CDS

CDS has been one of the principal drivers for HIT over the more than five decades during which the field has evolved (Greenes, 2015). Table 12.1 shows the principal trends introduced in each decade prior to the current one.

These foci have involved a combination of methods that have also been refined over the decade. Major categories of methodologies include:

a. *Information retrieval and question answering technologies* for assembling context-appropriate knowledge. This includes various search engines, as well as infobutton manager technologies (Del Fiol et al., 2012). The latter

TABLE 12.1 Major CDS Foci Introduced by Decade

Decade	Major Foci Introduced
1960s	Early fascination with the computer as a diagnostician with exploration of Bayesian, and then in the 1970s also artificial intelligence/expert systems methods.
	Collection of data on specific topics at particular points in the care process, as structured data entry or documentation templates
1970s	Use of alerts and reminders for key notifications and actions
1980s	A focus on guidelines and protocols for workup and management
1990s	Organizing specific actions or sets of actions, as orders and order sets for use at appropriate points in the care process
	Retrieval of context-appropriate information resources tailored to what the user is doing or is likely to want or need to know
2000s	More recent approaches to population management seeking to provide information about most appropriate actions in particular care settings and to identify subgroups (of patients or providers) that need specific attention
	Visualization and presentation approaches to support cognition and decision making, e.g., as trends, graphs, animations, relationship maps, or other forms—typically through new applications and extensions to the user interface of EHRs

have enabled knowledge resources to be assembled from disparate sources in response to prototypical queries appropriate for each anticipated context.

b. *Probabilistic modeling* of disease and findings distributions to predict diseases based on findings or to make prognostic estimates. Decision analytic methods, using both Bayesian probabilities of alternative outcomes, plus estimated utilities of alternative action pathways, to determine optimal choice for a patient's care plan.

c. *Heuristic strategies using artificial intelligence approaches.* The aim is to function as experts by emulating subject domain experts, capturing, and modeling the knowledge they use as they reason over patient data to make recommendations.

d. *Rules and guidelines.* Evidence-based literature review, critical appraisal, and meta-analysis, or derivation of knowledge from experts to inform the development of rules and guidelines for care processes. These involve *if ... then* statements with conditional logic expressions in the *if* portion and recommendations or actions in the *then* clause. A variation called an Event-Condition-Action (ECA) rule has definition of a triggering event. Branching sequences of rules can be created to form guidelines.

e. *Data analytics and machine learning to develop models of best practice* based on process and outcome, improvements, and to build predictive models (logistic regression, classification and regression trees, neural networks, and other modeling methods).

f. *Association models*, such as ontologies/taxonomies, and richer relations among concepts to organize information by categories or purposes.

CDS as Variations or Aspects of Guidelines

There are four main operational purposes for CDS:

1. to avoid errors and ensure safety;
2. to help make a decision—diagnostic, workup, treatment choice, or prognosis;
3. to be alerted or reminded of forthcoming tasks or changes in patient status that might otherwise escape attention;
4. to facilitate appropriate actions.

The fourth purpose is more subtle and less obvious than the first three and overlaps with them. If CDS is well-designed, it should facilitate user activities (rather than interrupt and impede them). The goal is "to make the right thing the easy thing to do." This can be achieved with documentation templates (for data entry as well as for reporting), order sets, and visualization displays that group information for ready analysis, to ensure that all the pertinent elements are considered, and to facilitate choice of actions. CDS gets a bad rap when it is interruptive (Ash et al., 2007) and causes "alert fatigue" from too many or not sufficiently patient-specific interruptions, and as a consequence, may be ignored by providers or simply turned off. But when it is designed to present and facilitate appropriate actions, it can be helpful. Designing this mode is difficult and is the subject of much current work on creating specialized apps and visualization modes for improving the user−computer interaction.

When we consider the four individual purposes identified above, they coincide with the idea of following a "guideline." In general, avoiding errors means conforming to accepted standards of practice. Making optimal decisions also implies following recognized best practices embodied in guidelines. Alerting or reminding may relate to either an explicit guideline that is the focus of current attention or others that are in operation for other purposes, such as when the patient is being managed for multiple conditions, and multiple care plans are in process. We will therefore use the notion of "guideline" to explore the various forms of CDS and how they relate to this notion.

Typically a care process unfolds over time. There is an initial presentation of a problem, and efforts are made to understand it and try to treat it. This usually happens iteratively. Some initial data are available, and using the hypothetico-deductive process (Elstein et al., 1978), a working set of possible explanations and hypotheses is identified. These suggest either the need for more data (testing) or desirability for treatment. Treatment of course generates more data as to its effectiveness, which causes the hypotheses to be updated or revised.

In line with this framework, CDS can be helpful at several stages in Fig. 12.1.

Note that the process begins with obtaining data, and the four steps iteratively repeat. Table 12.2 provides some notes on this process and why each step is getting more complicated and needing decision support.

Additional notes regarding how various CDS methods can fit into this framework are the following:

- As one initiates actions like testing and treating, order sets are useful.
- As one waits to continue the cycle, alerts and reminders advise when data are ready to evaluate.
- Advice on the possibilities at each step in the above can be provided by infobuttons.

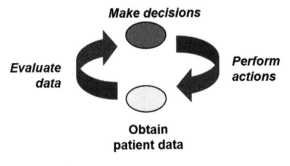

FIGURE 12.1   The basic patient care process.

TABLE 12.2 Basic Patient Care Process, Notes, and Complicating Factors

Patient Care Process Steps	Notes	Complicating Factors
Obtain patient data	Starts the process—both de novo data gathering and reviewing pertinent data from previous encounters	• More data – More sources, e.g., gene analyses, assays, imaging, and image processing – More data volume – More specialization – Only see part of picture
Evaluate data	Diagnostic process. Analyze data to form hypotheses	• More knowledge – Genes, proteins, cellular processes, pathways – Clinical trials, meta-analyses – Products, interactions, side effects
Make decisions	Workup/management decision process a. Acquire more data or b. Treat (also producing data)	• More pressures – Safety and quality initiatives – Cost-effectiveness foci – Liability concerns, compliance – Patient information, demands – Decreased time – Increased documentation
Perform actions	Activation or update of orders in a patient's overall plan of care a. Order tests b. Order or revise treatments c. Provide prognosis	• More pressures – Safety and quality initiatives – Cost-effectiveness foci – Liability concerns, compliance – Patient information, demands – Decreased time – Increased documentation
(Repeat)	The sequence unfolds in a series of cycles, and the overall progression and its possibilities could be portrayed as a guideline	

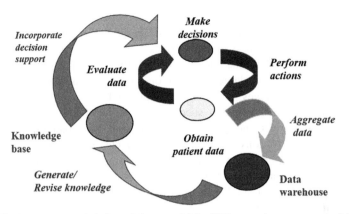

FIGURE 12.2 An expanded view of the potential for CDS to continue to renew and improve as part of a comprehensive data analytics framework based on ongoing capture of data from care processes and outcomes.

- Data analytics can assist in the evaluations needed in the first three steps, by indicating preferred choices based on match of patients to various subgroups. See also the expansion of the above process in Fig. 12.2 if one presumes that all data are being aggregated in a data warehouse. The data collected are used to help continue to improve both the knowledge base and the ongoing process of care.

Thus one can consider the above model as a concise representation of the possibilities for CDS.

THE CURRENT STATE OF CDS AND HOW WE GOT HERE

Discrete CDS Capabilities

Over many years, most of the effort in CDS development has been placed on individual knowledge artifacts that can be invoked or triggered at particular steps in the above iterative process, mainly because they corresponded to points that could be implemented, for example, in EHRs, given that the appropriate preconditions for their use existed. Most success has been achieved with order sets and with infobuttons, in that they have standardized representations and have been adopted by commercial knowledge providers who offer them in the marketplace. Alerts and reminders have continued to be a focus of major interest. But the latter have been plagued with implementation issues like poor interface with the user, interruption of workflow, insufficient patient specificity, and alert fatigue. Indeed, much of the hostility to EHRs in the academic, professional, and public press is owing to this.

We will have more to say about rules such as alerts and reminders later, because the need to adapt rules to many custom preferences in how they are triggered, who they are directed to, how the conditions are adapted to work-flows, and how notifications are processed is responsible for a huge customization effort. Even if best practice rules can be shared, this need for adaptation greatly adds to the cost of implementing, multiplies the variations that may be in place in a large organization, and makes tracking and updating, as knowledge changes, much more complicated.

Given the above, the modest success of CDS has been greatest in academic medical centers and larger institutions that have had the resources (and often the motivations) to do the necessary customizations and adaptations. There has been progress in developing standards for the data and knowledge representations needed, although most of the standards in use continue to evolve (Zhou et al., 2013). Deployment of CDS in community hospitals and practices has been much slower, due to resource limitations, inability to adapt to custom workflows, and frustration with the resulting implementations.

Progress in expansion of range of features and in widespread adoption has been incremental at best.

Guideline Modeling

We describe guidelines and variations to be discussed here as more comprehensive descriptions of optimal care processes that unfold over time, components of which have been discussed earlier. There has also been interest in this overall flow of the process and its iterative data gathering, evaluation, and action steps (cycles). This has been expressed in the form of formalized guidelines, protocols, care process models, care pathways, and workflow models (Peleg et al., 2003). By and large, despite the interest, the experience in introducing these into care have been less successful than the implementation of more discrete CDS capabilities. Several reasons may be cited for this (Greenes, 2014):

1. There is not universal agreement on what the various kinds of processes that unfold over time are, and how they should best be supported. This holds back the development of standard and interoperable models.
2. While an overall process model for managing diabetics, for example, may be able to be defined, it is typically not executable as such, because:
   a. Most guidelines are not defined in sufficient detail to apply to specific patients with variations in findings, disease status, concurrent diseases, treatments, and their side effects.
   b. The overall process might include initial workup, diagnosis, and subsequent management, and how to respond to new findings, treatment failures, side effects, and complications of the disease or other disease, and could cover many days, weeks, or years. Guidelines and care process models for some conditions, such as hypertension,

chronic heart failure, diabetes, etc., can also involve several venues, providers, and aspects of care—including office, emergency department, imaging center, lab, specialists, hospital, extended care facilities, and home—the latter involving data capture by patients with sensors and mHealth apps. This means several different systems must potentially have access to the same patient information and guideline model and be able to update it. Such is not the case generally—we will come back to this point.

c. Related to (b), many actors may be involved—including multiple providers with different roles, the patient, and other caregivers.

d. Despite their complexity for a specific disease, guidelines may be considered somewhat idealized, in that they typically do not account for the many other concurrent problems a patient may have and which may cause the guideline to become less relevant for any particular patient.

e. Even if a patient is eligible for a guideline at one point in time, the guideline may not remain applicable as the patient's status evolves.

f. The provider is typically in control of the process, not the computer, so the guideline is not typically "executing." In other words, physicians do not tend to use guidelines to carry out the management of a patient's care but rather to consult or refer to them at difficult decision points. Guidelines are available to be inspected in terms of particular states/nodes that may pertain and recommendations that may be appropriate in those states.

g. Steps that are pertinent can be either invoked by the user or eligibility at those entry points can be determined in order to trigger a notification to the user. The latter is what occurs with alerts and reminders and with the generation of order sets or data entry templates for use at particular junctures in the care process.

So a *guideline* is typically a human-readable framework for managing a care process, and computer implementation is done by identifying selected parts of it such as nodes/states that can be automatically identified and actions that can be triggered at such points (such as data entry or order set invocation). Guideline representation languages and formalisms have been developed for computer-based modeling, but have not converged on a standard, partly because of the problems with how one envisions execution, due to considerations such as the above. Some effort has been made to build execution models nonetheless (Wang et al., 2003; Tu et al., 2007).

Some exceptions where guideline-type processes are intended for execution can include: Guidelines expressed more deterministically and prescriptively rather than as advice may be used by providers in lesser positions of autonomy such as physician assistants or nurse practitioners to manage processes that are somewhat routine. Or physicians in training may be more reliant on the step-by-step guidance.

The term *protocol* is used to describe the flow when procedures are highly deterministic such as dialysis or cancer chemotherapy protocols, for which strict adherence is required, especially if the patients are enrolled in clinical trials. Failure to follow a clinical trial protocol, in fact, will cause the patient to need to drop off the trial.

Care pathways are attempts to predict tasks and resources for specific times following an event like surgery, such as postoperative day 1 tasks and expectations, day 2, etc.

Care process models are more loosely defined as guidelines with touch points where they can be implemented. Mayo Clinic defines these as connecting to evidence bases, having human-readable descriptions, and having refinements indicating how and at what points they should be integrated into care and into the computer systems to support the care (Cook et al., 2016).

Workflow models are an attempt to focus on the coordination of the multiple processes and actors involved in carrying out specific tasks.

Formalization of Guidelines

There have been a series of efforts over the years to try to formalize computer-based models of clinical guidelines, i.e., computer-interpretable guidelines (CIGs) (Peleg et al., 2003). But, because of the various subpoints of item 2 in the list above, there has not been sufficient enthusiasm of the developer community to converge on a standard model. Also, the authoring of guidelines and recommendations is often done in a rather imprecise fashion, partly because of lack of formal models being available and partly because it is sometimes artificial or forced to be too specific, given the likelihood that the guideline may not fit the patient very closely, or where the evidence for a particular action is not very strong.

Guideline development has ranged from (1) tools for markup of narrative guidelines to identify key elements of the source, provenance, evidence strength, and specific recommendations, to (2) tools for constructing flow charts that can be browsed and linked to details, to (3) guideline sequences that can be formally executed by user–computer interaction, possibly retrieving some information from the EHR, to (4) tools for integrating parts of CIGs into EHRs to provide trigger points or settings in which appropriate decision choices or actions/order sets are suggested.

A Mismatch With the Status Quo

The evolution of CDS has been in part driven by changes in:

a. healthcare structure and organization,
b. the way healthcare is financed,

c. the technology for implementation of EHR systems and other IT and communication functionality.

Initial efforts in the United States (beginning in the 1960s) were focused on HIT in individual hospitals (largely academic medical centers) or in some cases clinics or practices, had limited data, and had local optimization goals. As healthcare organizations expanded to include networks of practices and hospitals (beginning in the 1980s), a more comprehensive view evolved, in which the need to recognize interactions of problems, provider actions, treatments, and other aspects of care increased. Technology was initially focused on the individual organization and later on the enterprise, with health information exchange (HIE) as a way to achieve some level of interoperability at the edges, and across enterprises (beginning in the mid-2000s). In other nations with more central and coordinated health systems, such as the United Kingdom and Sweden, progress has tended to be greatest in the ambulatory setting and less so in hospitals, but with higher degree of standardization of systems deployed. In developing nations with limited infrastructure, the approach is beginning with the connectivity to the various care providers and clinics. Thus evolution of a fully connected system differs greatly among nations.

The Current Milieu—a Time of Disruption and Transformation

Today's focus in the United States is largely on a *learning health system*, moving toward emphasis on *wellness and early intervention* in disease, is having a dramatic effect on healthcare structure and organization, healthcare financing, and the technology needed. To a large extent this reflects goals and approaches of many nations, but the ability to realize this in the United States is largely impeded by the history of disparate systems and silos of care that have been in place for many decades. In the United States, we now have activities aimed at supporting Accountable Care Organizations (ACOs), service line optimization in organizations, and increased focus on the need to coordinate and communicate care across different venues and participants (Rochester-Eyeguokan et al., 2016).

Care is also becoming more patient-centric—not only in terms of patient engagement, but also as a source of data and action through the use of home sensors, medication administration devices, exercise and other activities, and reporting of results to mobile devices and cloud communication. An insightful look into where this trend is ultimately headed is offered by Topol (2015).

Financing based on pay-for-value rather than quantity of care (fee-for-service) and on bundled payments for categories of patients is requiring an entirely different level of need for coordination, communication, tracking, and analytics. This is bringing about a rapid increase in recognition of the urgency for CDS and a shift in emphasis to the kinds of CDS needed to provide guidance across the care continuum and tracking of process and outcomes across these venues.

It is important to note that the requirements for the new holistic focus on the care continuum are not well mapped to existing EHR systems, which have evolved over 20−40 (or more) years with a more limited purview of both organizational elements and data, effectively creating silos. Systems being used for EHRs are generally proprietary, with restricted access externally—due to organizational policy, HIPAA requirements, and vendor EHR technology incompatibilities, with data in proprietary formats. The knowledge resources used for decision support typically built into each of the organization's EHR systems is also in proprietary format. If multiple systems are involved and data are coming from more than one of them, there is no shared knowledge base, and the knowledge must be replicated and continually updated in each of the systems. The situation is compounded for most patients who obtain care for different problems in different venues, and typically not part of a single healthcare delivery enterprise.

We thus have a dilemma that much of the knowledge needed for decision support is not transparent to subject matter experts, since it is embedded in the proprietary EHR systems' own knowledge bases, and is not consistent across the different venues in which a patient receives care. Similarly, the data on which decision support must be based to produce recommendations are also not readily available across venues.

There is currently much discussion about the need for interoperability, but the term has multiple meanings, and there is little consensus on how to achieve it. Most work is occurring at the levels of:

1. Terminology agreement for clinical concepts, converging on SNOMED, LOINC, and RXNORM for clinical, laboratory, and medication terms, and ICD10 for billing.
2. Simple interfaces for retrieving data of specific types (such as medications or laboratory results) from EHRs through new interface technology known as the Health Level Seven (HL7, 2016) FHIR specification (HL7 FHIR, 2016) and point-to-point structured HL7 messaging.
3. HIE through summary records in a format known as CCDA (D'Amore et al., 2012).

Implications for CDS

In addition to depending on terminology and messaging/data exchange standards development, CDS has evolved its own standards, with some level of agreement. Most success has been in format for infobutton managers, to be able to describe interfaces to external information resources in a standard way. Order sets have standard formats facilitating import of them from knowledge content vendors and other shared sources. Alerts and reminders have had various formats proposed. The main standard is Arden Syntax (Jenders et al., 1998), although it is not widely used except in some proprietary implementations. Many models for guidelines and protocols have been proposed over the decades, but there is not an accepted standard for these.

Against the backdrop of the above, we will now focus on the nature of the CDS landscape, why there has been limited uptake, and where the needs and opportunities are heading.

One aspect that is instructive to understand is that there have been three main modes of CDS invocation up to the present:

a. *User-imitated*: the user invokes the CDS when the need for it is recognized. Infobutton selection is one example of this. Calling up a consultation program or selecting a graphical visualization tool are other examples.

b. *Event-driven*: an asynchronous occurrence such as elapse of a designated amount of time or availability from the laboratory of a particular test result causes a decision rule to be evaluated, with possible generation of a notification to the user.

c. *Workflow-integrated*: decision support built into regular user tasks and sequences, for example, the automatic generation of an order set for post-operative hip replacement care, when the patient reaches the recovery room after the designated surgery, or an interactive message during order entry advising that the drug to be prescribed has a harmful interaction with another medication the patient is receiving.

All of these modes require careful integration with the clinical applications and underlying system to determine when and how they should be made available or automatically surfaced. Both event-driven and workflow-integrated approaches require substantial modification of the basic knowledge to account for considerations such as to whom the advice should be directed, when, in what form, and by what means, often referred to as the "Five Rights for CDS" the right information, at the right time, to the right person, in the right format, via the right channel (Sirajuddin et al., 2009). As a result of this, as we noted earlier, a single rule, for example, to remind physicians to evaluate a diabetic patient's HbA1c level at regular intervals might have many variations within an organization to accommodate the different workflows that are preferred in various clinics or by various providers or as a function of different kinds of staff support. This multiplies greatly both the time involved in adapting the basic piece of knowledge to an operational rule and also the number of variations of it that must be maintained and updated as knowledge changes over time (Greenes et al., 2010).

A new mode of delivering CDS is beginning to be recognized as having potential to overcome some of the limitations of the above three modes. We refer to this as:

d. *Context-based*: One can think of an analogy to a GPS navigator. The GPS software has knowledge of the user's mode of transportation, directional heading, actual position or state, and can provide highly context-appropriate advice about upcoming intersections, traffic, weather, detours, and local points of interest or relevance. This can be done to a large extent even without informing the GPS of the destination. Doing so enables even more specific guidance to be possible.

To achieve context-based delivery of appropriate CDS, the IT system would, like the context tracked by a GPS navigator, similarly have awareness of the problems of the patient, the specialties and areas of expertise of the provider, the kind of encounter being conducted, and the application being used, as well as the device used for interaction. Like the GPS, this would enable the system to provide context-appropriate findings, actions, and knowledge/recommendations to be considered. This could be made available in a noninterruptive, always-available fashion, with more important items highlighted or requiring acknowledgments in various ways. A potential benefit of this mode of CDS provision is that it would not require the kind of laborious integration into applications and workflows that the other modes for CDS do require. This is because the logic and triggering conditions would not need to be as highly customized, since simply indexing the knowledge artifact by its contextualizing attributes would allow it to be retrieved at the appropriate point in the care process. If an explicit goal or purpose is stated (such as conformance with a guideline), as with a GPS navigator, the advice could be even more specific.

As implied by the above, the context-based delivery of CDS is predicated on having a comprehensive set of data and a knowledge base of knowledge content tagged and identified by the various contexts in which they are appropriate to use. In 2013, work began, sponsored by the US Office of the National Coordinator for Health IT (ONC), on development of a new shared model for CDS knowledge artifacts known as the Health e-Decisions (HeD) model, which has become an HL7 standard and is continually being refined and extended (HeD, 2014) and is now further modified to what is known as the Clinical Quality Framework (CQF) to accommodate quality measures as well as CDS artifacts (CQF, 2016). Although the HeD/CQF model was largely intended to be a standard means for representing best practice knowledge to encourage sharing, its formal underlying model and metadata tagging system potentially enables the knowledge artifacts to be richly indexed by domain, type of knowledge, purpose, and other attributes. Extension of this metatag scheme to include setting and context attributes is a possible means for maintaining a repository of available knowledge for the context-based delivery we describe. This idea is very preliminary, but is an intriguing possibility.

Problems of Multiple Venues

Knowledge, even in the form of alerts and reminders, is often embedded in specific EHRs, and thus tends to be locked into proprietary rule editors of the EHR systems, and their execution environments, and also must be tailored to workflows and triggers and setting-specific factors (SSFs) of particular environments (Greenes et al., 2010, 2014). Hence it is not readily shared.

When we consider the larger goal of coordination for a patient, this involves one or more guidelines or care process models that unfold over multiple venues, actors, settings, and times, and possibly spanning across data and responsibilities of disparate EHR systems in the offices and institutions involved in providing care for that patient. This is another reason that knowledge used for CDS should not be locked up in individual EHRs, and thus impede coordination.

Other Modes of CDS

Large additional areas of activity are in the realm of patient engagement and in population health management. We will not cover these in detail because they are addressed in other chapters (see Chapter 15: Predictive Analytics and Population Health and Chapter 18: Social and Consumer Informatics).

Patient engagement in the form of mobile health, personal data entry, use of home sensors and monitors, and communication with the care team are active areas of pursuit.

Population management involves organizing data on cohorts, performing analytics, and tracking cohort processes and outcomes. Registries are a way to enroll patients in such cohort management. Another would be to enroll patients who are on a particular care pathway or protocol. One can then develop dashboards for tracking patients in various cohorts, at the practice, provider, or individual patient level, and use these to provide monitoring and alerting of needs for intervention. Combining this with mobile health, for example, can be done to track patients who are sent home on a congestive heart failure protocol for reminding patients about their medications, and processes for the recording (or direct input) of weight, blood pressure, and other parameters.

APPROACHES TO MANAGING THE KNOWLEDGE

The above brings us to the topic of managing all the knowledge that is created. We have identified many purposes and methods of CDS in the above discussion.

We can group them another way:

- *Category 1*: models of diagnostic decision making, decision analysis, data analytics/prediction, formalized rules such as alerts and reminders, infobuttons for providing context-specific reference materials, structured grouping of information such as in documentation templates for data entry and reporting and in order sets, and visualization models for presenting and viewing related information.

- *Category 2*: methods for describing whole processes rather than separate stages of them achieved by the above components. These include formal guideline models, protocols, care process models, care pathways, and workflow models.

Standards exist for most of the items in category 1—the more-or-less discrete CDS components. Exceptions include diagnostic systems, decision analysis, and data analytics. For the components, the standards are not perfect, and there is much flux about them, based on ongoing issues about the EHR interface model for retrieving needed data, information models for how to postcoordinate discrete data elements to create the clinical meaningful elements to be used, the terminologies to be used to represent the clinical element concepts and the value sets for their data fields (using coded terms from taxonomy schemes such as SNOMED, LOINC, and RXNORM for disease and finding, test, and medication elements, respectively), and the knowledge to manipulate them. How, for example, should an alert like an HbA1c testing rule be represented, including its triggering event, the conditions for when it should be considered true, and the actions to occur if it is true? A review of current standards for CDS is provided by Zhou et al. (2013).

Although there has been much research on the topic, standards do not largely exist for the items in category 2, the more integrative longitudinal types of CDS. There is a need to do so, and this remains an open agenda item.

The HeD model we mentioned above was an attempt, promoted by ONC, to get best practice knowledge to be sharable, in terms of three classes of category 1 CDS: decision rules, documentation templates, and order sets. The goal of the HeD model was to have a standard means for sharing of specific knowledge artifacts to be required as part of Meaningful Use Stage 3. Its successor, the CQF, is intended to add quality measures. But as designed, these frameworks do not get into workflow customization or how to insert them into the care processes. The main goal in creating the frameworks was to be able to distribute standard best practice modules or specify how services might be interfaced to EHRs to offer their CDS content. Short of having services that can be invoked, however, the only way to use it is through distribution in eXtensible Markup Language (XML) form to translate it into the internal form used by individual EHRs' knowledge content editors—for order sets, rules, documentation templates, etc. Any customization and mapping to workflow and how triggered, who it interacts with, when, and how notifications are communicated, and what happens in terms of acknowledgment, override, etc., are typically done in the host editing system. If CDS is to be delivered as a service, these workflow customizations still need to be done in determining how and when the CDS service should be invoked, to respond to requests for needed data, and to specify how the response should be handled.

The HeD framework and the successor CQF are model-driven, as we have mentioned, in that their elements are formally defined. With suitable extensions, CQF could model most of these setting-specific customizations. The US Veterans Administration (VA) is, for example, interested in possibly doing that. The author and colleagues built a first-version editor for HeD knowledge artifacts, as part of a research and development contract from ONC (Greenes et al., 2014). Because HeD was model-driven, this means that an HeD authoring/editing system could formally convert knowledge content into any well-formed alternate rendering, whether it is Java code, Arden Syntax, Drools, or other rules languages supported by particular EHRs, or output as XML. Presumably, if vendor systems would allow it, they could import such content into their own environments to execute the CDS, or they could invoke it as services at well-defined points. Such adoption by vendors has not yet occurred however.

The customizations for workflow and SSFs, as described in the work of Greenes et al. (2014), could be done in the host EHR systems but then they would be buried (as they are now) in those systems and not transparent when the need comes for reviewing and updating the knowledge by subject matter experts. As we have noted, this is also required even invoking CDS as an external service, in that the context information for customizing the workflow integration needs to be either done by the EHR or it needs to be communicated to the service. The CDS Consortium (Middleton, 2009) and the Morningside Initiative (Greenes et al., 2010), followed on by the SHARPc project (Greenes et al., 2014), have each proposed formal models for such refinement, which differ from each other but reinforce the point that such refinement is needed.

The main advantage, though, in doing all the customizations in a system like HeD or CQF—before translating them into proprietary rule management and execution environments in EHRs—is the aforementioned growing need to address coordination and communication of care across care venues—which will often entail a multiplicity of EHRs and other patient-centric data sources. This means that the knowledge being used by each EHR needs to not only be consistent (and synchronized) in each of them, but perhaps maintained in a central knowledge repository external to all of them. Whether it can be executed externally, e.g., through services, is still an open question. But the very need to identify all the places where an element of knowledge is used, and all the SSF-based variations of it, is essential. Knowledge evolves frequently and is expected to do so. In fact, many organizations have policies that all CDS be reviewed at regular, e.g., 2-year, intervals. Current knowledge, such as about new therapies or more cost-effective practices, need to be incorporated as soon as it is shown to be effective.

Another potential advantage of the HeD/CQF model is that, because it models and represents the knowledge artifacts formally, each component is of well-defined syntactic type, and its semantics are typically known, such as the medical/problem domain to which it relates, its function, such as drug dose check, medication interaction check, or abnormal laboratory result condition, and other attributes. It also can have a variety of metatags indicating source, provenance, version, adaptation source, etc. To the extent it is used to incorporate SSF customization, these settings and contexts could also be indexed.

The tools now available for knowledge management are pitifully inadequate. There are systems for document management, for example, maintaining repositories of evidence-based clinical recommendations in human-readable form. The Care Process Model approach of Mayo Clinic seeks to attach to a document management system a set of formalized refinements of the human-readable knowledge, more formal descriptions of how various care processes are to be done, the points at which they should be introduced into clinical systems, e.g., as triggers, or in the form of alerts or order sets, and connections to the various subsystems that implement them (Cook et al., 2016).

Ideally a knowledge management system would also maintain all the versions and customizations introduced for workflow and SSF adaptation. But this requires more than document management. It requires formal editing tools with templates and syntactic and semantic constraints enforcing representation consistency and appropriateness. Knowledge editing and management tools for healthcare have in some cases been adapted from those in other domains like software development, and a few organizations (notably, Intermountain Health and Partners Healthcare) have spent considerable effort building their own custom knowledge management environments. Despite initial efforts to do this, their work in each case has been somewhat curtailed when the organizations have gone from home-grown EHR systems to commercial ones. There are very few healthcare organizations that are able to devote much effort to this, or they rely on commercial and sometimes proprietary solutions.

To the extent we move to a multivenue care coordination and communication model for a learning health system emphasizing wellness and pay for value, the more such a communal repository of best practice knowledge is needed, and the more it must be able to curate and manage updates and synchronize those with all places that they are executed.

As noted above, we could envision the context-based mode of CDS using an enhanced HeD/CQF repository as the source of knowledge to provide in appropriate settings. Given the metatags and indexing possible with HeD or CQF, a process maintaining and aware of the context of a clinical encounter (who the patient is and what problems he/she has, the identity, specializations, and expertise of the provider, the setting and purpose of the encounter, and the application context) could use those descriptors to automatically retrieve and make available the CDS options that are appropriate. This would be infinitely easier than needing to customize the CDS modules in every proprietary system or figure out how to pass all the context parameters to a CDS service.

WHERE WE NEED TO GO

Current approaches to implementing CDS focused on the parts that have worked and have become somewhat standardized will certainly continue. But judging from the mostly incremental progress that has occurred over many years, with no significant acceleration as a result of EHRs having become broadly adopted, and with growing frustration with EHRs and pushback, it is unlikely that continuing only the current approaches will achieve broad-based adoption and in-depth use of CDS.

To summarize, this is largely because of the following factors:

- The fragmented HIT systems/EHRs with their own CDS capabilities
- Proprietary knowledge editors and execution engines
- Limited (siloized) data on which they operate
- Lack of transparency of the knowledge embedded in disparate systems, making it hard to view what exists or to update it
- The need for highly specific workflow and SSF adaptations to make CDS acceptable, which only few health systems can afford to do, and which compound the knowledge management and update problem
- Lack of ability to do external knowledge management in a way that supports subject matter experts, and to integrate that into EHRs either through import by them or invocation of external CDS services by them.

These factors do not seem likely to change much without significant disruption in the health ecosystem. However, such disruption may indeed be upon us over the next decade. Looking ahead, we can project what a future HIT system might need to look like (Fig. 12.3), if disruptive forces do indeed bring about the expected changes in the health/healthcare delivery system:

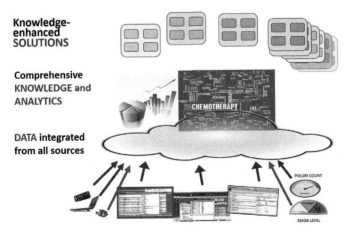

FIGURE 12.3 A multitiered architecture that may evolve to support the needs of a learning health system.

- Evolution to a multitiered HIT environment, in which existing EHRs and other data sources including personal data clouds and environmental and public health data, provide the data in a standardized interoperable format, at the lowest layer
- A standardized, normalized, knowledge-enriched data layer constructed from the underlying data sources, shared by multiple players in the health system, with appropriate security and access control capabilities
- An application layer (apps considered as mobile or desktop) that assembles data in highly specific ways suitable to users and contexts to facilitate appropriate visualization, cognitive understanding, and decision making
- Management of the underlying knowledge in a shared repository of best practice knowledge that is both human-readable and has formalized representations that are executable
- Use of context to not only facilitate app-specific assembly of information but also to identify CDS needed at particular points in healthcare so they can be surfaced as appropriate (the GPS model applied to healthcare)
- Data analytics to provide population management support for identifying cohorts at risk or in need of specialized care processes, to provide feedback to users, and to generate new knowledge about best practices.

The above may be wishful thinking—that the health system will evolve in such a fashion. Currently the author is engaged with Mayo Clinic and with the VA system, both of which are seeking to build aspects of the above. Other organizations are also working in this area. The Healthcare Services Platform Consortium (HSPC, 2015) has espoused such goals, and organizations such as Intermountain Healthcare, LSU Health, Regenstrief Institute, Vanderbilt, Partners Healthcare, and some vendors are beginning to work together to explore joint activities. The further development and maturation of ACOs and approaches to service line integration which require not only data analytics but persistent data and comprehensive care plans for guiding appropriate actions are demanding this.

CONCLUSIONS

CDS has been an active area of pursuit for more than five decades. Many approaches have been explored and much has been learned. But the field is plagued with difficulties in scaling from successful implementations of particular methods in one institution to broad adoption. This has been in part due to the different goals of various approaches but also because of the difficulties in achieving standardization and interoperability. Part of the reason for the latter is the proprietary nature of individual EHR systems, but another is the need for considerable customization into the workflows and processes of the users in those systems.

We looked at guidelines as an overall framework for how patients are diagnosed, worked up, and managed over the course of their diseases. While implementation of guidelines themselves has not been successful generally, the components of guidelines that are relevant at particular points in the care process include if-then rules, models of decision making, order sets, and other knowledge artifacts that themselves can be integrated into care, albeit with the need for customization mentioned.

The importance of guidelines or care process models is greater as we consider a more holistic, patient-centered view of health/health care. Transformations of health care into a learning health system are gathering momentum, with a focus on wellness and more proactive early intervention for disease, and for continuity and coordination of care across care venues and engaging the patient more directly, and tracking process and outcomes actively.

Our EHR infrastructure is not set up for this new mode, and issues of overall knowledge management and incorporation into the different systems and components of the care process are large challenges. This is giving rise to some new thinking about best ways to deliver CDS and to overcome some of the barriers of the past.

The need is great and the opportunity is large, but the direction we will collectively move is still not established. It is important for stakeholders to recognize both the capabilities and barriers that now exist and the new directions that show promise and to be involved in helping to chart these directions.

REFERENCES

Ash J.S., Sittig D.F., Campbell E.M., et al., 2007. Some unintended consequences of clinical decision support systems. AMIA Annual Symposium Proceedings 2007. pp. 26−30.

Bodenheimer, T., Sinsky, C., 2014. From triple to quadruple aim: care of the patient requires care of the provider. Ann. Fam. Med. 12, 573−576.

Cook D., Pencile L., Sorensen K., et al., 2016. Usage of electronic online care process models for learning at the point of care. Proceedings, AMIA 2016 Annual Symposium. Chicago, IL. American Medical Informatics Association.

CQF, 2016. Clinical Quality Framework Initiative. Available at http://wiki.siframework.org/ Clinical + Quality + Framework + Initiative.

D'Amore, J.D., Sittig, D.F., Ness, R.B., 2012. How the continuity of care document can advance medical research and public health. Am. J. Public Health 102, e1−4.

Del Fiol, G., Huser, V., Strasberg, H.R., et al., 2012. Implementations of the HL7 Context-Aware Knowledge Retrieval ("Infobutton") Standard: challenges, strengths, limitations, and uptake. J. Biomed. Inform. 45, 726−735.

Elstein, A.S., Shulman, L.S., Sprafka, S.A., 1978. Medical Problem Solving: An Analysis of Clinical Reasoning. Harvard University Press, Cambridge, MA.

Greenes, R., Bloomrosen, M., Brown-Connolly, N.E., et al., 2010. The Morningside Initiative: collaborative development of a knowledge repository to accelerate adoption of clinical decision support. Open Med. Inform. J. 4, 278−290.

Greenes, R.A., 2014. Clinical Decision Support, 2nd ed.: The Road to Broad Adoption. Elsevier, New York.

Greenes, R.A., 2015. Evolution and revolution in knowledge-driven health IT: a 50-year perspective and a look ahead. In: Riano, D., Lenz, R., Miksch, S., et al., Knowledge Representation for Health Care. AIME 2015 International Joint Workshop, KR4HC/ProHealth 2015. Springer, Pavia, Italy, pp. 3–20.

Greenes, R.A., 2016. Health information systems 2025. In: Weaver, C., Ball, M., Kim, G., et al., Healthcare Information Management Systems: Cases, Strategies, and Solutions, 4 ed. Springer, New York, pp. 579–600.

Greenes, R.A., Sottara, D., Haug, P.J., 2014. Authoring and editing of decision support knowledge. In: Zhang, J., Walji, M. (Eds.), Better EHR: Usability, Workflow and Cognitive Support in Electronic Health Records. University of Texas, Houston, TX.

HeD, 2014. Health eDecisions Homepage. Available at http://wiki.siframework.org/Health + eDecisions + Homepage.

HL7, 2016. Health Level Seven International. Available at http://www.hl7.org/.

HL7 FHIR, 2016. FHIR DSTU2. Available at http://www.hl7.org/fhir/summary.html.

HSPC, 2015. The Healthcare Services Platform Consortium. Available at https://healthservices.atlassian.net/wiki/display/HSPC/Healthcare + Services + Platform + Consortium.

Jenders R.A., Huang H., Hripcsak G., et al., 1998. Evolution of a knowledge base for a clinical decision support system encoded in the Arden Syntax. Proc AMIA Symposium. pp. 558–562.

LHS, 2015. Continuous improvement and innovation in health and health care. The Learning Health System Series: Roundtable on Value and Science-Driven Health Care. National Academy of Medicine (NAM).

Middleton, B., 2009. The clinical decision support consortium. Stud. Health Technol. Inform. 150, 26–30.

Peleg, M., Tu, S., Bury, J., et al., 2003. Comparing computer-interpretable guideline models: a case-study approach. J. Am. Med. Inform. Assoc. 10, 52–68.

Rochester-Eyeguokan, C.D., Pincus, K.J., Patel, R.S., et al., 2016. The current landscape of transitions of care practice models: a scoping review. Pharmacotherapy 36, 117–133.

Sirajuddin, A.M., Osheroff, J.A., Sittig, D.F., et al., 2009. Implementation pearls from a new guidebook on improving medication use and outcomes with clinical decision support. Effective CDS is essential for addressing healthcare performance improvement imperatives. J. Healthc. Inf. Manag. 23, 38–45.

Topol, E.J., 2015. The Patient Will See You Now. Basic Books.

Tu, S.W., Campbell, J.R., Glasgow, J., et al., 2007. The SAGE Guideline Model: achievements and overview. J. Am. Med. Inform. Assoc. 14, 589–598.

Wang D, Peleg M, Bu D, et al., 2003. GESDOR—a generic execution model for sharing of computer-interpretable clinical practice guidelines. AMIA Annual Symposium Proceedings. pp. 694–698.

Zhou, L., Hongsermeier, T., Boxwala, A., et al., 2013. Structured representation for core elements of common clinical decision support interventions to facilitate knowledge sharing. Stud. Health Technol. Inform. 192, 195–199.

Chapter 13

Mobile Health

Karandeep Singh[1] and Adam B. Landman[2]

[1]*University of Michigan Medical School, Ann Arbor, MI, United States,* [2]*Harvard Medical School, Boston, MA, United States*

INTRODUCTION

Health systems across the world are seeking to achieve the Institute of Healthcare Improvement's Triple Aim: improve the health of populations, improve the experience of care (quality and satisfaction), while reducing costs (Berwick et al., 2008). While there is not a single, simple way to achieve the Triple Aim, health information technology (HIT) will be an important component of solutions (Institute of Medicine, 2012). Traditionally, HIT has focused on the implementation and optimization of electronic health records (EHRs). Mobile phones are becoming increasingly common, with sophisticated capabilities at lower costs. According to a 2016 Pew Research report titled "Smartphone Ownership and Internet Usage Continues to Climb in Emerging Economies," a global median of 88% of adults own either a smartphone (43%) or a cellphone that is not a smartphone (45%) (Fig. 13.1). Over 70% of adults in South Korea, Australia, Israel, the United States, and Spain have smartphones. Cellular phones include the ability to make phone calls and send SMS (short message service) messages; while smartphones add a computer processor, operating system, ability to run applications (apps), browse the Internet, and send/receive electronic mail (e-mail). Given this level of penetration of smartphones and their capabilities, smartphones are an increasingly important medium to reach and engage patients in both developed and developing countries. Mobile Health (mHealth) is the use of mobile and wireless communication technologies to improve healthcare delivery, outcomes, and research (Healthcare Information and Management Systems Society, 2012). In this chapter, we will focus on smartphone-based mHealth solutions (text messaging campaigns, apps, wearable sensors) and describe their history, the app marketplace, the evidence-base for mHealth, and future care delivery models enabled by mHealth.

Key Advances in Clinical Informatics. DOI: http://dx.doi.org/10.1016/B978-0-12-809523-2.00013-3

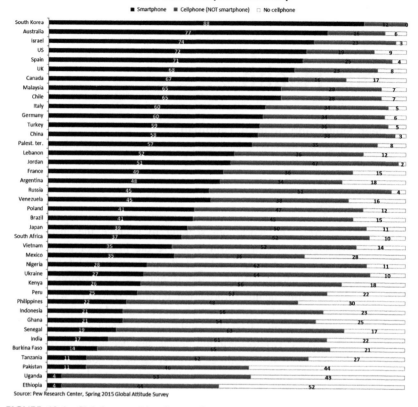

FIGURE 13.1 Global ownership of smartphones. *Reproduced from Pew Research Center, 2015. Smartphone ownership and internet usage continues to climb in emerging economies but advanced economies still have higher rates of technology use. http://www.pewglobal.org/2016/ 02/22/smartphone-ownership-and-internet-usage-continues-to-climb-in-emerging-economies (accessed 30.01.17).*

A BRIEF HISTORY OF mHEALTH

The history of mHealth can be firmly divided into the pre-app and post-app era. While this chapter primarily focuses on the post-app era, a brief foray into the history of the term is useful to understand how its meaning has evolved over time. The term "mHealth" was coined by Robert Istepanian, who defined it as "emerging mobile communications and network technologies for healthcare" (Istepanian et al., 2006). The emphasis on "mobile" differentiates mHealth from telehealth and telemedicine, which also use telecommunications technologies but are not necessarily mobile. In the early 2000s, mobile communications referred primarily to text (or SMS) messaging, mobile phone calls, and to a lesser extent, data exchange over cellular networks. According to Pew Research, 53% of Americans owned a cellular (cell) phone in 2000 (Pew Research Center, 2016a). Being able to call and message others without being tethered to a landline proved to be revolutionary for patients and healthcare providers. For patients, text messaging enabled real-time two-way exchange of health information. Patients could log their blood sugars (Kwon et al., 2004; Ferrer-Roca et al., 2004) and asthma symptoms (Anhøj and Møldrup, 2004) and in some instance receive real-time feedback from healthcare providers. Physicians could consult one another both in the hospital and in remote areas of the world.

In 2006, Istepanian foresaw a future where mHealth would power "personalized predictive healthcare," "mobile on-demand home health care," and "virtual mobile hospitals." These grand challenges would require dramatic improvements in phone hardware incorporating technologies such as wireless and Bluetooth connectivity, global positioning system (GPS), improved data download and upload speeds, and a way to install and update software on mobile phones. Each of these advances took place sequentially beginning with Apple's first iPhone going on sale June 29, 2007 and then the first Android phone arriving on October 22, 2008. While text messaging continued to be recognized under the umbrella of mHealth when the US Federal Communication Commission released its Broadband Plan in 2010, its importance as a communication and care tool had decreased considerably (Federal Communications Commission, 2010). In 2016, the Pew Research Center reported US cell phone ownership had risen to 95% and 77% of Americans owned a smartphone (Pew Research Center, 2017). Internationally, smartphone ownership ranged from 4% in Ethiopia and Uganda to 88% in South Korea in Spring 2015 (Pew Research Center, 2016b).

Unlike cell phones, smartphones were more widely Internet-connected, had fully functional web browsers, and had access to an increasing number of sensors, including GPS, compass, a video camera, a touch screen, rotation (accelerometer), brightness, and a pedometer. Smartphones used these abilities in software packaged with the operating system, but other developers did not have access to these functions in the first generation of devices. Both Apple and Google recognized the need to enable developers to take advantage of these features, so Apple's iOS App Store and Google's Android Market were born in 2008, with Google transitioning its apps to the more broadly placed Google Play Store. The app stores created a one-stop shop for buying and selling apps, and their reach extended globally. For the first time in the history of mHealth, developers could take ideas related to personalized or mobile healthcare, package them in a single location, and distribute them to millions of users.

In 2013, an IMS Health Informatics report characterized apps in the "Health and Wellness" and "Medical" sections of the iOS App Store and identified 23,682 apps that were genuine healthcare apps, of which 7407 were provider-facing, and 16,275 were patient-facing (i.e., intended for use by patients) (IMS Institute for Healthcare Informatics, 2013). While use of health apps was estimated to be quite low at the time, the report envisioned how rising evidence supporting the use of apps would place them firmly into the mainstream of medical care (Fig. 13.2). By 2015, the number of iOS apps had more than doubled, and the total number of consumer-facing iOS and Android health apps exceeded 165,000 (IMS Institute for Healthcare Informatics, 2015). Most recently, in October 2016, Research 2 Guidance estimated there were 259,000 mHealth apps available on major app stores (Research 2 Guidance, 2016).

The growing role of apps in healthcare has been fueled by a public that is increasingly empowered in self-care and the care of family members (see Chapter 18: Social and Consumer Informatics), advances in technology, rising interest from start-up companies, and by healthcare organizations' need to deliver value-based care. In a survey of US adults conducted by Pew Research in 2013, 69% reported keeping track of at least one health indicator such as weight, diet, exercise routine, or symptom (Pew Research Center, 2013). Individuals with chronic illnesses were more likely to track their health information and more likely to report that tracking affected a decision about how to treat an illness or condition as compared to healthy adults. Although a minority of tracking involved the use of technology, this finding demonstrates that the proliferation of health apps is mostly a response to existing consumer needs. One limitation of apps has been that they typically

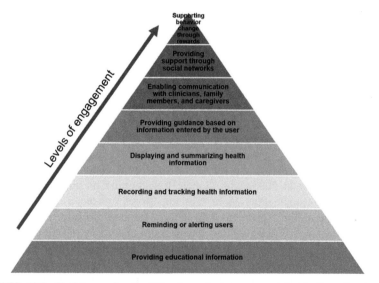

FIGURE 13.2 Useful app functionalities by patient engagement level. *Reproduced with permission from Singh, K., Drouin, K., Newmark, L.P., Rozenblum, R., et al., 2016a. Developing a framework for evaluating the patient engagement, quality, and safety of mobile health applications. Issue Brief (Commonw. Fund) 5, 1–11.*

require manual data entry. Advances in wearable and connected devices have largely lifted this barrier, allowing consumers of fitness bands to track exercise pattern and scales to automatically sync weight data. Awareness (70%) and ownership (15%) of wearable devices are at all-time highs and likely to expand further (The Nielsen Company, 2014).

THE MARKETPLACE OF HEALTH APPS

The marketplace of health apps is diverse in its focus, functionality, and quality (see Chapter 16: An Apps-Based Information Economy in Healthcare). The *Patient Adoption of mHealth* report by the IMS Institute of Healthcare Informatics issued in 2015 provides a detailed overview of the health app marketplace for patients. The primary focus of this section is on patient-facing apps, though we will give some attention to provider-facing apps as well.

While the sheer number of apps is impressive, available download data (for Android) demonstrates that 36 apps account for nearly half of all mHealth app downloads, and 900 apps (representing 12% of the mobile health apps on Android) account for over 90% of the downloads. On the other hand, 40% of apps have fewer than 5000 downloads. The variation in app popularity does not appear to be explained by the cost because 90% of the apps on the marketplace are free to download. Wellness apps account for about two-thirds of available apps, and disease and health management apps account for the remaining third. Mental health and diabetes apps comprise nearly half of all disease-specific apps.

Apps can support the ability of patients to participate in their care through a broad set of functionalities. Just as consumers may find different nonhealth-related apps to be useful, the same is true of health apps. Patients who are not very engaged in their own care may benefit from health education, reminders, and the ability to record and track their health. These functions do not require smartphone apps, though apps provide a new platform through which to store and deliver information. Moderately engaged patients who are already participating in tracking their health may benefit from being able to visualize and summarize their health information, receive guidance on next steps, and communicate with family members and healthcare providers. Finally, the most engaged patients may benefit from peer-to-peer support networks or ongoing motivational challenges that can be delivered through "gamification"—that is, using elements of game design, like competition or point scoring, to make an activity more fun. While gamification is the subject of an entire biomedical journal (*JMIR Serious Games*); research about gamifications effectiveness remains sparse (Brown et al., 2016; Boyle et al., 2017). Elements of gamification have been incorporated into apps such as Mango Health (developer: Mango Health) and Wizdy Pets (developer: LifeGuard Games, Inc.). Mango Health is a medication reminder app that rewards users with points for tracking medication adherence; weekly raffles for real-world prizes favor individuals with the most points. Wizdy Pets is a game in which users take care of an asthma-afflicted pet dragon; when the dragon needs to breathe fire, users must help the dragon properly set up and administer an inhaler in order to manage the asthma.

There are few one-size-fits-all solutions when it comes to apps, and the types of apps that are useful to patients are likely to vary with their degree of engagement in self-management (Fig. 13.3). Additionally, there are categories of apps whose presence is growing but whose user base is tough to quantify: patient health record apps, telemedicine apps, and health coaching apps linked to a healthcare provider or system. These apps are often downloadable by anyone, but require an account linked to a healthcare system in order to function. Apps that link patients with providers may enhance care in ways that apps intended for use by only patients cannot. For example, a patient with uncontrolled high blood pressure using the Twine Health app (developer: Twine Health) can receive advice about medication changes directly from his or her physician's office through the same app used to enter blood pressure. Often a health coach, disease-specific educator, or population health manager uses the app under a physician's supervision to monitor patient symptoms and provide feedback. Because of their tight integration with the health record and ability to communicate electronically with a healthcare team member, these apps are likely to be highly useful to patients.

We analyzed a subset of 137 popular and highly regarded iOS and Android chronic disease management apps to evaluate app functionality and

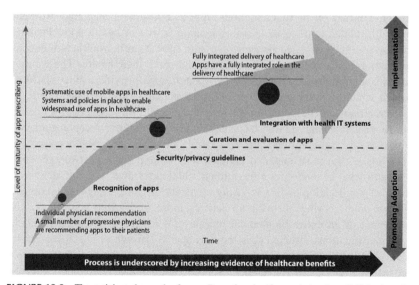

FIGURE 13.3 The anticipated growth of apps. *Reproduced with permission from IMS Institute for Healthcare Informatics, 2013. Patient apps for improved healthcare: from novelty to mainstream.*

quality (Singh et al., 2016b). We discovered variations in functionality by disease (Fig. 13.4). Nearly all of the apps for each disease included functionality to record, display, and summarize user-entered information and many apps provided educational information, reminders, or alerts. In comparison, very few apps focused on providing guidance based on user-entered information, support through social networks, or behavior change through rewards, which are functionalities likely to be useful for highly engaged patients.

We also discovered significant variations in apps' quality depending on who is assessing them. A high-quality app would be expected to be well-liked and usable by consumers and rated as useful by clinicians. We compared the consumer app store ratings to clinical utility as rated by clinicians and usability as rated by nonclinicians and found poor correlations between the three measures (Fig. 13.5). These differences suggest in some cases there is a tension between the forces guiding the marketplace as compared to those guiding patient care.

Provider-facing apps represent approximately a third of the health app market (IMS Institute for Healthcare Informatics, 2013) and are targeted to the healthcare system or healthcare providers. PricewaterhouseCoopers further classified provider-facing apps into four categories: emergency response,

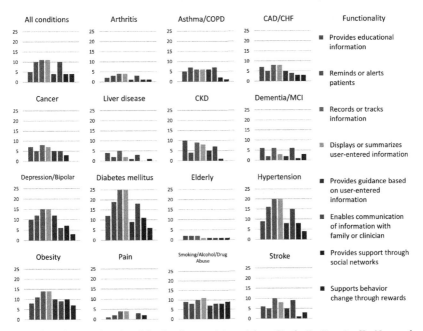

FIGURE 13.4 App functionalities by disease. *Adapted from Singh, K., Drouin, K., Newmark, L.P., et al., 2016b. Many mobile health apps target high-need, high-cost populations, but gaps remain. Health Aff. 35 (12), 2310–2318.*

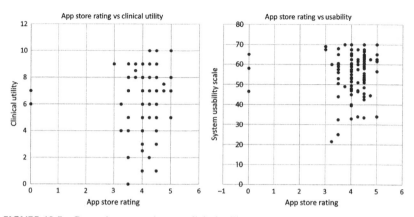

FIGURE 13.5 Comparing app ratings to clinical utility to usability. *Adapted from Singh, K., Drouin, K., Newmark, L.P., et al., 2016b. Many mobile health apps target high-need, high-cost populations, but gaps remain. Health Aff. 35 (12), 2310–2318.*

healthcare practitioner support, healthcare surveillance, and healthcare administration. Emergency response applications facilitate the response to emergencies and disaster situations (PricewaterhouseCoopers, 2012). For example, Twiage enables prehospital providers to securely communicate electronically with the emergency department while the patient is in transit (Twiage, 2016). Healthcare practitioner support constitutes the majority of provider-facing apps and includes apps that provide medical information, training, and updates to providers, such as drug dosing guides (ePocrates), textbooks, and medical calculators. Healthcare surveillance includes apps that help healthcare workers in resource-poor settings collect patient information and communicate with physicians. This category also includes apps enabling mobile access to the EHR. Healthcare administration includes apps that improve the efficiency of administrative and back-end processes, such as apps that provide appointment reminders and help fill last minute appointment cancellations.

Since a handful of apps dominate the marketplace, the vast majority of health apps are not widely used or widely known. To date, Apple and Google have not attempted to curate health apps by disease, functionality, or clinical utility—this means that high-quality apps are hard to find. Medical professional societies have generally taken a hands-off approach to app curation as well, so patients and providers have needed to lean heavily on proprietary tools such as IMS Health's AppScript (https://www. appscript.net). AppScript categorizes and scores health apps taking into account patient and provider input, and each app is accompanied by a "Prescribe" button that providers can use to send app download instructions to patients.

EVIDENCE SUPPORTING THE USE OF mHEALTH APPS

There is enormous potential for mHealth apps to improve health and health-care delivery. However, out of the 259,000 apps available on the iTunes (iOS) and Google Play (Android) app stores, very few have robust evidence of their effectiveness (Powell et al., 2014). Some of these apps may even be misleading or dangerous. Dr. James Madara, the American Medical Association (AMA) Chief Executive Officer, called apps and digital health products the "digital snake oil of the early 21st century" (American Medical Association, 2016).

Rigorous evaluation to understand an mHealth app's impact is needed, but is uncommon and often uses weaker study designs. A systematic review of randomized controlled trials of apps for self-management of diabetes mellitus, cardiovascular disease, and chronic lung diseases from 2005 to 2016 included only nine studies; three app-only interventions demonstrated a significant improvement in outcomes (Whitehead and Seaton, 2016). A scoping review of studies on apps for chronic disease management for high-need, high-cost populations that we conducted found that the studies were cross-sectional, had small sample size, were performed by app developers, and lacked a clinical outcome (Singh et al., 2016c).

Reviews of mHealth apps may supplement formal studies, providing patients and clinicians with information on the quality of apps. In a vast marketplace, such reviews can guide patients toward apps that are well-suited to their health conditions and serve the needs of their care team as well. Commercial ventures to certify health apps, such as Happtique, have thus far failed (Dolan, 2013). The National Health Service in England also failed at a curated app library (Hall, 2015). Efforts to curate health apps led by non-profit organizations are in progress. In addition to providing ratings, RANKED Health—a project run by the Hacking Medicine Institute—explains how patients should use individual apps (RANKED Health, 2017). Xcertia is the name of a recently formed collaboration among the American Heart Association, AMA, DHX Group, and the Healthcare Information and Management Systems Society to issue guidelines on health apps (Xcertia, 2017). App reviews may be of variable quality and may lack interrater reliability (Powell et al., 2016), so readers must remain vigilant. The Mobile App Rating Scale is a promising standardized, reliable tool to assess mobile health apps, but would benefit from additional validation (Stoyanov et al., 2015). In addition to standardized and validated app evaluation frameworks, further work is needed to identify a sustainable model for accurate, reliable, up-to-date mHealth app reviews. Importantly funding sources that do not have an inherent conflict of interest with the reviews must be identified. Second, more robust research on individual mHealth apps and meta-analyses of categories of mHealth apps are needed to inform the reviews. Finally, dissemination methods to provide high-quality reviews to the public are needed.

OTHER KEY ISSUES WITH mHEALTH APPS

Apps have begun to play a role in how we manage our own health due to advances in smartphone operating systems, sensors, and wearables. Apple, Google, and Samsung have each incorporated elements of health into their respective phones. Apple was the first to introduce a feature that replaces "medical identification bands" with important health information that is immediately accessible from a smartphone's lock screen, including information about medical conditions, allergies, medications, and emergency contact information. Android has a limited version of this feature that allows users to specify emergency contacts. Apple's HealthKit, Google's Fit, and Samsung's S Health serve as hubs for sensors and wearables to store information. Apple has since developed the ResearchKit platform to enable medical researchers to collect data via smartphones and CareKit to enable the development of apps that streamline healthcare delivery.

As the sensors on smartphones have become more sophisticated, apps have begun to compete with medical devices that have traditionally been covered and paid for by insurance companies. Interestingly, three-quarters of the top 20 most expensive apps—all of which cost more than $150—support augmentative alternative communication (AAC) for autism, an area where apps compete directly with dedicated AAC devices that may be more expensive (IMS Institute for Healthcare Informatics, 2015).

Privacy and security remain significant barriers to unlocking the full potential of mobile apps. One review found that just over 30% of health apps had privacy policies (Sunyaev et al., 2015). In our analysis of a subset of popular chronic disease-management apps, we found that 64% apps reviewed contained a privacy policy (Singh et al., 2016c), which is better than the prior review but still less than ideal. This finding is made more concerning by the fact that we found that nearly half of apps that enabled data sharing did so using e-mail, an insecure method not well-suited for sharing of potentially sensitive health information.

Increasingly, powerful apps have the potential to benefit health and to cause harm. The Food and Drug Administration (FDA) has used the risk-based approach it uses to assure safety and effectiveness for devices to guide the regulation of mobile apps (US Department of Health and Human Services Food and Drug Administration, 2015). In this framework, apps that resemble medical devices are regulated, informational apps are not regulated, and a number of apps with in-between functionality are regulated at the discretion of the FDA. In 2013, the FDA sent its first apps-related inquiry to the developer of an app that could be used to interpret a urine test, asking the developer why its app had not been cleared by the FDA (Edney, 2013). The FDA's regulatory role is currently in flux, as the 21st Century Cures Act passed in December of 2016 removed the FDA's ability to regulate wellness apps. The Federal Trade Commission (FTC) has also taken an active role in regulating mobile apps primarily related to false advertising in the areas of automated melanoma detection and "brain training" as a way to prevent dementia. Though such apps may carry liability issues if linked to patient harm, as of this printing, we are not aware of any lawsuits by health app users against app developers. In April of 2016, the FTC created an interactive tool to help developers determine which laws may apply to their apps based on the app's functionalities (Federal Trade Commission, 2016).

LOOKING AHEAD—HOW mHEALTH WILL POWER THE FUTURE

In 2030, a 60-year-old woman with type 1 diabetes will eat a meal at a restaurant while enjoying light conversation. A necklace camera will send a picture of her meal to her phone, which will analyze the contents of the plate, the woman's location, and triangulate the menu item she selected to determine the glucose content of her meal. A notification will pop-up on her smartphone, which she will view on her smartwatch: "Would you like to take 4 units of pre-meal insulin?" She will swipe right, and the units will be administered through an implanted insulin pump. After her meal, she will return home to her apartment.

Three months later, she will develop a urinary tract infection, become confused, and develop rising sugar levels. Her smartphone, noting the rising sugars, will ask her: "Are you okay? I noticed your sugars are going up." Confused and unable to react, she will not respond. Her phone will first reach out to her emergency contact—her daughter—with a brief message: "Your mother isn't answering her phone and her sugars are rising." Concerned, her daughter will write back: "Show me what's going on." The phone will transmit its video feed to the daughter, and she will spot her mother lying on the couch and looking ill. "Call an ambulance," she will instruct the phone. The phone will quietly transmit the sensor data along with the GPS and altimeter coordinates to the local ambulance company— glucose 450, patient lying on the couch based on video feed, in New York City, building 108, 8th floor, daughter aware.

The capabilities needed to enable smartphones to participate actively in the care of patients with chronic disease exist today. However, barriers ranging from interoperability and the willingness of health systems to engage with streams of mHealth data may stifle the ability of apps to take personalized healthcare delivery to the next level. Interoperability has been a particularly challenging problem to solve both from a policy and a technology standpoint. From a policy standpoint, healthcare exchange standards have been intentionally flexible to ensure broad adoption. However, this flexibility has led to varying degrees of interoperability. A newer and more rigid standard called SMART on FHIR is currently being developed for the purpose of enabling apps to securely exchange information with EHRs (SMART Health

IT, 2017) (see Chapter 16: An Apps-Based Information Economy in Healthcare). FHIR is promising to improve data sharing between apps and EHRs but still in early phases of the technology life cycle; Duke recently reported integrating provider- and patient-facing apps with their Epic EHR using FHIR (Bloomfield et al., 2017).

CONCLUSIONS

While it is clear that mobile apps are likely to play an important role in engaging patients in self-care, market forces may push the development of apps in a direction that leaves behind certain chronically ill populations. This phenomenon has been observed in the pharmaceutical industry, where it is often not profitable to target drugs to people with rare disorders. How the role of apps evolves over time will depend on the willingness of medical professional societies and clinical experts to engage with app developers and regulators to align the needs of all of the stakeholders. Mobile phones and smartphones already serve as the primary infrastructure for health interventions in developing nations but are often delivered via community workers because of low rates of smartphone ownership. As smartphone adoption rises globally, we expect there to be significant gains in public health efforts in developing nations. As interoperability improves (see Chapter 5: Interoperability), we also expect apps targeting clinicians to grow in number and create opportunities to enhance clinical workflow and patient safety by placing relevant and patient-specific knowledge at the clinician's fingertips. It is not uncommon for adoption of new technologies to lag in the healthcare domain compared to other sectors. Given the widespread reach of smartphones, moving slowly to incorporate mHealth into everyday health and healthcare would represent one of the largest missed opportunities in clinical informatics.

DISCLOSURES

Adam Landman is a senior editor with RANKED Health, which is a non-profit that reviews digital health apps.

REFERENCES

American Medical Association, 2016. AMA CEO Madara outlines digital challenges, opportunities facing medicine. http://www.ama-assn.org/ama/pub/news/news/2016/2016-06-11-a16-madara-address.page (accessed 28.06.16).

Anhøj, J., Møldrup, C., 2004. Feasibility of collecting diary data from asthma patients through mobile phones and SMS (short message service): response rate analysis and focus group evaluation from a pilot study. J. Med. Internet Res. 6 (4), e42.

Berwick, D.M., Nolan, T.W., Whittington, J., 2008. The triple aim: care, health, and cost. Health Aff. 27 (3), 759–769.

Bloomfield, R.A., et al., 2017. Opening the Duke electronic health record to apps: implementing SMART on FHIR. Int. J. Med. Inform. 99, 1−10.

Boyle, S.C., et al., 2017. PNF 2.0? Initial evidence that gamification can increase the efficacy of brief, web-based personalized normative feedback alcohol interventions. Addict. Behav. 67, 8−17.

Brown, M., et al., 2016. Gamification and adherence to web-based mental health interventions: a systematic review. JMIR Ment. Health 3 (3), e39.

Dolan, B., 2013. Happtique suspends mobile health app certification program. http://mobihealthnews.com/28165/happtique-suspends-mobile-health-app-certification-program (accessed 28.06.16).

Edney, A., 2013. iPhone urinalysis draws first inquiry of medical apps. http://www.bloomberg.com/news/articles/2013-05-23/iphone-urinalysis-draws-first-fda-inquiry-of-medical-apps (accessed 28.06.16).

Federal Communications Commission, 2010. National Broadband Plan.

Federal Trade Commission, 2016. Mobile health apps interactive tool. https://www.ftc.gov/tips-advice/business-center/guidance/mobile-health-apps-interactive-tool (accessed 28.06.16).

Ferrer-Roca, O., et al., 2004. Mobile phone text messaging in the management of diabetes. J. Telemed. Telecare 10 (5), 282−285.

Hall, K., 2015. Shonky securo-nightmare NHS apps library finally binned. http://www.theregister.co.uk/2015/10/13/nhs_apps_library_shelved/ (accessed 28.06.16).

Healthcare Information and Management Systems Society, 2012. Definitions of mHealth. http://www.himss.org/definitions-mhealth (accessed 28.06.16).

IMS Institute for Healthcare Informatics, 2013. Patient apps for improved healthcare: from novelty to mainstream.

IMS Institute for Healthcare Informatics, 2015. Patient adoption of mHealth.

Institute of Medicine, 2012. Best care at lower cost: the path to continuously learning health care in America. http://www.iom.edu/Reports/2012/Best-Care-at-Lower-Cost-The-Path-to-Continuously-Learning-Health-Care-in-America.aspx (accessed 06.11.14).

Istepanian, R.S.H., Laxminarayan, S., Pattichis, C.S. (Eds.), 2006. M-Health. Springer, Boston, MA.

Kwon, H.-S., et al., 2004. Development of web-based diabetic patient management system using short message service (SMS). Diabetes Res. Clin. Pract. 66 (Suppl. 1), S133−S137.

Pew Research Center, 2013. Tracking for health. http://www.pewinternet.org/2013/01/28/tracking-for-health/ (accessed 06.01.16).

Pew Research Center, 2016a Device ownership over time. http://www.pewinternet.org/data-trend/mobile/device-ownership/ (accessed 20.06.16).

Pew Research Center, 2016b. Smartphone ownership and internet usage continues to climb in emerging economies. http://www.pewglobal.org/2016/02/22/smartphone-ownership-and-internet-usage-continues-to-climb-in-emerging-economies/ (accessed 30.01.17).

Pew Research Center, 2017. Demographics of mobile device ownership and adoption in the United States. http://www.pewinternet.org/fact-sheet/mobile/ (accessed 29.01.17).

Pew Research Center, 2015. Smartphone ownership and internet usage continues to climb in emerging economies but advanced economies still have higher rates of technology use. http://www.pewglobal.org/2016/02/22/smartphone-ownership-and-internet-usage-continues-to-climb-in-emerging-economies (accessed 30.01.17).

Powell, A.C., Landman, A.B., Bates, D.W., 2014. In search of a few good apps. JAMA 311 (18), 1851−1852.

Powell, A.C., et al., 2016. Interrater reliability of mHealth app rating measures: analysis of top depression and smoking cessation apps. JMIR mHealth uHealth 4 (1), e15.

PricewaterhouseCoopers, 2012. Touching lives through mobile health: assessment of the global market opportunity.

RANKED Health, 2017. Curated health apps and devices. http://www.rankedhealth.com/ (accessed 30.01.17).

Research 2 Guidance, 2016. mHealth App Developer Economics 2016: the current status and trends of the mHealth app market.

Singh, K., Drouin, K., Newmark, L.P., Rozenblum, R., et al., 2016a. Developing a framework for evaluating the patient engagement, quality, and safety of mobile health applications. Issue Brief (Commonw. Fund) 5, 1−11.

Singh, K., Drouin, K., Newmark, L.P., et al., 2016b. Many mobile health apps target high-need, high-cost populations, but gaps remain. Health Aff. 35 (12), 2310−2318.

Singh, K., Drouin, K., Newmark, L.P., Filkins, M., et al., 2016c. Patient-facing mobile apps to treat high-need, high-cost populations: a scoping review. JMIR mHealth uHealth 4 (4), e136.

SMART Health IT, 2017. SMART on FHIR. http://docs.smarthealthit.org/ (accessed 30.01.17).

Stoyanov, S.R., et al., 2015. Mobile app rating scale: a new tool for assessing the quality of health mobile apps. JMIR mHealth uHealth 3 (1), e27.

Sunyaev, A. et al., 2015. Availability and quality of mobile health app privacy policies. J. Am. Med. Inform. Assoc. 22 (e1), e28−e33.

The Nielsen Company, 2014. iHealth: how consumers are using tech to stay healthy. http://www.nielsen.com/us/en/insights/news/2014/ihealth-how-consumers-are-using-tech-to-stay-healthy.html (accessed 06.01.16).

Twiage, 2016. Twiage − Accelerating Life-Saving Care. http://www.twiagemed.com (accessed 30.01.17).

US Department of Health and Human Services Food and Drug Administration, 2015. Mobile medical applications: guidance for industry and Food and Drug Administration Staff.

Whitehead, L., Seaton, P., 2016. The effectiveness of self-management mobile phone and tablet apps in long-term condition management: a systematic review. J. Med. Internet Res. 18 (5), e97.

Xcertia, 2017. Pioneering mHealth app guidelines. http://www.xcertia.org/ (accessed 30.01.17).

Chapter 14

A Sociotechnical Approach to Electronic Health Record Related Safety

Dean F. Sittig[1] and Hardeep Singh[2,3]
[1]The University of Texas Health Science Center at Houston, Houston, TX, United States,
[2]Michael E. DeBakey Veterans Affairs Medical Center, Houston, TX, United States,
[3]Baylor College of Medicine, Houston, TX, United States

THE NEED FOR A SOCIOTECHNICAL APPROACH TO ELECTRONIC HEALTH RECORD SAFETY

The modern information and communications technology (ICT)-enabled healthcare delivery system epitomizes a complex adaptive sociotechnical system in which individuals' actions in relation to the electronic health record (EHR) are based on often unpredictable physical, psychological, or social rules rather than constraints imposed by someone, or something, that is "in-charge" (Rouse, 2008). Individuals' needs, desires, and actions are often not homogeneous even within their work groups which often leads to goals and behaviors that are in conflict. For instance, most doctors would prefer to dictate, write, or type a concise, highly technical note that describes a patient's problem(s) and their treatment plans and then have another individual enter that into the patient's EHR. Patients desire easy to understand information on their medical problems and a clear explanation of what they should do. The finance department wants a concise structured progress note for accurate billing. The pharmacy wants an unambiguous medication order consistent with formulary. Payers who offer quality-based incentive programs want clear evidence that key performance metrics, or the reasons for exclusion, have been met. Taken together, these different perspectives make clinical documentation modules overly complex, difficult, and time-consuming to use and error prone.

Over the past 10 years, many healthcare delivery systems around the world have added an enormously complex set of interconnected, often externally developed, software applications that together create an EHR. These

Key Advances in Clinical Informatics. DOI: http://dx.doi.org/10.1016/B978-0-12-809523-2.00014-5

EHRs were designed to improve the quality and safety of the care delivered while also reducing its cost. Addressing the safe and effective design, development, implementation, and use of state-of-the-art EHRs requires a sociotechnical approach since focusing on the technology alone is not adequate.

Briefly, the sociotechnical approach regards individual system components (e.g., the people, their culture, organizational structures and processes, and technology) as a single, complex adaptive system in which the technology changes the way some people organize and conduct their work and the way some people change the technology. In other words, the various human and technical components cannot be separated and analyzed in isolation, rather they must be examined together as a whole, complete functioning system.

Using these principles, we developed an eight-dimension sociotechnical model, which includes all key elements and stakeholders in the ICT-enabled healthcare system, for the safe and effective implementation and use of EHRs. We routinely use this model to study these complex adaptive sociotechnical systems (see Fig. 14.1 for illustration and Table 14.1 for a brief overview of the eight dimensions).

One key aspect of the sociotechnical approach is that one cannot analyze any of these interrelated dimensions in isolation, rather one must always recognize their dependence and interaction with each other. As with any complex, adaptive system, these interacting components result in nonlinear,

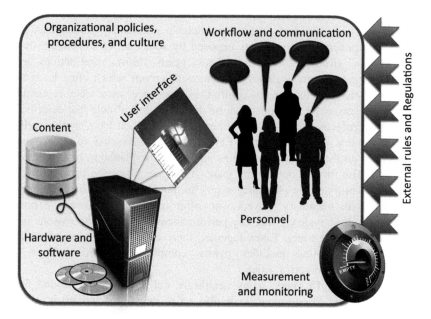

FIGURE 14.1 Diagram of the eight-dimension sociotechnical model.

TABLE 14.1 The Eight-Dimension Sociotechnical Model

Dimension	Description
Hardware and software	Computing infrastructure used to support and operate clinical applications and devices
Clinical content	The text, numeric data, and images that constitute the "language" of clinical applications, including clinical decision support (CDS)
Human–computer interface	All aspects of technology that users can see, touch, or hear as they interact with it
People	Everyone who is involved with patient care and/or interacts in some way with healthcare delivery (including technology). This includes patients, clinicians and other healthcare personnel, IT developers and other IT personnel, and informaticians
Workflow and communication	Processes to ensure that patient care is carried out effectively, efficiently, and safely
Internal organizational features	Policies, procedures, the physical work environment, and the organizational culture that govern how the system is configured, who uses it, and where and how it is used
External rules and regulations	Official Journal of the European Union project tender rules, US Federal or state rules (e.g., Centers for Medicare & Medicaid Services Physician Quality Reporting Initiative, Health Insurance Portability and Accountability Act, and Meaningful Use program) and billing requirements that facilitate or constrain the other dimensions
Measurement and monitoring	Measurement and monitoring required for evaluating both intended and unintended consequences through a variety of prospective and retrospective, quantitative and qualitative methods

emergent, and dynamic behaviors (e.g., small changes in one aspect of the system lead to small changes in other parts of the system under some conditions, but large changes at other times) that from the outside can appear random or chaotic.

WHAT IS "EHR-RELATED SAFETY"?

Over the past decade, the risks that define the intersection of patient safety and EHRs have become clearer. In 2007, Weiner et al. defined the term "e-iatrogenesis" as "patient harm caused at least in part by the application of health information technology" (Weiner et al., 2007). This recognizes the relationship of e-iatrogenic events to virtually any part of a comprehensive EHR-enabled healthcare system and includes both "traditional" errors (i.e., ones that are similar in nature to those that occurred during the paper-based medical record era), as well as new kinds of errors with no exact analog in the paper-based era. Furthermore, because EHRs are integrated with most aspects of care delivery, a wide variety of heterogeneous safety concerns can occur, often in temporally or physically separated circumstances. Various types of EHR-related errors have been identified in several landmark studies (Magrabi et al., 2012, 2015; Meeks et al., 2014b).

In an attempt to illustrate the wide variety of EHR-related safety events, we developed a taxonomy to increase understanding of how to address EHR-related safety concerns. We use the term "EHR-related safety concern" to broadly include patient safety events that reached the patient (regardless of whether harm occurred) as well as near misses and unsafe conditions. This terminology of "safety concern" is consistent with current Agency for Healthcare Research and Quality (AHRQ) common error reporting format standards (Clancy, 2010). In an effort to describe EHR-related safety concerns comprehensively to guide identification, measurement, and improvement efforts in this area, we describe five major types of EHR-related safety concerns in Table 14.2.

The current version of the AHRQ common format standards does not adequately capture the breadth of EHR-related safety concerns defined in Table 14.2. Only the first two or three types of EHR-related safety concerns have been widely encountered and reported, therefore, the current EHR-error reporting taxonomies focus on these types of concerns. Additionally, the failure to have a particular EHR safeguard in place would often not likely to be a reportable error per se.

Errors could be multifactorial and involve several sociotechnical dimensions. For example, Horsky et al. identified a traditional EHR-related error (i.e., similar errors occurred prior to the introduction of EHRs), due to a poorly designed EHR, involving a medication dosing error that occurred in part over "confusing on-screen laboratory results review, EHR usability difficulties, user training problems, and suboptimal clinical system safeguards that all contributed to a serious dosing error" (Horsky et al., 2005). Similarly, Wright et al. identified a series of new EHR-related errors that occurred due to malfunctions in the CDS functionality within an in-house developed EHR due in part to changes in data codes and field definitions, software upgrades, inadvertent disabling or editing of decision support logic rules, and malfunctions of external systems upon which the CDS system relied (Wright et al., 2016). Some of these errors involved increased inappropriate firing of alerts, which could lead to "alert fatigue" resulting in clinicians deciding to ignore other potentially important alerts. Other errors resulted in alerts not firing for many patients. None of these resulted in patient harm.

TABLE 14.2 Five Major Types of EHR-Related Safety Concerns

Category	Definition	Potential Consequences
EHR fails	EHR fails during use or is otherwise not working as designed (Kilbridge, 2003). The safety concern is directly attributable to the health information technology (HIT).	Clinicians are forced to work without their normal set of tools and their safety net.
Poorly designed EHR	EHR is working as designed, but the design does not meet the user's needs or expectations (i.e., bad design) (Horsky et al., 2005). HIT is a contributing factor to the safety concern.	Clinicians are forced to perform extra clicks, copy and paste information from other locations, which may contribute to loss of their train of thought or situational awareness.
Incorrect EHR usage	EHR is well designed and working correctly, but was not configured, implemented, or used in a way anticipated or planned for by system designers and developers (Koppel et al., 2008). These events are related to use of HIT (i.e., rather than HIT itself) and may be referred to as configuration errors, "work-arounds" or incorrect usage.	Clinicians are forced to use work-arounds which often lead to other unforeseen safety problems.
EHR interface issues	EHR is working as designed, and was configured and used correctly, but interacts with external systems (e.g., via hardware or software interfaces) so that data is lost or incorrectly transmitted or displayed (Spencer et al., 2005; Schreiber et al., 2017). These events are inevitable due to the interactive complexity of tightly coupled systems. They are often referred to as EHR system interface safety concerns (Perrow, 1984).	Even though a clinician may have done everything correctly, the end result is an error that may be difficult or impossible for him/her to detect. Similarly, the person receiving the request or instruction has very little chance in recognizing an error has occurred since they are physically and often temporally separated from the patient.
EHR features not available	Specific EHR safety features or functions were not implemented or not available (Bobb et al., 2004).	Clinicians are forced to rely on their own knowledge, skill, and ability to remember a dizzying array of facts, figures, and processes while simultaneously thinking about difficult life-altering decisions.

On the other hand, some of these adverse events cannot be attributed to either human or EHR errors, but are rather comparable to other unavoidable adverse outcomes from appropriately used non-EHR-related medical interventions (e.g., an unexpected reaction to the first administration of a medication (Subramaniam et al., 2013)). It is inevitable that a well-designed EHR which is used appropriately may also contribute to an undesirable adverse outcome.

ADDRESSING EHR-RELATED SAFETY CONCERNS

Improving patient safety in EHR-enabled healthcare systems requires measurement of safety concerns at the intersection of EHRs and patient safety. Many EHR-related adverse events, near misses, and unsafe conditions are difficult to define and detect for several reasons as illustrated above. Few strategies currently exist to systematically detect and correct EHR-related safety issues, and most users are unaware of recommended practices for safe EHR implementation and use. In response to the fundamental conceptual and methodological gaps related to both defining and measuring EHR-related patient safety, we developed a new framework, the Health IT Safety (HITS) measurement framework, to provide a conceptual foundation for HIT-related patient safety measurement, monitoring, and improvement (Fig. 14.2).

The first domain of the framework focuses on safety risks that are unique and specific to the use of technology (Kilbridge, 2003) (e.g., unavailable or

Health Information Technology Safety measurement framework (HITS framework)

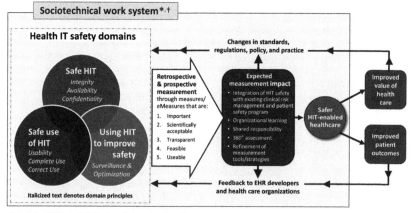

* Includes eight technological and non-technological dimensions.
† Includes external factors affecting measurement such as payment systems, legal factors, national quality measurement initiatives, accreditation, and other policy and regulatory requirements.

FIGURE 14.2 Diagram of the HITS framework. *Used with permission from Singh, H., Sittig, D.F., 2016. Measuring and improving patient safety through health information technology: the Health IT Safety Framework. BMJ Qual. Saf. 25 (4), 226–232.*

malfunctioning hardware or software). Poorly designed EHR software would be included in this domain. For example, if the login component of an EHR was only available 91% of the time, users would be unable to login to the EHR for over 2 h each day (Meyer, 2015). Wang et al. (2016) found that the time for clinicians to review potassium and hemoglobin test results were five (10.3 h vs 48 h) and six (13.4 h vs 50.4 h) times longer when the computer system was unavailable even for a relatively short time. Similarly, if a computer-generated graph of the patient's laboratory test results, collected at unequally-spaced points in time, were displayed on a graph using equally-spaced data points, a clinician could easily misinterpret the slope (i.e., rate of change in the patient's physiologic status) (Sittig et al., 2015b). Occasional failures of electromechanical devices such as desktop computers, servers, routers, and tablet computers caused by either natural (e.g., weather storms or earth quakes) or man-made disasters (e.g., ransomware attacks (Sittig and Singh, 2016)) are inevitable. Likewise, failures in either the application or network operating system or application software that create EHR functionality are also bound to include errors. Kilbridge (2003) described a network failure that lead to a 4-day downtime in one hospital. Although the hospital was able to stay open, many of their key workflows including order entry and results reporting were unavailable forcing numerous work-arounds. Furthermore, the potential consequences of an EHR hardware or software failure become increasingly important as a single instance of an EHR is deployed across multiple facilities within a geographically-distributed healthcare delivery system (McCann, 2013).

The second domain of the framework addresses EHR-related safety issues resulting from a failure to use or the misuse of appropriate technology (Sittig and Singh, 2011). Partial use of the EHR system to order tests or medications is risk-prone, since clinicians are required to check at least two sources to find all patient orders. Failure to document the clinician's thoughts and actions, or misuse of copy—paste functionality (Tsou et al., 2017) is also more widespread than one would expect due to the idiosyncratic billing rules that reimburse clinicians based on what is documented rather than patient outcomes. In parallel with these use failures, many clinicians misuse the technology due to either inadequate training or physical limitations of the software. For example, we and others have documented numerous types of "work-arounds." These work-arounds represent a significant source of EHR-related safety concerns and can lead to inappropriate medication administration (Koppel et al., 2008), duplication of work (Saleem et al., 2011), delays in addressing abnormal test results (Menon et al., 2016), and delays in entering data into the patient's record (Ser et al., 2014).

The final domain of the framework describes uses of EHR systems to monitor safety events and identify potential safety issues before they can harm patients (Jha and Classen, 2011). Most healthcare organizations today rely on user-initiated incident reports to identify EHR-related safety concerns

(Wright et al., 2016). It has been well documented that these incident reports detect only a small proportion of all safety-related events. Therefore, we advocate for the use of automated surveillance methods that rely on use of computer programs to automatically detect these easily overlooked and often underreported errors of omission, such as patients who experience errors during surgery (Mull et al., 2013), patients who experience delays in diagnostic evaluation (Murphy et al., 2014), or who do not receive follow-up for abnormal laboratory or radiology tests (Singh et al., 2009a, 2010c). These EHR-based trigger methods can also be used to detect errors of commission such as adverse drug events (Nwulu et al., 2013), postoperative complications (Griffin and Classen, 2008), and errors related to misidentification of patients (Adelman et al., 2012). All healthcare organizations should begin to leverage EHRs for purposes of improving detection of common EHR-related errors, to monitor for high-priority safety events and to more reliably track trends over time.

CONCEPTUAL APPROACHES USEFUL FOR IMPROVING EHR SAFETY

As healthcare organizations around the world are rapidly moving toward EHR-enabled patient care, it is critically important that we: (1) refine the science of measuring EHR-related patient safety; (2) make EHR-related patient safety an organizational priority by securing commitment from organizational leadership and by refocusing the organization's clinical governance structure to facilitate measurement and monitoring; and (3) develop the local organizational infrastructure and environment, as well as encourage national efforts to help in detecting, fixing, and learning from EHR vulnerabilities.

The HITS framework is based in large part on the principles of Continuous Quality Improvement (CQI). CQI is "a philosophy that encourages all healthcare team members to continuously ask the questions, 'How are we doing?' and 'Can we do it better?'" (HITRC, 2013). Our framework also addresses a third element "How can we do better?"

For example, if an organization identifies delays in following up on certain abnormal laboratory test results, our framework would help remind those responsible for addressing this problem to think about solutions across their entire sociotechnical work system, which includes the eight dimensions of our sociotechnical model, rather than just encouraging clinicians to be more vigilant in their work. Using such a systems approach might lead the organization to review their entire test result ordering, resulting, and reporting system (Singh et al., 2009b) to develop several strategies. For instance, they could: (1) develop a system to track and report on missed test results (Smith et al., 2013); (2) revise their internal organizational policies to standardize their service-level agreements on communicating test results to patients and providers (Singh and Vij, 2010a; VHA, 2015); and (3) develop and disseminate a set of best practices for management of abnormal test results to clinicians (Singh et al., 2010b).

The HITS framework (Fig. 14.2) is set within the sociotechnical work system described above and helps organizations to focus their attention on measurement and improvement of safety concerns. It encourages best practices for measurement including both retrospective as well as more proactive data collection methods. Additional measurement methods, such as proactive risk assessments (see following section for an example), could provide a more comprehensive picture of the extent and seriousness of current risks. Regardless of the timing of the measurement activities, good measures should be impactful (i.e., important to measure and report), scientifically acceptable (i.e., reliable, valid), feasible (i.e., clinically, technically, and financially), usable (i.e., easily extracted from existing EHRs), and transparent (i.e., reviewable by all stakeholders) (NQF, 2010).

RECOMMENDATIONS TO IMPROVE EHR SAFETY AND SYSTEM RESILIENCE

As previously described, identifying and fixing EHR-related safety concerns is inherently complex and requires a multifaceted, sociotechnical approach. In an effort to help healthcare organizations of all sizes, using different EHR configurations (e.g., locally- and remotely-hosted) to assess their readiness to meet the enormous challenges associated with using their EHR, we created the SAFER (Safety Assurance Factors for EHR Resilience) guides (ONC, 2014). Nine SAFER guides (Table 14.3) were developed using the eight-dimension sociotechnical model and all are useful for proactive risk assessment at the organizational level.

Each of the guides consists of a short checklist of 10−25 recommendations that reflect basic tenets of safe and effective EHR implementation and use. In addition, there is a worksheet for each recommendation (see Fig. 14.3) that provides its rationale along with several examples that help illustrate how that recommendation could be implemented. The recommendations are organized according to the HITS framework and additionally according to six key principles (Box 14.1) that describe how organizations should treat clinical and administrative data as well as their EHRs (Sittig et al., 2014b).

Use of the SAFER guides (see example in Fig. 14.3) is intended to stimulate adoption of the recommended practices by both EHR developers as well as those responsible for the configuration, implementation, and maintenance of the systems at the local site. Organizations could use the SAFER guides at regular intervals (i.e., before initial implementation, if possible, and then

TABLE 14.3 EHR-Related Structures and/or Processes Addressed by SAFER Guides

Name of Guide	Description of Each Guide
High-priority practices	The subset of processes determined to be "high risk" and "high priority," meant to broadly cover all areas that have a role in EHR safety.
Computerized provider order entry with decision support	Processes pertaining to electronic ordering of medications and diagnostic tests and aiding the clinical decision-making process at the point of care.
Test results reporting and follow-up	Processes involved in delivering test results to the appropriate providers.
Clinical communication	Communication processes in three high-risk areas: consultations or referrals, discharge-related communications, and patient-related messaging between clinicians.
Patient identification	Processes related to creation of new patients in the EHR, patient registration, retrieval of information on previously registered patients, and other patient identification processes.
Contingency planning	Processes and preparations that should be in place in the event that the EHR experiences a hardware, software, or power failure.
System configuration	Processes required to create and maintain the physical environment in which the EHR will operate, as well as the infrastructure related to the hardware and software that are required to run the EHR.
System interfaces	Processes that enable different hardware devices and software applications to be connected both physically and logically so they can communicate and share information.
Organizational responsibilities	The organizational activities, processes, and tasks that people must carry out to ensure safe and effective EHR implementation and continued operation.

Source: Used with permission from Sittig, D.F., Ash, J.S., Singh, H., 2014b. The SAFER guides: empowering organizations to improve the safety and effectiveness of electronic health records. Am. J. Manag. Care 20 (5), 418—423.

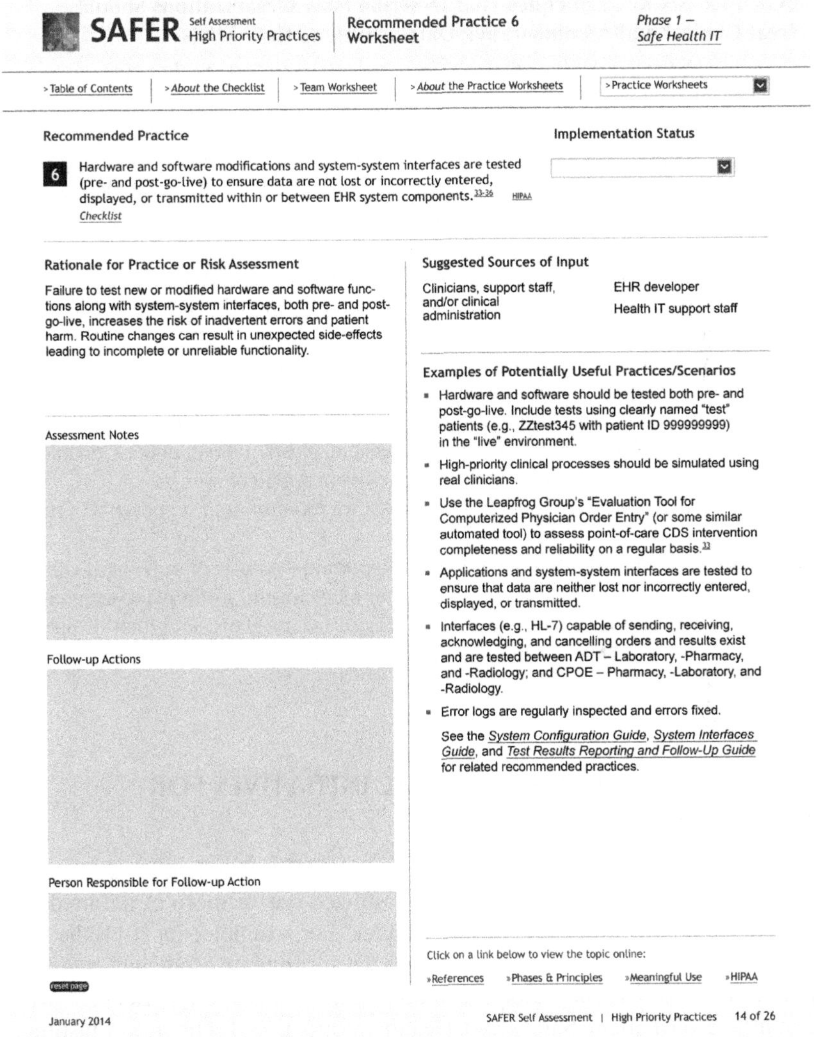

SAFER Self Assessment High Priority Practices | Recommended Practice 6 Worksheet | *Phase 1 – Safe Health IT*

›Table of Contents | ›*About* the Checklist | ›Team Worksheet | ›*About* the Practice Worksheets | ›Practice Worksheets

Recommended Practice

Implementation Status

6 Hardware and software modifications and system-system interfaces are tested (pre- and post-go-live) to ensure data are not lost or incorrectly entered, displayed, or transmitted within or between EHR system components.[33-36] HIPAA *Checklist*

Rationale for Practice or Risk Assessment

Failure to test new or modified hardware and software functions along with system-system interfaces, both pre- and post-go-live, increases the risk of inadvertent errors and patient harm. Routine changes can result in unexpected side-effects leading to incomplete or unreliable functionality.

Assessment Notes

Follow-up Actions

Person Responsible for Follow-up Action

reset page

Suggested Sources of Input

Clinicians, support staff, and/or clinical administration

EHR developer

Health IT support staff

Examples of Potentially Useful Practices/Scenarios

- Hardware and software should be tested both pre- and post-go-live. Include tests using clearly named "test" patients (e.g., ZZtest345 with patient ID 999999999) in the "live" environment.
- High-priority clinical processes should be simulated using real clinicians.
- Use the Leapfrog Group's "Evaluation Tool for Computerized Physician Order Entry" (or some similar automated tool) to assess point-of-care CDS intervention completeness and reliability on a regular basis.[33]
- Applications and system-system interfaces are tested to ensure that data are neither lost nor incorrectly entered, displayed, or transmitted.
- Interfaces (e.g., HL-7) capable of sending, receiving, acknowledging, and cancelling orders and results exist and are tested between ADT – Laboratory, -Pharmacy, and -Radiology; and CPOE – Pharmacy, -Laboratory, and -Radiology.
- Error logs are regularly inspected and errors fixed.

  See the *System Configuration Guide*, *System Interfaces Guide*, and *Test Results Reporting and Follow-Up Guide* for related recommended practices.

Click on a link below to view the topic online:

»References »Phases & Principles »Meaningful Use »HIPAA

January 2014 SAFER Self Assessment | High Priority Practices 14 of 26

FIGURE 14.3 A screen shot showing one of the recommendations from the High Priority SAFER guide.

annually or whenever significant changes to the system are made) to help them understand their progress toward a safe and effective EHR system (Sittig et al., 2014a). Use of the SAFER guides by both EHR developers and healthcare organizations could contribute significantly toward improving the safety and effectiveness of the modern, EHR-enabled healthcare system.

BOX 14.1 Six Key Principles That Describe How Organizations Should Treat Clinical and Administrative Data as well as Their EHRs.

Domain 1/Safe HIT: Address Safety Concerns Unique to EHR Technology

- Data Availability—EHRs and the data or information contained within them are accessible and usable upon demand by authorized individuals.
- Data Integrity—Data or information in EHRs are accurate and created appropriately and have not been altered or destroyed in an unauthorized manner.
- Data Confidentiality—Data or information in EHRs are only available or disclosed to authorized persons or processes.

Domain 2/Using HIT Safely: Optimize the Safe Use of EHRs

- Complete/Correct EHR Use—EHR features and functionality are implemented and used as intended.
- EHR System Usability—EHR features and functionality are designed and implemented so that they can be used effectively, efficiently, and to the satisfaction of the intended users to minimize the potential for harm. For information in the EHR to be usable, it should be easily accessible, clearly visible, understandable, and organized by relevance to the specific use and type of user.

Domain 3/Monitoring Safety: Use EHRs to Monitor and Improve Patient Safety

- Safety Surveillance, Optimization, and Reporting—As part of ongoing quality assurance and performance improvement, mechanisms are in place to monitor, detect, and report on the safety and safe use of EHRs, and then to optimize the use of EHRs to improve quality and safety.

NATIONAL AND INTERNATIONAL INITIATIVES FOR EHR-RELATED SAFETY

Over the last 5 years there have been several EHR-related safety initiatives in the United States. Many of these have utilized one or more of the models, tools, or strategies described in this chapter. For example, in 2011 the US National Academy of Medicine (formerly the Institute of Medicine) released "Health IT and Patient Safety: Building Safer Systems for Better Care" (Committee on Patient Safety and Health Information Technology; Institute of Medicine, 2011). Briefly, this report adapted our eight-dimension sociotechnical model of safe and effective EHR implementation and use to help explain the complex interrelationships between the people, hardware and software, processes, organization, and the external environment involved in creating safer systems for better health care.

Following this seminal report, in 2015 the Joint Commission released its report entitled, "Investigations of Health IT-related Deaths, Serious Injuries, or Unsafe Conditions" which adapted the eight-dimension sociotechnical model to create a coding system for the contributing and causal factors identified during their investigation (Castro, 2015). Their main conclusion was that HIT was most often a latent factor that may not be readily apparent when the adverse event is first identified. But once identified as a potential contributing factor, a multidisciplinary team should work to try and understand how the technology could have contributed to the event within the context of the underlying clinical workflows.

More recently, the National Quality Forum (NQF) released a landmark report "Identification and Prioritization of Health IT Patient Safety Measures" (NQF, 2016). This report focuses on the necessity of developing and validating measures to help identify the types, scope, and prevalence of HIT-related safety issues and to assess how well healthcare providers, EHR vendors, and government agencies are preventing or mitigating HIT-related safety concerns. Building on the HITS framework as a conceptual foundation, the NQF committee identified nine key facets of the EHR-enabled healthcare system upon which researchers should focus their efforts (Table 14.4).

Several international studies and initiatives on HIT or EHR safety are underway. Magrabi et al. (2013) compared the patient safety initiatives for HIT across six countries. They identified 27 different initiatives related to HIT that predominantly focused on software systems for health professionals. Similarly, Kushniruk et al. (2013) reviewed national efforts to improve HIT safety in three countries. They identified several different approaches including: developing national standards related to usability and interface design, certifications for various types of software artifacts, directives from regulatory bodies that clarify the definitions of various types of software, educational initiatives aimed at clinicians, along with encouraging research into safer design and implementation methods.

POLICY EFFORTS FOR IMPROVING EHR-RELATED SAFETY

Our frameworks and strategies call for better mechanisms to share lessons in safety (see Chapter 7: Public Policy and Health Informatics). Despite the aforementioned national and international efforts, much remains to be done for learning and disseminating lessons in safety. We previously proposed a national oversight mechanism to promote EHR safety, similar to that proposed in the USA's National Academy of Medicine report (Singh et al., 2011). However, policy efforts in this area have been slow. For example, the US Office of the National Coordinator for Health IT (ONC) has repeatedly (Fiscal Years 2016 and 2017) requested funds to establish a National Health IT Safety Center (Collaboratory) but this has not been funded yet. This includes some of the components we previously identified (Sittig et al., 2015a) but not others, such as random inspections and investigations (ONC, 2015). Here is an overview of what was proposed:

1. Collaboration to support development of targeted, sociotechnical solutions to HIT-related safety issues identified through evidence, as well as

TABLE 14.4 Key Facets of the EHR-Enabled Healthcare System Requiring Safety-Related Measures

Facet of the EHR-Enabled Healthcare System	Potential Safety-Related Measures
Clinical decision support	<ul><li>% of alerts that occur at the right time, for the right person, in the right context, and are useful</li><li>Instances of inappropriate alert overrides (number of patient allergic reactions/number of overrides that would be harmful)</li><li>% of alerts for situations that do not warrant alerting</li><li>% of medication orders that generate an alert</li><li>SAFER guide metrics on CDS; Leapfrog simulator results</li><li>Monitoring of CDS logic for agreement with current best evidence</li></ul>
System interoperability	<ul><li>% of diagnostic test results not available, transmitted, or displayed for the correct clinician or patient</li><li>Total number and % of completed transactions between any two systems</li></ul>
Patient identification	<ul><li>% of potential duplicate records in EHR database</li><li>% of retract-and-reorder events</li><li>% of medications administered following barcode scanning</li><li>% of incorrect patient ID alerts in barcode medication administration</li></ul>
User-centered design and use of testing, evaluation, and simulation to promote safety across the HIT lifecycle	<ul><li>End user involvement in EHR lifecycle (design, development, implementation, use, evaluation)</li><li>Assessments of EHR usability during all phases of the lifecycle for the purpose of increasing patient safety</li><li>% of records using scribes or dictation systems</li><li>% of users that are tested in an EHR simulator</li><li>% of users that successfully complete a given task</li></ul>
System downtime (data availability)	<ul><li>% of system availability (uptime)</li><li>System response time</li><li>Number of times that downtime procedures are activated/month</li><li>Number of downtime drills/year</li></ul>

Feedback and information sharing	• Timely vendor notifications are sent to all organizational users following identification of software, hardware, or other issues that materially affect patient safety • Mean time from safety issue reports until vendor provides solutions • Vendors share comparative user experiences across organizational users • Software and hardware license and purchase agreements permit shared learning of comparative user experiences (e.g., screen shots) to promote shared accountability for safety
Timely and high-quality documentation	• Mean time from electronic request for information to provision of information • % of patients in the hospital who have at least one progress note per day written within an established time period • % of diagnostic tests (i.e., laboratories, imaging) with documented timely follow-up • Mean time from diagnostic test result availability to clinician acknowledgment and patient notification of result • Discharge and transition note quality • % of charts with allergies in free text versus in structured or designated fields
Patient engagement	• % of patient acknowledgment of diagnostic test results via patient-facing technology • % of patients who suggest corrections to the EHR information • % of patients viewing, annotating, and editing the medical record • % of patients with viewable progress notes via patient portal • % of clinician responses to patient-generated electronic communication within 48 h
HIT-focused risk-management infrastructure	• EHR safety metrics shared with the governing board • Number of successful (inappropriate access) network penetration tests • Number of open patient orders (i.e., not acted upon by clinicians) after a set period • Mean time to fix serious EHR errors • % of risk managers who have received continuing education units from the Joint Commission's safe HIT module

Source: Adapted from NQF Report on Identification and Prioritization of Health IT Patient Safety Measures.

the dissemination, pilot testing, adoption, and evaluation of these solutions.

2. Strengthening and augmenting existing ways to identify and classify HIT-related safety events, as well as identify ways to encourage better reporting of HIT-related events by improved training, developing easier reporting mechanisms that are better integrated into workflows, and reduction of other barriers to reporting.

3. Production of yearly reports summarizing current evidence of HIT safety from a broad range of sources, including Patient Safety Organizations, medical liability insurers, academic researchers, EHR vendors, provider organizations, and hospital accreditors. The Center should also conduct targeted examinations of specific issues—such as safety-related concerns in EHR user interfaces and usability—and identify approaches to addressing these issues.

4. Develop new educational resources and training materials to build HIT-related competencies and server as a clearinghouse for HIT-related safety.

In June 2016, the American Medical Association (AMA) went several steps further in suggesting that the Health IT Safety Center should actually collect data on EHR-related safety risks and passed a resolution at their annual meeting calling for the United States "congress to create a National Health IT Safety Center that can implement an effective EHR safety program designed to reduce EHR-related patient safety risks through collection, aggregation and analysis of data reported from EHR-related adverse patient safety events and near misses" (AMA, 2016).

On the international front, several research teams have begun reviewing patient safety data collected via survey, observation, and national error reporting systems. While these reporting systems were not all solely for HIT-related concerns, there are important lessons to be learned from their review. For example, Meeks et al. (2014a) "conducted a secondary analysis of interview data from a 30-month longitudinal, prospective, case study-based evaluation of EHR implementation in 12 National Health Service hospitals" in the United Kingdom. They found that "patient safety improvement activities as well as patient safety hazards change as an organization evolves." Similarly in Finland, Palojoki et al. (2016b) found in a review of patient safety incidents reported by 23 hospitals over a 2-year period, that the complex interactions between humans and computers were the most frequently reported error types. Furthermore, in a survey of 2864 respondents from across Finland, Palojoki et al. (2016a) found that extended EHR unavailability was reported as a high level of risk by almost half of respondents. Similarly, Magrabi et al. (2015) found that 92% of 1606 events reported to the United Kingdom's National IT program were associated with technical rather than human factors.

Taken together, governments around the world have begun to address the issue of EHR-related patient safety. In the future, we expect that policy makers and regulators will continue to increase their focus on this topic and bring additional oversight of these complex, sociotechnical interactions that will be driving improvements in healthcare quality, access, and cost reductions in the 21st century.

CONCLUSIONS

EHRs are one of the most important tools the modern healthcare delivery system has to improve patient safety. However, as EHRs become bigger, more complex, and interconnected, our ability to identify EHR-related safety concerns and fix them becomes more challenging. Furthermore, the rapid adoption of EHRs by healthcare organizations requires close monitoring and rapid responses to newly identified safety concerns. By using the sociotechnical models, frameworks, strategies, and tools described in this chapter, healthcare organizations should be able develop techniques and strategies to identify, mitigate, and prevent future EHR-related safety events from harming patients. Concomitantly, we encourage governments around the world to use sociotechnical approaches to guide policies and procedures to improve the safety of their EHR-enabled healthcare delivery systems.

REFERENCES

Adelman, J.S., Kalkut, G.E., Schechter, C.B., Weiss, J.M., Berger, M.A., Reissman, S.H., et al., 2012. Understanding and preventing wrong-patient electronic orders: a randomized controlled trial. J. Am. Med. Inform. Assoc. Available from: http://dx.doi.org/10.1136/amiajnl-2012-001055.

American Medical Association (AMA), 2016. AMA Throws Support Behind Development of a National Health IT Safety Center. Digitized Medicine. http://www.digitizedmedicine.com/2016/06/ama-throws-support-behind-development-of-a-national-health-it-safety-center.html.

Bobb, A., Gleason, K., Husch, M., Feinglass, J., Yarnold, P.R., Noskin, G.A., 2004. The epidemiology of prescribing errors: the potential impact of computerized prescriber order entry. Arch. Intern. Med. 164 (7), 785−792.

Castro, G., 2015. Investigations of Health IT-related Deaths, Serious Injuries, or Unsafe Conditions. The Role for the EHR in Patient Safety—What Does the Evidence Tell Us. https://www.healthit.gov/sites/default/files/safer/pdfs/Investigations_HealthIT_related_SE_Report_033015.pdf.

Clancy, C.M., 2010. Common formats allow uniform collection and reporting of patient safety data by patient safety organizations. Am. J. Med. Qual. 25 (1), 73−75.

Committee on Patient Safety and Health Information Technology, Institute of Medicine, 2011. Health IT and Patient Safety: Building Safer Systems for Better Care. National Academies Press, Washington, DC.

Griffin, F.A., Classen, D.C., 2008. Detection of adverse events in surgical patients using the trigger tool approach. Qual. Saf. Health Care 17 (4), 253−258.

Health Information Technology Research Center (HITRC), 2013. Continuous Quality Improvement (CQI) Strategies to Optimize Your Practice—Primer. The National Learning Consortium (NLC).

Horsky, J., Kuperman, G.J., Patel, V.L., 2005. Comprehensive analysis of a medication dosing error related to CPOE. J. Am. Med. Inform. Assoc. 12 (4), 377–382.

Jha, A.K., Classen, D.C., 2011. Getting moving on patient safety—harnessing electronic data for safer care. N. Engl. J. Med. 365 (19), 1756–1758.

Kilbridge, P., 2003. Computer crash—lessons from a system failure. N. Engl. J. Med. 348 (10), 881–882.

Koppel, R., Wetterneck, T., Telles, J.L., Karsh, B.T., 2008. Workarounds to barcode medication administration systems: their occurrences, causes, and threats to patient safety. J. Am. Med. Inform. Assoc. 15 (4), 408–423.

Kushniruk, A.W., Bates, D.W., Bainbridge, M., Househ, M.S., Borycki, E.M., 2013. National efforts to improve health information system safety in Canada, the United States of America and England. Int. J. Med. Inform. 82 (5), e149–e160.

Magrabi, F., Ong, M.S., Runciman, W., Coiera, E., 2012. Using FDA reports to inform a classification for health information technology safety problems. J. Am. Med. Inform. Assoc. 19 (1), 45–53.

Magrabi, F., Aarts, J., Nohr, C., Baker, M., Harrison, S., Pelayo, S., et al., 2013. A comparative review of patient safety initiatives for national health information technology. Int. J. Med. Inform. 82 (5), e139–e148.

Magrabi, F., Baker, M., Sinha, I., Ong, M.S., Harrison, S., Kidd, M.R., et al., 2015. Clinical safety of England's National Programme for IT: a retrospective analysis of all reported safety events 2005 to 2011. Int. J. Med. Inform. 84 (3), 198–206.

McCann E., 2013. Setback for Sutter after $1B EHR crashes. Healthcare IT News. http://www.healthcareitnews.com/news/setback-sutter-after-1b-ehr-system%20crashes.

Meeks, D.W., Takian, A., Sittig, D.F., Singh, H., Barber, N., 2014a. Exploring the sociotechnical intersection of patient safety and electronic health record implementation. J. Am. Med. Inform. Assoc. 21 (e1), e28–e34.

Meeks, D.W., Smith, M.W., Taylor, L., Sittig, D.F., Scott, J.M., Singh, H., 2014b. An analysis of electronic health record-related patient safety concerns. J. Am. Med. Inform. Assoc. 21 (6), 1053–1059.

Menon, S., Murphy, D.R., Singh, H., Meyer, A.N.D., Sittig, D.F., 2016. Workarounds and test results follow-up in electronic health record-based primary care. Appl. Clin. Inform. 7, 543–559.

Meyer R., 2015. The Secret Startup That Saved the Worst Website in America. The Atlantic. http://www.theatlantic.com/technology/archive/2015/07/the-secret-startup-saved-healthcare-gov-the-worst-website-in-america/397784/.

Mull, H.J., Borzecki, A.M., Hickson, K., Itani, K.M., Rosen, A.K., 2013. Development and testing of tools to detect ambulatory surgical adverse events. J. Patient Saf. 9 (2), 96–102.

Murphy, D.R., Laxmisan, A., Reis, B.A., Thomas, E.J., Esquivel, A., Forjuoh, S.N., et al., 2014. Electronic health record-based triggers to detect potential delays in cancer diagnosis. BMJ Qual. Saf. 23 (1), 8–16.

National Quality Forum (NQF), 2010. National Voluntary Consensus Standards for Public Reporting of Patient Safety Event Information. A Consensus Report. NQF, Washington, DC.

National Quality Foundation, 2016. Identification and Prioritization of Health IT Patient Safety Measures. http://www.qualityforum.org/WorkArea/linkit.aspx?LinkIdentifier = id&ItemID = 81710.

Nwulu, U., Nirantharakumar, K., Odesanya, R., McDowell, S.E., Coleman, J.J., 2013. Improvement in the detection of adverse drug events by the use of electronic health and prescription records: an evaluation of two trigger tools. Eur. J. Clin. Pharmacol. 69 (2), 255–259.

Office of the National Coordinator of Health IT (ONC), 2014. SAFER Guides. https://www. healthit.gov/safer/.

Office of the National Coordinator for Health IT (ONC), 2015. Health IT Safety Center Roadmap: Collaborate on Solutions, Informed by Evidence. http://www.healthitsafety.org/ uploads/4/3/6/4/43647387/roadmap.pdf.

Palojoki, S., Pajunen, T., Saranto, K., Lehtonen, L., 2016a. Electronic health record-related safety concerns: a cross-sectional survey of electronic health record users. JMIR Med. Inform. 4 (2), e13.

Palojoki, S., Mäkelä, M., Lehtonen, L., Saranto, K., 2016b. An analysis of electronic health record-related patient safety incidents. Health Informatics J.pii:1460458216631072.

Perrow, C., 1984. Normal Accidents: Living with High-Risk Technologies. Basic Books.

Rouse, W.B., 2008. Health care as a complex adaptive system: implications for design and management. Bridge Washington Natl Acad. Eng. 38 (1), 17.

Saleem, J.J., Russ, A.L., Neddo, A., Blades, P.T., Doebbeling, B.N., Foresman, B.H., 2011. Paper persistence, workarounds, and communication breakdowns in computerized consultation management. Int. J. Med. Inform. 80 (7), 466–479.

Schreiber, R., Sittig, D.F., Ash, J., Wright, A., 2017. Orders on file but no labs drawn: investigation of machine and human errors caused by an interface idiosyncrasy. J. Am. Med. Inform. Assoc. Available from: http://dx.doi:10.1093/jamia/ocw188.

Ser, G., Robertson, A., Sheikh, A., 2014. A qualitative exploration of workarounds related to the implementation of national electronic health records in early adopter mental health hospitals. PLoS One 9 (1), e77669.

Singh, H., Thomas, E.J., Mani, S., Sittig, D., Arora, H., Espadas, D., et al., 2009a. Timely follow-up of abnormal diagnostic imaging test results in an outpatient setting: are electronic medical records achieving their potential?. Arch. Intern. Med. 169 (17), 1578–1586.

Singh, H., Wilson, L., Petersen, L.A., Sawhney, M.K., Reis, B., Espadas, D., et al., 2009b. Improving follow-up of abnormal cancer screens using electronic health records: trust but verify test result communication. BMC Med. Inform. Decis. Mak. 9, 49.

Singh, H., Vij, M.S., 2010a. Eight recommendations for policies for communicating abnormal test results. Jt Comm. J. Qual. Patient Saf. 36 (5), 226–232.

Singh, H., Wilson, L., Reis, B., Sawhney, M.K., Espadas, D., Sittig, D.F., 2010b. Ten strategies to improve management of abnormal test result alerts in the electronic health record. J. Patient Saf. 6 (2), 121–123.

Singh, H., Thomas, E.J., Sittig, D.F., Wilson, L., Espadas, D., Khan, M.M., et al., 2010c. Notification of abnormal lab test results in an electronic medical record: do any safety concerns remain? Am. J. Med. 123 (3), 238–244.

Singh, H., Classen, D.C., Sittig, D.F., 2011. Creating an oversight infrastructure for electronic health record-related patient safety hazards. J. Patient Saf. 7 (4), 169–174.

Sittig, D.F., Singh, H., 2011. Defining health information technology-related errors: new developments since to err is human. Arch. Intern. Med. 171 (14), 1281–1284.

Sittig, D.F., Singh, H., 2016. A socio-technical approach to preventing, mitigating, and recovering from ransomware attacks. Appl. Clin. Inform. 7, 624–632.

Sittig, D.F., Ash, J.S., Singh, H., 2014a. ONC issues guides for SAFER EHRs. J. AHIMA 85 (4), 50–52.

Sittig, D.F., Ash, J.S., Singh, H., 2014b. The SAFER guides: empowering organizations to improve the safety and effectiveness of electronic health records. Am. J. Manag. Care 20 (5), 418−423.

Sittig, D.F., Classen, D.C., Singh, H., 2015a. Patient safety goals for the proposed Federal Health Information Technology Safety Center. J. Am. Med. Inform. Assoc. 22 (2), 472−478.

Sittig, D.F., Murphy, D.R., Smith, M.W., Russo, E., Wright, A., Singh, H., 2015b. Graphical display of diagnostic test results in electronic health records: a comparison of 8 systems. J. Am. Med. Inform. Assoc. 22 (4), 900−904.

Smith, M., Murphy, D., Laxmisan, A., Sittig, D., Reis, B., Esquivel, A., et al., 2013. Developing software to "track and catch" missed follow-up of abnormal test results in a complex sociotechnical environment. Appl. Clin. Inform. 4 (3), 359−375.

Spencer, D.C., Leininger, A., Daniels, R., Granko, R.P., Coeytaux, R.R., 2005. Effect of a computerized prescriber-order-entry system on reported medication errors. Am. J. Health Syst. Pharm. 62 (4), 416−419.

Subramaniam, S.R., Cader, R.A., Mohd, R., Yen, K.W., Ghafor, H.A., 2013. Low-dose cyclophosphamide-induced acute hepatotoxicity. Am. J. Case Rep. 14, 345−349.

Tsou, A.Y., Lehmann, C.U., Michel, J., Solomon, R., Possanza, L., Gandhi, T., 2017. Safe practices for copy and paste in the EHR. Systematic review, recommendations, and novel model for health IT collaboration. Appl. Clin. Inform. 8 (1), 12−34.

Veterans Health Administrative (VHA) Directive 1088, 2015. Communicating Test Results to Providers and Patients. http://www.va.gov/VHAPUBLICAtIONs/ViewPublication.asp?pub_ID = 3148.

Wang, Y., Coiera, E., Gallego, B., Concha, O.P., Ong, M.S., Tsafnat, G., et al., 2016. Measuring the effects of computer downtime on hospital pathology processes. J. Biomed. Inform. 59, 308−315.

Weiner, J.P., Kfuri, T., Chan, K., Fowles, J.B., 2007. "e-Iatrogenesis": the most critical unintended consequence of CPOE and other HIT. J. Am. Med. Inform. Assoc. 14 (3), 387−388.

Wright, A., Hickman, T.T., McEvoy, D., Aaron, S., Ai, A., Andersen, J.M., et al., 2016. Analysis of clinical decision support system malfunctions: a case series and survey. J. Am. Med. Inform. Assoc.pii: ocw005. Available from: http://dx.doi.org/10.1093/jamia/ocw005.

Chapter 15

Predictive Analytics and Population Health

Peter S. Hall and Andrew Morris

The University of Edinburgh, Edinburgh, United Kingdom

WHAT IS PREDICTIVE ANALYTICS AND WHY IT IS IMPORTANT

Predictive analytics is the use of statistics, epidemiology, data mining, machine learning, and artificial intelligence techniques to identify the likelihood of future events based on historical data. Information sources relevant to predictive analytics can include clinical, demographic, and social data. This is combined with information from past treatment exposure and outcomes measurement as well as genomic and other biological data. Such information may be derived from the medical research literature and publications as well as original databases.

The development of predictive analytics is essential as a basis for the development and implementation of contemporary models of healthcare and is relevant to both individuals and populations. For individual patients, the quest for precision medicine (also sometimes referred to as personalized or stratified medicine) aspires to provide a tailored account of an individual's risk of disease combined with a quantified estimate of the expected benefits and harms of treatment or intervention (see also Chapter 11: Bioinformatics and Precision Medicine). The aim of predictive analytics is therefore to determine the risk of developing certain conditions such as asthma, diabetes, cardiovascular disease, and cancer, or of response to treatment. In an environment of person-centered medicine, the role of the clinician is to provide this information to facilitate empowered decision making by patients.

Beyond the individual patient, the responsibility of a health system is to implement policy that maximizes population health; in this context the aim of predictive analytics is to integrate, and analyze healthcare, clinical and socioeconomic data to inform the provision of services, coordinate care within the community, and give patients the best possible chance of

Key Advances in Clinical Informatics. DOI: http://dx.doi.org/10.1016/B978-0-12-809523-2.00015-7

217

maintaining a healthy lifestyle as well as avoiding emergency hospital admissions, with the aim of improving healthcare while reducing cost.

Predictive analytics is a subject of intense activity; in the last 6 months alone approximately twice the number of papers on using prediction in healthcare have been published compared with the entire 1990s. Additionally, numerous new "analytics" commercial organizations and established health IT vendors are harnessing innovative data sources and marketing new analytics processing techniques as tools to support precision medicine and population health management. This chapter reviews the origins and scientific basis for prediction in healthcare, discusses the current status and the likely challenges if predictive analytics are to be widely adopted across health systems.

HISTORICAL APPROACHES TO PREDICTION IN HEALTHCARE

At its historical beginnings, prediction was based on a physician's experience combined with the knowledge gained through education. The availability of data and the development of analytical methods has made empirical estimation possible as the basis of informed decision making. Risk prediction is undertaken for the two main purposes of prognostication of outcome, including the risk of a future clinical event or disease, and prediction of the effect of a treatment on the outcome of interest. The latter may be extrapolated from prognosis using an estimate of relative risk reduction (Fig. 15.1) or in relation to a predictive biomarker (Fig. 15.2).

Well known and widely used examples of prognostic models are seen in cardiovascular disease with the Framingham risk score (Anderson et al., 1991) which estimates the 10-year risk of developing cardiovascular disease

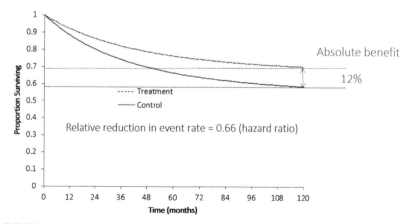

FIGURE 15.1 Estimating an absolute benefit from treatment by combining an estimate of prognosis with a relative treatment effect.

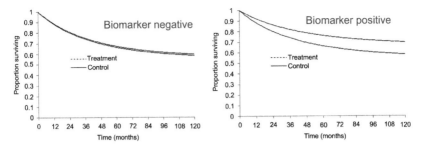

FIGURE 15.2 A predictive biomarker divides a population into subgroups characterized by a differential response to a treatment.

using a Weibull parametric survival regression model; and in breast cancer with the Nottingham prognostic (Haybittle et al., 1982) index, which predicts the 5-year survival probability of a woman with newly diagnosed early breast cancer using a Cox proportional hazards regression model.

STATISTICAL AND ANALYTICAL METHODS

The usual basis for prediction is a regression model, the exact type of which will depend on the outcome of interest and may rely on a parametric, semi-parametric, or nonparametric specification. Common examples are logistic models for binary outcomes or Cox models for time-to-event outcomes.

It is important to employ transparent and reproducible methods in the development and validation of a predictive model and there are many texts that describe this process in detail. Broadly, they can be broken down into the following:

1. Data cleaning, coding, and imputation of missing data: This is the process of turning a messy source database into something useable by the analyst.
2. Identification and selection of predictor or independent variables (e.g., patient age), referred to as model development should follow a predefined objective process for including or discarding variables: This creates a shortlist of information that will be used for prediction.
3. Model fitting and diagnostics, to ensure the most appropriate model has been chosen and that it does not lead to spurious predictions: This ensures the predictive model is technically robust.
4. Internal and external validation involves confirmation that the model performs as promised, in terms of its predictive ability, both in within the original dataset in which it was developed and in alternative "external" datasets: This tests the ability of the model to accurately predict outcomes or seeks to refine it in relevant settings.

5. Visualization and implementation in decision context with "impact evaluation": In order for a model to be used at the front line of clinical decision making it requires to be implemented in a form that is accessible and useable in a routine practice or policy-making context.

It is also important to note that the identification of a predictive biomarker as one of the independent variables requires inclusion of an interaction term between the biomarker and a treatment indicator, usually requiring a randomized comparison between treatment and no treatment in the datasets used for development and validation.

CURRENT CHALLENGES AND A NEED TO DO BETTER

Progression from development to external validation and corroboration is a vital step, but often undertaken with insufficient rigor. The availability of data suitable for external validation has been a historical limitation, but new opportunities for the use of large routinely collected healthcare datasets provide an opportunity to overcome such limitations. The *QRISK2* (Hippisley-Cox et al., 2008) model for cardiovascular event prediction is an example of where this has been done for cardiovascular risk prediction: Development used the QRESEARCH database which includes over 1.5 million patients from 355 randomly selected family health centers in the United Kingdom. Validation was undertaken in a similarly sized independent dataset.

A significant limitation of existing predictive tools is due to a disparity between the patient populations included in the development and validation datasets. These are frequently derived from clinical trials, research databases, or local case series which are small in sample size and homogenous in patient characteristics. The outcomes and treatment effects may vary between such highly selected populations and the population of eventual routine use. Patients or patient subpopulations may differ in terms of comorbidity, frailty, ethnicity, and socioeconomic status. It is also likely that were populations evolve over time model performance will deteriorate. The thorny issue of accounting adequately for heterogeneity in model performance through systematic case-mix variation can be dealt with by the application of appropriate methods in large routine datasets (Kang et al., 2013). The prospect of universal or "big data" presents the enticing possibility to provide a solution by reestimation or calibration of a model for a specific population of interest.

Counteracting the advantages of large routinely collected healthcare data model development and validation are the problems that arise from collection in routine rather than in a highly controlled research context. Predictor variables may be systematically missing and, although multiple imputation techniques may provide a solution, they rely on an assumption that data is missing at random, which is rarely a robust assumption. Data quality may vary, not because of inaccuracy, but because of the use of nonstandard definitions of diagnoses and outcomes or an absence of recording of diagnoses that are not pertinent to a specific care episode.

The availability of large, individual patient-level meta-analysis of randomized controlled trials curated by international cooperative groups has advanced our ability to estimate relative treatment effects to a high level of accuracy across populations. For example, the international Early Breast Cancer Trialists Collaborative Group (EBCTCG) (Peto et al., 2012) has honed our measurement of the improvements in survival seen with chemotherapy after surgical treatment of early breast cancer. But there is a key problem limiting the applicability of these results in the clinic which is again the disparity between the characteristics of the patients included in the contributing clinical trials and the characteristics of a real-world population. The availability of routinely collected healthcare datasets, such as cancer registries and death registries, provide an opportunity for the reestimation of these prognostic models in complete and representative populations. An example is the "NHS Predict" tool which has been estimated in a UK population and validated in UK and Canadian representative cohorts.

NHS Predict (Wishart et al., 2012) is an example of where prognostication has been made possible based on real-world populations, but the estimation of treatment benefit from chemotherapy in NHS Predict is still based on the relative risk reduction seen in the clinical trials in the EBCTCG meta-analysis. The estimation of treatment effect from nonrandomized routinely collected data is not readily possible. There are, however, a toolkit of methods that may help us achieve this. These include statistical methods that are more sophisticated than simple regression to adjust for confounding variables such as propensity score matching or methods for quasi-experimentation such as the regression discontinuity (RD) design that takes opportunist advantage of situations where a natural experiment exists. The RD design has been widely used in social sciences and has been heralded as a simple to implement and transparent method for providing "real-world" effects of treatments, but is underused in healthcare. RD applies when participants are assigned to an intervention using a cut-off value (or threshold) of a continuous assignment variable, e.g., a risk score or test result. The treatment effect is estimated by comparing outcomes in individuals who lie just below the cut-off with those just above it; under several assumptions, any discontinuity in the outcome at the cut-off can be attributed to treatment (Lee and Lemieux, 2010; Cook, 2008).

NEW OPPORTUNITIES AND BIG DATA

To date, predictive analytics has mainly used research datasets and classical statistical regression methods described above. The large-scale adoption of electronic healthcare records (EHRs) at organizational level has, however, made possible a new era in predictive modeling that has not historically been possible. Our attempts to move toward a representative population sample may be fully realized if our datasets constitute the entire population of interest. There are now examples of EHRs providing comprehensive datasets that encompass the patients treated by individual institutions. These may exist at national level, such as the disease registries used in parts of Northern Europe, or at the level of often large healthcare provider organizations. Leading examples are seen in the private sector and include Aetna or other US systems such as the Veterans Affairs or Kaiser-Permanente in the United States, and the UK's National Health Services.

The ability to utilize new comprehensive health system datasets opens up an opportunity for a continuous feedback loop such that an evolving predictive model, made possible by rapid access and respecification of a model can rapidly influence health policy and service design in an evolving manner. This is one of the underlying requirements of what has been referred to by Friedman and colleagues as a learning health system (Friedman et al., 2014).

The challenges of using such large and routinely collected healthcare datasets are significant, arising from the primary purpose of data collection being either individual patient care or financial administration. Barriers include variation in data definitions, variable data quality or missing-ness, and hidden incentives for selective data capture. But the advantages, once these challenges are overcome, may be significant, including an ability to rapidly repeat analysis and update models in the context of a rapidly changing healthcare environment or population demographic. To realize this ambition, the underlying requirements must be met with regard to availability of data, quality of data, representativeness of data, and robust methods for translation into clinical utility.

NEW TOOLS FOR PREDICTIVE ANALYTICS

The approach to date cited in the above examples has a foundation in regression model development, Monte Carlo simulation, and other statistical methods. More recently, advances in information technology, machine learning, and statistics have made it possible for predictive tools to access and manipulate big data, often in real time opening up opportunities to anticipate issues with unprecedented precision. Other industries including financial services, retail, airline, and human resources have seen an exponential increase in the use of data to drive new business models. But nowhere is the potential of this new era of opportunity more apparent, exciting, and challenging than it is in healthcare.

Access to comprehensive large healthcare datasets makes available large numbers of predictor and outcome variables that traditional methods for building and validating regression models struggle to handle, leading to overfitting, unrealistic distributional assumptions, unstable estimation, and consequent inaccurate or sometimes bizarre inference. In response to this, analysts are turning to alternative methods such as those developed within the field of artificial intelligence and machine learning (see Chapter 11: Bioinformatics and Precision Medicine). Such approaches identify repeated reliable patterns in data that lend themselves for prediction in individual cases, rather than fit statistical models and parameters for those models that we hope will generalize to the population. Such pattern recognition algorithms offer the additional advantage that information may be extracted from healthcare data that would otherwise go unnoticed.

A FOUNDATION FOR PRECISION MEDICINE

These advances provide a necessary foundation for the step change in healthcare promised by the new paradigm of precision medicine. For clinical decision making to be truly personalized, the clinician and patient require a prediction of their risk and benefits from treatment to be unique to them specifically, as distinct to a prediction for a population of patients similar to themselves. The capability of large healthcare datasets as the foundation of this, empowered by predictive analytics, will only become complete if molecular data is added to phenotypic variables.

This is arguably the most advanced in the United States in light of the digital maturity of health systems increasing significantly over the last 5 years yielding vast amounts of data. While some progress is being made in "big data" techniques to predict individual outcomes like postoperative complications, hospital readmissions, and diabetes risk, big data remains largely a promise rather than a proven reality, in the routine delivery of healthcare. A *Harvard Business Review* article (Parikh and Obermeyer, 2016) defined four contextual, technical, and socio-ethnographic criteria that will define the successful uptake of predictive analytics into routine healthcare.

1. *Determine the clinical decision.* Despite the abundance of data, it is essential that any predictive algorithm has clinical utility that adds value or redefines the clinical workflow or treatment pathway. For example, the Parkland Health and Hospital System in Dallas, TX, developed a validated EHR-based algorithm to predict readmission risk in patients with heart failure. The pathway was altered—patients deemed at high risk for readmission receive evidence-based interventions, including education by a multidisciplinary team, follow-up telephone support within 2 days of discharge to ensure medication adherence, an outpatient follow-up appointment within 7 days, and coordinated primary care review. In a prospective study, the algorithm-based intervention reduced readmissions by 26%. The success was derived from specific population and the prediction was mapped to a discrete clinical intervention (Amarasingham et al., 2013).

2. *Harness the data from EHRs.* Algorithms that utilize greater amounts of clinical data have greater accuracy and potential clinical applications— the importance of data quality and completeness however are vital.

3. *Focus on key decision points.* Predictive analytics can allow clinicians to steer high-cost interventions to those high-risk patients who actually need them.
4. *Integrate—do not force—analytics into the existing workflow.* It is important that predictive tools are proportionate to clinical need—healthcare professionals are bombarded with best evidence, guidelines, and data. It is therefore vital that any decision support or predictive tool fits well into a workflow. If excess prompts/interventions are made at the clinical encounter, evidence suggests that the tools are ignored or disabled (Dikomitis et al., 2015).

COMMERCIAL DEVELOPMENTS

The last 5 years have witnessed an exponential increase in the number of start-up companies that are harnessing machine learning and advanced analytics in an attempt to redefine the healthcare market. This applies to both structured and unstructured data and with a clinical and omics focus. For example, deep learning has been applied to image analysis to make medical diagnostics faster, more accurate, and more accessible. Teams from MIT spun out Ginger.io which has a focus on supporting people with depression and anxiety. It uses a combination of smartphone technology, data science, and clinical services to create a personalized, affordable way to deliver mental healthcare. Likewise, Aetna and Cerner have developed data science platforms that underpin Accountable Care Act support and population management toolset. Likewise, there are multiple data science start-ups that integrate real-time clinical and genomics data to identify patient risks, interpret complex data, and highlight the best evidence-based interventions, such as Syapse (http://www.syapse.com/), Flatiron health (https://www.flatiron.com/), and NantHealth (http://nanthealth.com/).

CONCLUSIONS

In conclusion, the focus and interest in predictive analytics in healthcare is becoming complex, sophisticated, and now attracting a lot of commercial interest. However, our quest for personalized medicine and predictive analytics will be hampered if healthcare professionals and health systems cannot apply such algorithms to realize value for patients and populations in everyday clinical care. This will also be dependent on large, quality-assured, deeply phenotyped data assets, ideally derived in real time from EHR systems. Currently the level of digital maturity of health systems internationally is highly variable, which will hamper adoption. In addition, in the quest for the triple aim, health systems must prioritize the key clinical situations/service planning business needs where enhanced analytics can be useful, and be confident about the performance, validation and applicability within the local healthcare context.

REFERENCES

Amarasingham, R., et al., 2013. Allocating scarce resources in real-time to reduce heart failure readmissions: a prospective, controlled study. BMJ Qual. Saf. 22 (12), 998−1005, http://www.ncbi.nlm.nih.gov/pubmed/23904506 (accessed 18.02.17).

Anderson, K.M., et al., 1991. Cardiovascular disease risk profiles. Am. Heart J. 121 (1 Pt 2), 293−298, http://linkinghub.elsevier.com/retrieve/pii/000287039190861B (accessed 17.02. 17).

Cook, T.D., 2008. "Waiting for life to arrive": a history of the regression-discontinuity design in psychology, statistics and economics. J. Econom. 142 (2), 636−654.

Dikomitis, L., Green, T., Macleod, U., 2015. Embedding electronic decision-support tools for suspected cancer in primary care: a qualitative study of GPs' experiences. Prim. Health Care Res. Dev. 16, 548−555, http://www.ncbi.nlm.nih.gov/pubmed/25731758 (accessed 18.02.17).

Friedman, C., et al., 2014. Toward a science of learning systems: a research agenda for the high-functioning Learning Health System. J. Am. Med. Inform. Assoc. 22 (1), 43−50, https://academic.oup.com/jamia/article-lookup/doi/10.1136/amiajnl-2014-002977 (accessed 17.02.17).

Haybittle, J.L., et al., 1982. A prognostic index in primary breast cancer. Br. J. Cancer 45 (3), 361−366, http://www.ncbi.nlm.nih.gov/pubmed/7073932 (accessed 18.02.17).

Hippisley-Cox, J., et al., 2008. Performance of the QRISK cardiovascular risk prediction algorithm in an independent UK sample of patients from general practice: a validation study. Heart 94 (1), 34−39, http://www.ncbi.nlm.nih.gov/pubmed/17916661 (accessed 24.02.17).

Kang, J., Brant, R., Ghali, W.A., 2013. Statistical methods for the meta-analysis of diagnostic tests must take into account the use of surrogate standards. J. Clin. Epidemiol. 66 (5), 566−574.

Lee, D.S., Lemieux, T., 2010. Regression discontinuity designs in economics. J. Econ. Lit. 48 (2), 281−355, http://pubs.aeaweb.org/doi/10.1257/jel.48.2.281 (accessed 18.02.17).

Parikh R.B., Obermeyer Z.B.D., 2016. Making predictive analytics a routine part of patient care. Harvard Business Review. https://hbr.org/2016/04/making-predictive-analytics-a-routine-part-of-patient-care (accessed 18.02.17).

Peto, R., et al., 2012. Comparisons between different polychemotherapy regimens for early breast cancer: meta-analyses of long-term outcome among 100,000 women in 123 randomised trials. Lancet 379 (9814), 432−444, http://www.sciencedirect.com/science/article/pii/S0140673611616255 (accessed 27.01.14).

Wishart, G.C., et al., 2012. PREDICT Plus: development and validation of a prognostic model for early breast cancer that includes HER2. Br. J. Cancer 107 (5), 800−807, http://www.nature.com/doifinder/10.1038/bjc.2012.338 (accessed 18.02.17).

FURTHER READING

Miner, L., et al., 2014. Practical Predictive Analytics and Decisioning Systems for Medicine, first ed. Elsevier, ISBN: 9780124116436.

Friedman, C., et al., 2014. Toward a science of learning systems: a research agenda for the high-functioning learning health system. J. Am. Med. Inform. Assoc. 22 (1), 43−50.

Chapter 16

An Apps-Based Information Economy in Healthcare

Kenneth D. Mandl[1,2], Joshua C. Mandel[1,2] and Pascal B. Pfiffner[1]

[1]*Boston Children's Hospital, Boston, MA, United States,* [2]*Harvard Medical School, Boston, MA, United States*

INTRODUCTION

Scenario 1. Integration

A company—Medtastic—has designed an elegant medications-management app which needs access to a current medications list from the EHR. While the user-interface design is considered exemplary, and it market-tests extraordinarily well with end users, the company finds that the necessity of a long, involved sales cycle at each possible customer site is draining their venture capital funds. They had hoped to exercise the 80:20 rule and focus on integration with the top five vendors, but: (1) only two of those are willing to entertain the proposition; (2) the technical teams of those vendors are not enthusiastic about prioritizing the work; (3) the Medtastic technical team has found that there is such variation across different instances of each brand of EHR— because of versioning and extensive local customization—that installation of their app requires significant support from IT staff at each customer site. The process is not turnkey in any way.

Mandel et al. (2016)

Classical Electronic Health Records

Electronic health record (EHR) vendors may have an incentive to develop apps and integrate them for use solely within their own EHR system. But this may not be the best outcome for many functions, such as automated dissemination of triage protocols in a spreading epidemic or genomic test interpretation, where authoritative rapid dissemination of practice is in the patient's best interest (Mandl and Kohane, 2012).

Key Advances in Clinical Informatics. DOI: http://dx.doi.org/10.1016/B978-0-12-809523-2.00016-9

Application Programming Interfaces

Over the past decade, the concept of "Application Programming Interfaces" (APIs) has gained currency across the technology industry, including the consumer Web (e.g., Google, Facebook, Twitter) and finance (e.g., Square, Stripe). APIs open the clinical encounter to third-party information technology (IT) innovation and redefine clinical interoperability in terms of substitutability—apps that can be added to or deleted from EHRs as easily as from a smartphone. Fundamentally, APIs allow third-party developers to work in a well-supported integration environment, without the need to reverse-engineer existing systems or "shoehorn in" new integration points. APIs come with support from vendors including Service Level Agreements that set expectations for system uptime and for advance notification of any breaking changes (e.g., deprecation of functionality that developers may rely on). Fostering third-party apps creates a market where innovations compete with each other for purchase and use by providers (and patients/members of the public), thus reducing dependency on updates and specific functions made by an EHR vendor.

Integration Without APIs

Organizations increasingly rely on APIs as a means to extend the functionality of products with new workflows and tighter integration with other ecosystem components. But how is API-based integration fundamentally different than the kind of healthcare data integration strategies we have seen for decades? Here, we will briefly compare API-based integration with two alternatives:

1. *Compared with reverse engineering, user impersonation, and screen scraping.* In the absence of an API, deep integration between systems is still possible. For example, by reverse engineering network protocols, one of the authors of this chapter (JCM) was able to build automated data extraction on top of an otherwise rigid legacy EHR. More generally, any operation that can be performed by a human user of a system can be scripted and automated to simulate interaction; as such, techniques including "user impersonation" (where a software program drives inputs by simulating the keystrokes, taps, or mouse clicks of a human user) and "screen scraping" (where a software gleans information by interpreting pixel or text data from the screen) are a common fallback when some particular workflows step requires automation. There are two key challenges with this approach: (1) reverse engineering is a painstaking process that requires dedicated engineering time to develop an understanding of the underlying system and (2) reverse-engineered solutions depend on implementation details that may not be supported long term by the vendors of the underlying systems (indeed, vendors may incidentally break compatibility, or may actively change the behavior of systems to break

compatibility if they do not approve of a particular "adversarial" integration attempt).

2. *Compared with "messaging systems" like HL7 Version 2.* Longstanding healthcare data integration techniques like HL7 V2 messaging interfaces can be viewed as APIs in their own right: They are thoroughly-documented, standards-based, officially supported integration points designed to facilitate connections between software components. In this chapter, however, when we refer to APIs we will focus on contemporary Web-based APIs that model healthcare data as state within a system. This approach has been formalized by Roy Felding with the concept of Representational State Transfer (REST, https://www.ics.uci.edu/~fielding/pubs/dissertation/top.htm). With REST APIs, rather than relying on complex routing behavior to determine where ad-hoc messages should be delivered and how they should be acted upon, a data server exposes an API that provides developers with a concrete understanding of how to create, update, or retrieve data in the form of "resources" that are well-identified and can be clearly modeled and versioned.

APIs IN HEALTHCARE

SMART Health IT; Substitutable Apps

In 2009, Mandl and Kohane proposed that EHRs should be reimagined as "iPhone-like" platforms supporting a selection of "substitutable" modular third-party applications (apps) (Mandl and Kohane, 2009). Opening EHRs up to third-party apps brings the full power of the Web to patient interactions, including external services and data. Mobile phone platforms such as iPhone and Android lower the barrier to app development by providing a software platform with a published interface to a set of core services such as camera, address book, geolocation, and cell and wireless networks. A smartphone platform functionally separates the core system from the apps, and the apps are substitutable. Thus, for example, a consumer can download a calendar reminder system, reject it, and replace it with another one, without losing calendar data or undertaking a complicated migration process. Through substitutable apps, the iPhone and Android platforms now support myriad capabilities that the original platform designers never imagined. By defining an API that consistently presents well-specified data, we proposed to (1) enable purchasers, users, and administrators of platform-based systems to be able to install and subsequently substitute apps from different vendors without software programming and (2) create a broad market for app developers across multiple systems, including EHRs, patient-facing apps, and health information exchanges (Fig. 16.1).

In 2010, the Office of the National Coordinator for Health Information Technology funded the SMART (Substitutable Medical Applications, Reusable Technologies) Health IT project as a part of the Strategic Health IT

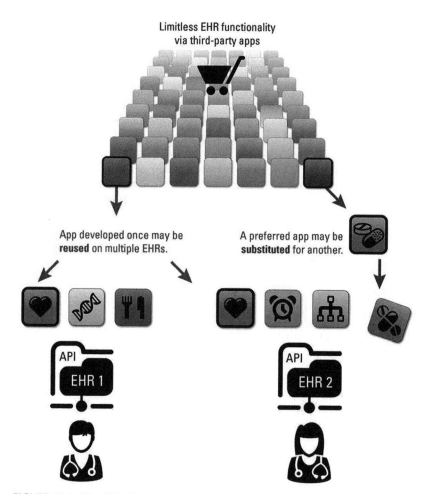

FIGURE 16.1   The SMART ecosystem. . With a uniform public API for healthcare data, a third-party app written once can run anywhere. The app can be reused on multiple electronic medical records and other forms of health IT. And just as on a smartphone, one app can be readily substituted for another. *Adapted from Mandl, K.D., Mandel, J.C., Kohane, I.S., 2015. Driving innovation in health systems through an apps-based information economy. Cell Syst. 1, 8–13.*

Advanced Research Projects Program, an appropriation under the Health Information Technology for Economic and Clinical Health Act, enacted as part of the American Recovery and Reinvestment Act of 2009.

The SMART API—Technical Specifications

SMART uses the HTML5 markup language and the JavaScript programming language for User Interface integration and the OAuth standard to delegate authorization. The API relies on a RESTful paradigm, which allows to access data via uniform resource locators (URL) and a predefined set of operations. Our first release, now dubbed SMART Classic, employed the Resource Description Framework to model and expose semantically precise data payloads, but in 2014 we transitioned to Fast Health Interoperabilty Resources (FHIR). The emerging FHIR standard was spearheaded by Grahame Grieve (Grieve, 2011), who eventually secured an agreement that HL7 publish it under an open license. FHIR focused on providing an API for healthcare that was "simple and easy to implement and manage," inspired by contemporary Web APIs used widely throughout multiple industries. The earliest version of FHIR defined data models to support laboratory result exchanges, and by 2012, a growing community had begun to participate in expanding FHIR's scope. SMART constrains FHIR with specific profiles relying on subsets of the "best-of-the-breed" coding systems including *Systematized Nomenclature of Medicine—Clinical Terms, Logical Observation Identifiers Names and Codes*, and RxNorm. SMART focused on coding systems that are helpful for developers also and aligned with national requirements for Meaningful Use. This API-oriented approach hides internal implementation details from SMART apps and their developers.

Case Study. Customization and Visualization. *At Boston Children's Hospital in 2011, a point-of-care blood pressure tracking interface for the EHR remained unaddressed on the development queue for several years. A multidisciplinary team of cardiologists, hospital IT leaders, and SMART-team informatics faculty and software developers created the BP Centiles app and a container enabling it to run in the Cerner deployment at the Hospital, in patient context. The app was designed to graph the blood pressure of a patient over time, adjusted for by age and percentile. A SMART "container" is any Health IT system that exposes the SMART API, and supports "substitutable" apps. The premise of substitutability is that users of a given container can add or delete any approved SMART apps; and conversely, application developers can anticipate that their app will run unmodified against any SMART container. Since Boston Children's Hospital used the Cerner Millennium product, the challenge at hand was to adapt the SMART API to that proprietary EHR system, creating a "SMART container" around the Cerner system. Enabling SMART applications to run on Boston Children's Hospital's Cerner EHR required implementing a substantial portion of the SMART API, providing apps with user-interface integration and an authenticated API to retrieve a subset of medical data (demographics and vital signs). The SMART technical team took advantage of two existing API tools that Cerner already exposed: MPages and Millennium Objects. The MPages API allowed the app to be launched when users clicked a button within the frame of the Cerner EHR. The result was an adaptor or "proxy layer" to (1) support an app-launch workflow initiated from within the Cerner PowerChart EHR, in the context of a single patient and without*

additional lookup or authentication steps by the clinician user; and (2)
expose demographics and vital signs through the SMART API. Each time the
app launched, it would issue a SMART API query against the HER's proxy
endpoint; the proxy translated this query into a set of Web service calls to
the Millennium Objects server; and the Millennium Objects server in turn
translated the calls into a set of SQL database queries against Cerner's
underlying data tables. The resulting app launched in the container
(Fig. 16.2) has been used thousands of times by Boston Children's Hospital
clinicians.

Diffusion of SMART Technology

We subsequently released a SMART on FHIR version of the BP Centiles
app that has been reused in multiple deployments nationally. Deployment of
SMART apps is occurring along two pathways. For example, at Duke
Health, backend Epic services were used to implement the SMART API. But
also, many vendors now see that nurturing this app ecosystem is essential to

FIGURE 16.2 The BP Centiles app as deployed at Boston Children's Hospital. The
SMART BP Centiles app reads the child's relevant vitals and provides a weight and height
adjusted calculation of diastolic/systolic blood pressure centiles, a pop-up calculator, and a
graphical history of the young patient's blood pressure percentiles, enabling full screening at
each visit. It was first designed to run on the Cerner EHR at Boston Children's Hospital, in
patient context and has been used tens of thousands of times in the course of routine care.

both continuous improvement of EHR modules and addressing the myriad specialized needs of our complex healthcare delivery system. In fact, multiple EHR vendors, including Cerner and Epic, and health systems have come together in a project called "Argonaut" to implement the SMART API into their products. Also, there are regulatory forces promoting SMART adoption. Inspired by SMART, the 2015 EHR Certification and Meaningful Use 3 Final Rules require an API for patients to download and transmit their data—a promising approach to replacing patient portals which rarely provide access beyond a view.

And there is a growing ecosystem of apps (Mandl et al., 2015). The SMART App Gallery offers an integrated online resource for testing, publishing, and discovering FHIR apps. The App Gallery features substitutable apps compatible with the SMART on FHIR API, an open specification seeing international adoption. Integrated with the App Gallery, the open-source SMART Sandbox lets developers simulate a commercial EHR and patient population when creating and testing apps, driving the availability of high-quality FHIR-based applications. Together, these tools power a "Live Demo" button that allows users to explore apps in real time as part of the evaluation process; such transparency is unusual in an environment where healthcare apps are often marketed through sales calls with vendor-led remote demonstrations, and it exceeds the functionality afforded by the Apple App and Google Play stores.

Interoperable Decision Support

Beyond substitutability, a second keystone of interoperability is adoption of a standardized, service-oriented architecture (SOA) for clinical decision support (CDS) (see Chapter 12: Clinical Decision Support and Knowledge Management). An SOA separates CDS rules from the EHR itself. Instead, the actionable knowledge is hosted on a rules engine made accessible by a third party. Hence when an organization like the Centers for Disease Control and Prevention (CDC) issues a new immunization recommendation or a new algorithm for emergency department triage in the midst of a public health emergency, that recommendation or rule can be added, deleted, or updated once and in one place, rather than repeatedly at each of thousands of EHR installations.

Scenario 2. Third-Party Decision Support

A 3-year old is brought to a pediatric clinic for her annual well-child visit.

When the pediatrician opens the well-child documentation template in Epic, the "Immunization App" automatically launches inside of the patient context. The app transmits a FHIR representation of the immunization history to a third-party rules engine, hosted by the CDC. The rules engine encodes all of the Advisory Committee on Immunization Practice recommendations and returns recommendations tailored to the patient's age and history. The recommended immunizations needed next are returned to the app and displayed in the app.

To facilitate this type of interaction, we are leading an effort that extends SMART called CDS Hooks, working with EHR vendors and the CDS providers. Our approach is to define specific trigger points in a user's interaction with the EHR, such as "opening a patient's chart," "prescribing a new medication," or "reviewing a set of orders." At these points in the workflow, we can trigger a notification (called a Web hook) to external services, which can respond with advice, in the form of "cards." Each card can display a message to a clinician, offer an alternative suggestion with a button to take the proposed action, or offer a link to launch a SMART app that the clinician might want to run.

Modular Extensions to the FHIR Ecosystem

Availability of a standardized API also plays an important role as "enabler" of novel, previously unforeseen use cases. Most existing data obtained from external sources (e.g., whole exome sequencing, fitness trackers and other devices, and patient reports) are handled through applications disconnected from health system data and clinician workflows. APIs such as SMART on FHIR essentially create a pipeline for bringing externally sourced data and services into the health system to augment the clinical encounter or the experience of the patient at home. This tectonic shift toward 21st century IT, which mirrors changes sweeping across other industries, will change the experience of physicians and patients, dramatically increasing return on EHR investments. Examples include the following:

> *Clinical Trials.* Clinical trials are facilitated by collecting patient data from mobile phones after obtaining consent. Programming frameworks such as "ResearchKit" for iPhone and "ResearchStack" for Android aid in creation of these apps but, by design, do not provide facilities to transmit this data back to the health system. Now, frameworks such as the Consent, Contact, and Community framework for Patient Reported Outcomes (C3-PRO) take advantage of the tapestry of toolkits and APIs emerging around FHIR and the SMART ecosystem and enable the use of FHIR in such research apps.
>
> *Surveys and Device Data.* In order to manage congestive heart failure, patient reports and daily weights are key elements not captured by the EHR. For disease management, patients and clinicians should be automatically alerted to weight gain. By encoding patient-reported outcomes in FHIR format (Pfiffner et al., 2016) a software developer can make use of existing frameworks such as Apple's ResearchKit or CareKit to rapidly

build an app that can download surveys from FHIR compliant servers, display them to the user, and return responses safely. Developers can leverage the C3-PRO backend to export data in FHIR format. As EHRs and other forms of Health IT extend the SMART API to enable granular "write" functions back into the EHR, these apps will be increasingly integrated into clinician and patient workflows.

Genomics. Genomic sequences will likely never be integrated into the core databases of existing EHR products. Instead, multiple systems need to interact to bring genomic medicine to the point of care. The SMART Genomics API (Alterovitz et al., 2015) is a means to classify and package genomic information for use in the clinical realm (Alterovitz et al., 2015). With the influx of data supporting personalized medicine, a need arises to accommodate the use of this information in the EHR by point-of-care providers. Since a clinical API has been defined and supported by the SMART clinical initiative, it is natural to model the use of genomic data in a way similar to this established method.

CHALLENGES

Facilitating the full apps ecosystem requires overcoming several challenges. First of all, clinical data must become accessible to innovators who are not doctors doing medical research, public health specialists, or clinicians doing quality improvement. Efforts at robust de-identification often limit the utility of the data—for example, dates are removed making temporal algorithms difficult to develop and validate.

Another barrier is the regulatory and legal surround. For example, because apps may receive and store data from HIPAA covered entities, business associate agreements will be needed between app purveyors and health systems.

A particularly thorny challenge is assessment of app quality. Are black box algorithms correct? Are they validated against the populations to which they are applied, and revalidated over time? Are data stored secure? Are API calls for data limited to only data needed for the app? Is the user experience clear and conducive to safe and efficient workflows?

CONCLUSIONS

A key property that APIs enable in healthcare is the substitutability of apps accessing core health system data—in other words that third-party apps can be added to or deleted from EHRs as easily as on an iPhone. Substitutability will underpin a tailored end-user experience—contrasting with today's one-size-fits-all approach, in which rheumatologists and urologists share the same EHR functions. When third-party apps are fostered, a market can emerge where innovations compete with each other for purchase and use by institutions, providers, and patients. This market-based information economy will sharply reduce customer's dependency on updates, innovations and specific functions made by an EHR vendor. Innovators can deploy their products within a façade of HTML5 Web apps or native smartphone apps to the EHR, which vests the end user with the full power of the Web, including data mashups, visualizations, and computation.

REFERENCES

Alterovitz, G., Warner, J., Zhang, P., Chen, Y., Ullman-Cullere, M., Kreda, D., et al., 2015. SMART on FHIR Genomics: facilitating standardized clinico-genomic apps. J. Am. Med. Inform. Assoc. 22, 1173—1178.

Grieve, G., 2011. Resources for health: a fresh look proposal. Health Intersections: Healthcare Interactivity by Graham Grieve [Online]. http://www.healthintersections.com.au/?p = 502.

Mandel, J.C., Kreda, D.A., Mandl, K.D., Kohane, I.S., Ramoni, R.B., 2016. SMART on FHIR: a standards-based, interoperable apps platform for electronic health records. J. Am. Med. Inform. Assoc. 23, 899—908.

Mandl, K.D., Kohane, I.S., 2009. No small change for the health information economy. N. Engl. J. Med. 360, 1278—1281.

Mandl, K.D., Kohane, I.S., 2012. Escaping the EHR trap—the future of health IT. N. Engl. J. Med. 366, 2240—2242.

Mandl, K.D., Mandel, J.C., Kohane, I.S., 2015. Driving innovation in health systems through an apps-based information economy. Cell Syst. 1, 8—13.

Pfiffner, P.B., Pinyol, I., Natter, M.D., Mandl, K.D., 2016. C3-PRO: connecting ResearchKit to the health system using i2b2 and FHIR. PLoS One 11, e0152722.

Part III

Future Developments

Chapter 17

Cloud-Based Computing

Mehrdad A. Mizani
Middle East Technical University, Ankara, Turkey

INTRODUCTION

Cloud computing is an umbrella term to describe a collection of technologies and models that constitute a new paradigm of information technology (IT). It represents a model to deliver ubiquitous and on-demand computing resources and self-service Internet-based infrastructure (Mell and Grance, 2011; Kuo, 2011). In cloud computing, resources such as processor, storage, and networking are accessed based on users' demand (Zhang et al., 2010). Its supporting technologies started in 1960s and cloud computing as we know it today emerged in the early 2000s (Moura and Hutchison, 2016). Since then, it has gained popularity for personal usage, enterprises, and research purposes. Amazon Web Services, Google Docs, and Salesforce are examples of popular cloud-based services.

Traditional health information systems (see Chapters 2 and 3: Inpatient Clinical Information Systems; Outpatient Clinical Information Systems) are mostly maintained in-house by healthcare organizations (e.g., healthcare service provider, hospital, clinic). These systems are based on client−server architecture where high-capacity servers make data and applications accessible to client computers used by physicians, nurses, administrators, and patients. This architecture requires the health organization to invest in hardware infrastructure, network, database servers, application development, user licenses, system maintenance, security, backup, and, most importantly, staff training. Current IT solutions in healthcare have several challenges, which include high cost, inefficient connectivity, and inadequate disaster recovery (Bamiah et al., 2012). The upfront cost of the infrastructure is high. There is uncertainty involved in sociotechnical aspects, acceptance by staff, and the capability of systems to provide the required/desired functionalities. During normal workload, many systems are underutilized whereas in times of high activity (e.g., high number of patients or cyber-attacks) system with their limited capacity may become unable to respond to requests.

Key Advances in Clinical Informatics. DOI: http://dx.doi.org/10.1016/B978-0-12-809523-2.00017-0

239

Cloud computing provides services to access functional infrastructures, application development platforms, databases, or software applications rapidly with minimum upfront costs. Also, cloud services facilitate a utility-based and flexible payment model enabling the users to pay exactly for what they use (Buyya et al., 2009). The engine for cloud computing is a huge and high-capacity data center with numerous computers, storage devices, high network capacity, and power and cooling systems (Mastelic et al., 2015). Cloud computing is highly scalable as powerful data centers enable the provision of virtually unlimited resources in case they are needed. The replication of systems in data centers makes cloud computing more reliable than traditional systems. Moreover, system updates, security measures, application updates, server maintenance, and many other operational processes are outsourced to the Cloud Service Provider (CSP).

There are many application areas of cloud computing in health informatics, both for clinical and secondary uses of health data. Examples are cloud-based electronic health record (EHR) systems (Lin et al., 2014) and bioinformatics research (Sandhu et al., 2016). Despite many benefits of cloud computing, concerns such as security, privacy, governance, and regulatory issues make healthcare organizations reluctant to move their systems to cloud environments. For example, sharing the responsibility of systems with the CSP leads to a loss of control over outsourced systems and complicates compliance with data protection regulations (Jansen and Grance, 2011).

This chapter begins by defining cloud computing in greater detail. It proceeds by describing opportunities and challenges of cloud computing in healthcare and finally, considers future trends.

CLOUD COMPUTING

Definition

The definition of the National Institute of Standards and Technology (NIST) for cloud computing has been largely adopted by cloud computing communities. According to NIST, cloud computing "is a model for enabling convenient, on-demand network access to a shared pool of configurable computing resources (e.g., networks, servers, storage, applications, and services) that can be rapidly provisioned and released with minimal management effort or service provider interaction" (Mell and Grance, 2011). Cloud computing is the name given to a collection of technologies such as distributed systems, grid computing, service-oriented architecture (SOA), and virtualization (Youseff et al., 2008). Fig. 17.1 shows the main enabling technologies of cloud computing. Table 17.1 identifies the essential characteristics of cloud computing (Mell and Grance, 2011), provides an explanation for each, and related concepts. Boxes 17.1 and 17.2 detail examples of cloud-based deployments in leading US hospitals.

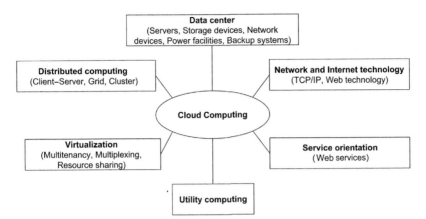

FIGURE 17.1   Main enabling technologies of cloud computing.

TABLE 17.1 The Essential Characteristics of Cloud Computing (Mell and Grance, 2011) and its Related Concepts

Characteristics	Explanation	Related Concepts
Broad network access	Cloud services are enabled over network using Internet technology and are accessed from workstations or mobile devices, usually via a Web browser.	Distributed computing, Internet technology, data center
Measured services	Resource utilization, for example, the amount of memory or network bandwidth, is measured in detail. Users only pay for the resources they use without any upfront payment and are able to monitor and control the utilization of resources (e.g., CPU, memory, disk space).	Utility-based computing, pay-as-you-go model, no upfront costs, monitoring
On-demand self-service	Users can utilize services they need in an on-demand basis without any human intervention from CSP. Required functionalities are provided as services via easy to use Web pages or Application Programming Interfaces (APIs).	Service orientation, API
Rapid elasticity	The capacity of provisioned services and systems can be scaled up and down on demand.	Scalability, reliability, virtually unlimited resources
Resource pooling	Virtualization technology allows the provision of several virtual machines (VMs) in the same physical machine. Resources, both physical and virtual, can be dynamically shared between cloud users independent of their geographic location.	Virtualization, resource sharing, multitenancy, multiplexing, energy consumption

BOX 17.1 Use Case: Cleveland Clinic

Cleveland Clinic is an academic hospital based in Cleveland, OH. The Clinic uses Amazon Web Services to run its Healthy Brain Initiative focusing on clinical assessment of brain health in elderly patients. Its cloud-based services enable patients and neurologists to access health information on brain health. To protect the security, privacy of personal data, and also to ensure compliance with HIPAA Rules, Cleveland Clinic uses ClearData cloud. ClearData provides cloud services for protecting patient data, providing comprehensive BAA, and ensuring adherence to security, privacy, and compliance standards. Using cloud services lowers the operational costs, and enables Cleveland Clinic to focus on activities that bring value to patients (Cleveland Clinic Case Study, 2017).

BOX 17.2 Use Case: Beth Israel Deaconess Medical Center (BIDMC)

BIDMC, a teaching hospital of Harvard Medical School, is based in Boston, MA. BIDMC will open their cloud application services in 2017. Previously, the cloud was used only for decision support databases and online educational materials. BIDMC uses cloud services for application hosting, storage and analytics from Amazon Web Services, G-Suite (e.g., Gmail, Docs, Drive, Calendar) from Google, community hospital EHR hosting from Dell/NTTData, and EHR/practice management for community ambulatory practices from AthenaHealth. The experience of using cloud services suggests that IT workload has not increased. Therefore, migrating to the cloud increases the free time of IT experts leading to better innovation and customer service. To tackle the challenge of network latency in accessing cloud services over the Internet, BIDMC uses a high-speed dedicated line to connect to Amazon Web Services (Halamka, 2017).

Distributed computing refers to connecting dispersed computing resources over the network in order to harness the collective power of the systems. The underlying distributed nature of the system is transparent to the end user to whom the system appears as a single supercomputer. Data center refers to a facility with large number of computers and storage devices interconnected with high-speed network lines (Mastelic et al., 2015). Data centers construct the physical infrastructure on which cloud computing services operate. These centers have specific security, power management, and cooling requirements.

The hardware systems in a cloud data center are shared among many users. This is realized by virtualizing, a basic enabling technology of cloud computing, which refers to running several software-based systems (VM) in a single physical machine. Virtualization abstracts the underlying infrastructure from the users to whom the VMs look like real physical systems. Several VMs of multiple users can share a single hardware (Buyya et al., 2009), which is called multitenancy (Ali et al., 2015). VM users can access the required functionalities provided by CSP as services like other utilities. In such on-demand and utility-based computing, the users pay for the resources they actually use.

Cloud Service and Deployment Models

NIST recognizes three cloud service models, namely, Software-as-a-Service (SaaS), Platform-as-a-Service (PaaS), and Infrastructure-as-a-Service (IaaS) (Mell and Grance, 2011). SaaS refers to services that provide access to software and applications mainly on Web browsers (e.g., Google Docs). User of SaaS simply utilizes the software functionalities while the responsibilities of software development, licensing, security, and upgrading belong to the CSP. PaaS refers to platforms for software and application development (e.g., Google App Engine and Salesforce). IaaS refers to provision of virtual resources such as VMs, storage devices, and network functionalities. One example is Amazon Web Services which provides IaaS in addition to other services such as storage, database, identity management, network, analytics, and Internet of Things (IoT), to name a few. The user of an IaaS has the illusion of having access to a physical computer with on-demand scalable resources. Users of IaaS pay for the actual utilized resources, such as CPU, memory, and disk space, and pay more if they need additional resources.

NIST also defines four cloud deployment models, namely, public, private, hybrid, and community models (Mell and Grance, 2011). A public cloud is accessible to any customer connected to the Internet (e.g., Google Drive) while underlying hardware and software are governed by the CSP. The systems hosting user data reside in the premises of the CSP, which may in some instances cause security, privacy, regulatory, and ownership concerns especially for healthcare data. Private cloud refers to provision of services on an infrastructure that is either hosted in-house or outsourced to a trusted third party. In both cases, the cloud infrastructure, platform, or software services are only accessible to the owner of the private cloud. Hybrid cloud is a mixture of distinct private and public clouds. In hybrid clouds, data and applications are divided between private and public clouds according to information management, policy, security, privacy, and regulatory requirements. Community cloud is accessible to a group of people or organizations with a common purpose and concerns, such as public sector organizations (O'Driscoll et al., 2013). Well-known examples of public and hybrid clouds are Amazon Web Services, Microsoft Azure, Google Cloud Platform, VMWare, and Rackspace. Many well-known cloud providers also provide private cloud capabilities. For example, users of the Amazon Web Services in some regions have the option of connecting to private servers via a dedicated line which is not connected to the Internet. Examples of other private cloud providers are Microsoft Private Cloud, VMWare, Rackspace, OpenStack, and Apache CloudStack.

Cloud Computing in Healthcare: Examples and Use Cases

A broad categorization of cloud computing research in healthcare are telemedicine and teleconsultation, medical imaging, public health and patient's self-management, hospital management and clinical information systems (CIS), therapy, and secondary uses of data (Griebel et al., 2015). Cloud computing is also used for EHR management (Lin et al., 2014; Fernández-Cardeñosa et al., 2012) and medical imaging systems (Silva et al., 2012; John and Shenoy, 2014). Well-known cloud providers such as Amazon and Microsoft can be used for healthcare applications. However, healthcare providers remain responsible for ensuring compliance with data protection regulations and reviewing Business Associate Agreement (BAA) as well as Service Level Agreement (SLA). Table 17.2 shows examples of cloud-based tools and services for healthcare.

Currently most of the applications of cloud computing are in the fields of big data and bioinformatics (e.g., Mahmud et al., 2016; Taylor, 2010; Jourdren et al., 2012; Dai et al., 2012; Onsongo et al., 2014; Krieger et al., 2017) (see Chapter 11: Bioinformatics and Precision Medicine). Another example is analyzing social networks to detect health-related incidents (Sandhu et al., 2016). Examples of other potential overlapping areas of healthcare and cloud computing are decision support system (Hussain et al., 2013), patient self-management (Lee et al., 2016), IoT (Suciu et al., 2015), and collaborative data sharing (Fabian et al., 2015).

TABLE 17.2 Examples of Cloud-Based Tools and Services for Healthcare

Cloud Services	Examples
EHR, practice management	AthenaHealth, PracticeFusion, CareCloud
Medical imaging	Tricefy, Nuance PowerShare Network, Terarecon, Purview ViVA
Bioinformatics	Genestack, InsideDNA
Security, privacy, and compliance	ClearData
Patient portal, patient engagement	Athenahealth, CareCloud, PatientFusion, iPatientPortal

BENEFITS OF CLOUD COMPUTING IN HEALTHCARE

Ubiquitous Access and Availability

The most important benefit of cloud computing in healthcare is ubiquitous access to health data and applications at any time, independent of geographical location, mostly using a Web browser. Availability of EHR and related services is crucial to prevent interruptions in healthcare processes, especially in emergency cases. The engine of a cloud environment is a data center with backup servers for additional capacity and data replication. Examples of systems in a data center are servers, storage devices, hard disks, network lines, and power facilities. Existing data replication mechanisms guarantee that huge amounts of data are replicated among distributed data centers to reduce access time and increase availability (Milani and Navimipour, 2016). All these factors decrease the risk of data loss or unavailability of online medical services.

Smooth IT Transition

In traditional in-house CIS, health organizations need to invest in expensive hardware infrastructure and mostly proprietary software licenses. The hardware requires setup time, extensive technical knowledge, transforming and migrating old data to the new system, sociotechnical change management, maintenance and updates, and many other costly and time-consuming procedures. The applications are tightly coupled with hardware and software systems (e.g., database servers). Any changes in underlying servers and systems necessitate the review and modification of applications and user interfaces which may result in unintended sociotechnical consequences. There is a risk that the new system will fail to provide required functionalities or be rejected by end users who are accustomed to old systems.

Conversely, in cloud computing, it is possible to establish fully functional VMs and servers (IaaS) for CIS, clinical software development platforms (PaaS), or application services (SaaS) for medical functionalities in a short time with minimum upfront costs and maintenance expenses (Zhang et al., 2010). The simplicity and low cost of setting up VMs in IaaS enables healthcare organizations to establish and try alternative test cloud services (e.g., EHR systems) in terms of functionalities and user acceptance, and to choose the most appropriate cloud for final transition.

High Scalability and Elasticity

Another benefit of cloud computing is its flexibility and scalability (Kuo, 2011). To reduce the risk of medical errors and delays, healthcare data and applications must be available at all times. This requirement must be maintained not only during normal system activities, but also amid high demands and occasional cyber-attacks. Examples of high peaks in demands are increasing service requests during epidemics and natural disasters.

In traditional IT, in order to respond to increasing resource-intensive requests, it is necessary to periodically upgrade the hardware. In times of high computing or network activity, the systems can become congested and unable to respond to the peaks in demand. In normal situations, most of the resources remain underutilized, negatively affecting the return on investment. In cloud computing, on the other hand, resources can be utilized and released dynamically based on demand (Zhang et al., 2010). It is possible to utilize more virtual resources in times of peak activity or cyber-attacks to respond to service requests. However, in times of normal activity, the resources can be reduced to the actual level needed. This prevents costly upgrades, under-utilization of expensive systems, or congestion of under-powered computing resources.

Service Models for Healthcare

Cloud services are supported by extensively tested APIs, backup, security measures, audit capabilities, and regular updating procedures. All these supporting features are provided and maintained by the CSP. With IaaS, cloud-enabled health organizations simply utilize cloud services without involving regular maintenance of hardware infrastructure. PaaS potentially provides necessary platforms for software development with specialized APIs for healthcare (Deng et al., 2012). Examples are general APIs for authentication, authorization, anonymization, and audit or specialized APIs for EHR access, picture archiving and communication systems, computerized decision support systems, and data analysis. Depending on PaaS capabilities, it will be possible to develop functional clinical applications in a short time. The most important benefit of SaaS is access to the services from the Web browser, which is a widespread, familiar, and lightweight interface. The SOA of the cloud makes it an ideal platform to respond quickly to technological changes (Choi et al., 2010). With SaaS, there is no need to know and control the underlying hardware or software infrastructure (Oh et al., 2015). Fig. 17.2 shows the service models of an example cloud-enabled healthcare organization.

Cloud Deployment Models for Healthcare

Public clouds, such as Amazon Web Services and Microsoft Azure, are easiest to deploy because the responsibilities of hardware, software, and service maintenance belong to the CSP. Due to high-scalable computing resources of cloud environment, its most common usage in healthcare is in resource-intensive applications such as big data analysis and bioinformatics. Security, privacy, and ownership issues are easier to control in private clouds which

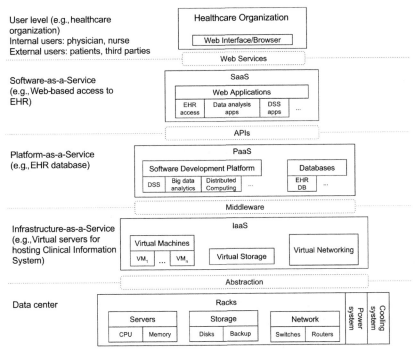

FIGURE 17.2 Service models of an example cloud-enabled healthcare organization.

facilitate hosting of sensitive data and applications only accessible to authorized users managed by internally controlled policies. It is also easier to move to private clouds if the underlying infrastructure is provided and maintained by a trusted third party.

Hybrid clouds can be used to counter the privacy concerns of pure public clouds. In a hybrid model, it is possible to isolate sensitive data and applications by hosting them in a private cloud. The public facing applications, such as patient access to CIS, will still be possible to provide. The fact that the user interface to the cloud is simply a Web browser also makes it easier to provide public access to CIS. Assuming there are adequate security measures in place, access to CIS encourages patient involvement in their healthcare processes through controlled sharing of EHRs. Community clouds can be used in healthcare domain to share data and to provide collaborative care. Community clouds make it possible to share data efficiently and securely with all parties (Sandhu et al., 2016). Currently there are not commonly accepted standards for interoperability between clouds. Open-source cloud systems encourage standardization and interoperability. They also facilitate the detection of security vulnerabilities, as the source code is open to the software development community, leading to extensive testing, and faster security incident response, compared to nonstandardized and proprietary systems.

Cloud and Large-Scale Data Analysis

The amount of health data gathered and processed is on the rise both in primary and secondary healthcare. Body area networks (e.g., wireless wearable sensors) produce gigantic amounts of data that need to be stored and analyzed (Kaur and Chana, 2014; Suciu et al., 2015). Single computers will not be able to store and process the volume and variety of healthcare data that will appear in near future. Cloud computing is a potentially ideal technology to store and analyze large amounts of data due to its virtually unlimited storage and computing capabilities. The underlying infrastructure of cloud computing is based on distributed architecture in huge data centers with high-capacity computers, large storage devices, high-speed network capabilities, and backup systems, making it an ideal platform for the application of big data analysis in healthcare. Currently the prominent areas of big data analytics in healthcare are bioinformatics and biomedical fields which require processing of large amounts of data (e.g., Mahmud et al., 2016; Taylor, 2010; Jourdren et al., 2012).

CHALLENGES AND CONCERNS SURROUNDING CLOUD COMPUTING IN HEALTHCARE

Despite aforementioned benefits of cloud computing, healthcare organizations are generally reluctant to move their data and applications to cloud environments. The prominent concerns are security, privacy, trust, ownership, governance, vendor lock-in, regulations, interoperability, and standardization (see Chapters 5, 6, and 12: Interoperability; Privacy and Security; Clinical Decision Support and Knowledge Management).

Security and Privacy

Security refers to protecting the confidentiality, integrity, and availability of data and applications. Security is the major concern in cloud computing in healthcare. Transferring data to the servers of the CSP and loss of control over parts of the system (i.e., data and applications) that reside on the cloud, raise serious security concerns. Resource pooling characteristics of cloud computing and multitenancy allow hosting VMs and databases of several customers on the same physical computer. Any vulnerabilities in the separation mechanisms of VMs can lead to unauthorized access to data. While customers' VMs are agnostic to this multitenant nature of the cloud, such shared environments pose risks to the confidentiality of data (Dillon et al., 2010) and patient privacy, especially in public clouds.

Data integrity including correctness, consistency, and coherency is an indispensable requirement to prevent medical errors. The accessibility of clouds over the Internet and multitenancy of cloud environments also raise concerns about accidental or intentional manipulation of health data leading to data deletion, corruption, or modification. Cryptographic methods are available to tackle such confidentiality and integrity concerns for both stored and in-transit data. However, many practical issues such as efficiency of cryptographic algorithms (e.g., homomorphic encryption) still remain to be solved (Ryan, 2013). Homomorphic encryption is a cryptographic method for sending encrypted data to the cloud which unlike normal encryption, enables CSPs to perform certain operations (e.g., mathematical calculations) on encrypted data. However, it is not fully applicable due to its inefficiency and slow performance (Ryan, 2013).

In order to provide high-quality care and to prevent interruptions, clinical health data and applications must be available in high speed. The availability of cloud services over the Internet also attracts cyber-attackers, especially for Distributed Denial of Service (DDoS) attacks which aim at making the servers unavailable to legitimate users. In a DDoS attack, the attacker compromises many computers over the Internet, called bot-nets or zombies, and initiates concurrent and high-volume service requests to the server being attacked. Hence, the victim server becomes unable to handle huge amounts of service requests, leading to the unavailability of services to legitimate users. In case of cloud computing, cyber-attack, technical faults, misconfiguration, and natural disasters might affect the services or data center, hence causing interruption in clinical applications and data access.

In order to mitigate security and privacy problems in the cloud, it is necessary to review the SLA to ensure two crucial components: firstly, the SLA should be in-line with the security policy of the healthcare provider; second, the CSP must have a well-defined auditing mechanism, disaster recovery plan, and incident response policy. Table 17.3 lists examples of technical mitigation methods for security and privacy issues in cloud-based healthcare.

Interoperability and Standardization

Cloud computing can potentially be used to connect health organizations for sharing data and services both for primary uses (e.g., collaborative care) or secondary uses (e.g., public health research). This is possible, for example, by having a community cloud, or an inter-cloud (Aazam et al., 2016), to connect health organizations. The most important challenge of realizing a cloud federation is the lack of commonly accepted standards to interconnect different clouds. Some clouds are based on proprietary standards making it extremely difficult for their users to migrate data and services to another CSP. This challenge is called vendor lock-in (Zhang et al., 2013). In healthcare, vendor lock-in restricts the healthcare organization to use databases and applications that are not easily adaptable to rapidly improving technologies. Another complicating issue is different data protection regulations in transborder data exchange (Seddon and Currie, 2013).

TABLE 17.3 Examples of Technical Methods to Mitigate Security and Privacy Issues in Cloud-Based Healthcare

General measures (Gonzalez et al., 2012)	Ensuring network security
	Ensuring the security of user interface, administrator interface, and APIs
	Ensuring virtualization security
Confidentiality	Using a hybrid cloud model with EHR stored in a private cloud
	Access control policies
	Dedicated servers and network lines
	Ensuring VM separation
Integrity	Cryptographic integrity checks
	Data versioning
	Replicating data to other geographical areas for backup
Availability	Backup server, storage media, network line, and power system
	Scaling up the capacity of servers in times of high demand
	Replicating data to closer data centers to the user
	Mitigating DDoS attacks

Governance, Regulations, and Trust

In traditional IT, the ownership of hardware, software, and data, along with protecting the security and privacy, belong to the healthcare organization. In cloud computing, particularly in public clouds, the hardware infrastructure containing data and applications is owned and maintained by the CSP. If no encryption is used, both stored and in-transit data will be visible to CSP administrators. The administrators in CSP are in charge of managing the physical or virtual resources, and potentially have access to data with full authentication permissions. Multitenancy may also expose data of one VM to another in case of security vulnerabilities and intentional cyber-attacks. The regulatory aspects of managing health data, especially sensitive and person-specific data, are more complicated in cloud computing than traditional IT. Moving data out of the premises of the health organization, and hosting them on the cloud, requires review of regulatory aspects of compliance and data protection (Schweitzer, 2012). In public clouds, the infrastructure may reside in another country with completely different data protection regulations. In case of community clouds, the regulations and policies governing the data management and sharing might be different among participating healthcare organizations.

Healthcare organizations must carefully and systematically review the SLA and conditions of the CSP (i.e., due diligence) to make sure that security and privacy issues are in accord with organizational policies and data protection regulations. It is also crucial to be able to monitor the security practices of the CSP, along with a comprehensive audit of the activities of legitimate users, administrators, and any unauthorized parties. However, access to internal security policies of the CSP and comprehensive access to audits and logs are often not easily granted in many public clouds and are subject to SLAs. As the boundary of data premise extends to the CSP, a lack of trust between the client (e.g., health organization) and the CSP may also become a barrier (Pearson, 2013).

According to the HIPAA Rules, a healthcare provider and "the CSP must enter into a HIPAA-compliant business associate agreement (BAA), and the CSP is both contractually liable for meeting the terms of the BAA and directly liable for compliance with the applicable requirements of the HIPAA Rules" (OCR, 2016). Popular cloud providers (e.g., Amazon Web Services, Microsoft Azure, Google Cloud Platform) provide a BAA to healthcare providers who want to use the cloud for protected health information. However, healthcare providers remain responsible for risk analysis, risk management, and careful review of the SLA to ensure its consistency with the BAA and the HIPAA Rules (OCR, 2016). According to the European Data Protection Supervisor a CSP is required to "guarantee by way of a contract with the customer or in binding corporate rules, that all transfers of information to non-EU jurisdictions will meet specific data protection requirements to provide adequate safeguards" (EDPS, Q&A, Cloud Computing, 2017).

FUTURE TRENDS

Current benefits of cloud computing in healthcare are ubiquitous access, high availability, scalability, smooth IT transition, and potential service and deployment models for healthcare. Currently, private clouds are the safest choice for healthcare organizations that are ready to move their systems to the cloud. More hybrid and community clouds in healthcare are likely to emerge with more efficient cryptography algorithms, standards for interoperability, and services for compliance with data protection regulations including HIPAA Rules and EU Data Protection Act.

In the area of service models, potential health-related services are SaaS for accessing EHR, medical imaging, and clinical decision support systems in cloud-based CIS. Although not yet commercially developed, specialized healthcare PaaS will also emerge. This will enable health organizations to easily develop clinical and health-related applications on the cloud, for example, with generic application templates (Jeferry et al., 2015). Services and APIs for secondary uses of health data also need to be developed.

Examples of such APIs and services are data aggregation, data joining, making synthetic datasets, de-identification, and anonymization. With the ease and low cost of setting up IaaS and relieving from the burden of maintaining hardware infrastructure, more healthcare organizations are likely to move their systems to secure IaaS.

Following the current trends, the volume and variety of health data will continue to increase. The volume of healthcare data is expected to reach 250 Exabytes in 2020 (Goli-Malekabadi et al., 2016). Advancements in the areas of the IoT (Botta et al., 2016) and wireless body area networks (Hassan et al., 2017) will produce vast amounts of data. Other sources of health and biology data are bioinformatics, biomedical, hospitals, patients, systematic disease surveillance, mobile health applications, and social networks. IaaS is becoming a mature cloud model, hence currently is an ideal model for big data analysis in healthcare. The future is also very likely to see specialized and sophisticated SaaS and PaaS for data analysis in healthcare domain. In the field of cloud-enabled big data analytics, standards and APIs will need to be developed to utilize and choose the best analytics solutions in the cloud (Assunção et al., 2015).

CONCLUSIONS

This chapter has introduced cloud computing and considered the benefits and challenges of cloud-enabled healthcare. The main benefits of cloud computing are ubiquitous access to EHRs, flexible payment models, availability, smooth IT transition, big data analysis capabilities, high scalability, and specialized service and deployment models for healthcare. Despite the many benefits of cloud computing, there are still major challenges remain to be addressed. Loss of control over the boundary of data storage results in serious security, privacy, ownership, and trust concerns. Many of these challenges are present in public clouds as they are publicly accessible over the Internet. Lack of commonly accepted standards also causes vendor-lock and difficulties in interoperability and the realization of community clouds. Currently private and hybrid clouds are better choices for cloud-enabled healthcare.

REFERENCES

Aazam, M., Huh, E.-N., St-Hilaire, M., 2016. Towards media inter-cloud standardization—evaluating impact of cloud storage heterogeneity. J. Grid Comput.1–19. Available from: http://dx.doi.org/10.1007/s10723-015-9356-5.

Ali, M., Khan, S.U., Vasilakos, A.V., 2015. Security in cloud computing: opportunities and challenges. Inform. Sci. 305, 357–383.

Assunção, M.D., Calheiros, R.N., Bianchi, S., Netto, M.A., Buyya, R., 2015. Big data computing and clouds: trends and future directions. J. Parallel Distrib. Comput. 79, 3–15.

Bamiah, M., Brohi, S., Chuprat, S., 2012. A study on significance of adopting cloud computing paradigm in healthcare sector. In: Cloud Computing Technologies, Applications and Management (ICCCTAM), 2012 International Conference. IEEE, pp. 65−68.

Botta, A., de Donato, W., Persico, V., Pescapé, A., 2016. Integration of cloud computing and internet of things: a survey. Future Gener. Comput. Syst. 56, 684−700.

Buyya, R., Yeo, C.S., Venugopal, S., Broberg, J., Brandic, I., 2009. Cloud computing and emerging IT platforms: vision, hype, and reality for delivering computing as the 5th utility. Future Gener. Comput. Syst. 25 (6), 599−616.

Choi, J., Nazareth, D.L., Jain, H.K., 2010. Implementing service-oriented architecture in organizations. J. Manage. Inform. Syst. 26 (4), 253−286.

Cleveland Clinic Case Study. <https://aws.amazon.com/solutions/case-studies/cleveland-clinic/> (accessed 30.01.17).

Dai, L., Gao, X., Guo, Y., Xiao, J., Zhang, Z., 2012. Bioinformatics clouds for big data manipulation. Biol. Direct. 7 (1), 1.

Deng, M., Nalin, M., Petković, M., Baroni, I., Marco, A., 2012. Towards trustworthy health platform cloud. In: Workshop on Secure Data Management. Springer, Berlin and Heidelberg, pp. 162−175.

Dillon, T., Wu, C., Chang, E., 2010. Cloud computing: issues and challenges. In: 2010 24th IEEE International Conference on Advanced Information Networking and Applications. IEEE, pp. 27−33.

EDPS, Q&A, Cloud Computing. <https://secure.edps.europa.eu/EDPSWEB/edps/EDPS/Dataprotection/QA/QA10> (accessed 30.01.17).

Fabian, B., Ermakova, T., Junghanns, P., 2015. Collaborative and secure sharing of healthcare data in multi-clouds. Inform. Syst. 48, 132−150.

Fernández-Cardeñosa, G., de la Torre-Díez, I., López-Coronado, M., Rodrigues, J.J., 2012. Analysis of cloud-based solutions on EHRs systems in different scenarios. J. Med. Syst. 36 (6), 3777−3782.

Goli-Malekabadi, Z., Sargolzaei-Javan, M., Akbari, M.K., 2016. An effective model for store and retrieve big health data in cloud computing. Comput. Methods Programs Biomed. 132, 75−82.

Gonzalez, N., Miers, C., Redígolo, F., Simplício, M., Carvalho, T., Näslund, M., et al., 2012. A quantitative analysis of current security concerns and solutions for cloud computing. J. Cloud Comput. 1, 11. Available from: http://dx.doi.org/10.1186/2192-113X-1-11.

Griebel, L., Prokosch, H.U., Köpcke, F., Toddenroth, D., Christoph, J., Leb, I., et al., 2015. A scoping review of cloud computing in healthcare. BMC Med. Inform. Decis. Mak. 15 (1), 1.

Halamka, J., 2017. Life as a Healthcare CIO: Our 2017 Priorities. <http://geekdoctor.blogspot.com.tr/2017/01/our-2017-priorities.html> (accessed 30.01.17).

Hassan, M.M., Lin, K., Yue, X., Wan, J., 2017. A multimedia healthcare data sharing approach through cloud-based body area network. Future Gener. Comput. Syst. 66, 48−58.

Hussain, M., Khattak, A.M., Khan, W.A., Fatima, I., Amin, M.B., Pervez, Z., et al., 2013. Cloud-based smart CDSS for chronic diseases. Health Technol. 3 (2), 153−175.

Jansen, W., Grance, T., 2011. SP 800-144. Guidelines on Security and Privacy in Public Cloud Computing. National Institute of Standards & Technology, Gaithersburg, MD.

Jeferry, K., Kousiouris, G., Kyriazis, D., Altmann, J., Ciuffoletti, A., Maglogiannis, I., et al., 2015. Challenges emerging from future cloud application scenarios. Proc. Comput. Sci. 68, 227−237.

John, N., Shenoy, S., 2014. Health cloud-Healthcare as a service (HaaS). In: Advances in Computing, Communications and Informatics, ICACCI 2014 International Conference. IEEE, pp. 1963−1966.

Jourdren, L., Bernard, M., Dillies, M.A., Le Crom, S., 2012. Eoulsan: a cloud computing-based framework facilitating high throughput sequencing analyses. Bioinformatics 28 (11), 1542–1543.

Kaur, P.D., Chana, I., 2014. Cloud based intelligent system for delivering health care as a service. Comput. Methods Programs Biomed. 113 (1), 346–359.

Krieger, M.T., Torreno, O., Trelles, O., Kranzlmüller, D., 2017. Building an open source cloud environment with auto-scaling resources for executing bioinformatics and biomedical workflows. Future Gener. Comput. Syst. 67, 329–340.

Kuo, M.H., 2011. Opportunities and challenges of cloud computing to improve health care services. J. Med. Internet Res. 13 (3), e67.

Lee, P., Liu, J.C., Hsieh, M.H., Hao, W.R., Tseng, Y.T., Liu, S.H., et al., 2016. Cloud-based BP system integrated with CPOE improves self-management of the hypertensive patients: a randomized controlled trial. Comput. Methods Programs Biomed. 132, 105–113.

Lin, C.W., Abdul, S.S., Clinciu, D.L., Scholl, J., Jin, X., Lu, H., et al., 2014. Empowering village doctors and enhancing rural healthcare using cloud computing in a rural area of mainland China. Comput. Methods Programs Biomed. 113 (2), 585–592.

Mahmud, S., Iqbal, R., Doctor, F., 2016. Cloud enabled data analytics and visualization framework for health-shocks prediction. Future Gener. Comput. Syst. 65, 169–181.

Mastelic, T., Oleksiak, A., Claussen, H., Brandic, I., Pierson, J.M., Vasilakos, A.V., 2015. Cloud computing: survey on energy efficiency. ACM Comput. Surv. 47 (2), 33.

Mell, P.M., Grance, T., 2011. The NIST Definition of Cloud Computing. National Institute of Standards and Technology, Special Publication 800-145. < http://nvlpubs.nist.gov/nistpubs/Legacy/SP/nistspecialpublication800-145.pdf > (accessed 10.05.17.).

Milani, B.A., Navimipour, N.J., 2016. A comprehensive review of the data replication techniques in the cloud environments: major trends and future directions. J. Netw. Comput. Appl. 64, 229–238.

Moura, J., Hutchison, D., 2016. Review and analysis of networking challenges in cloud computing. J. Netw. Comput. Appl. 60, 113–129.

OCR, Cloud Computing, 2016. HHS.gov., Guidance on HIPAA & Cloud Computing. <https://www.hhs.gov/hipaa/for-professionals/special-topics/cloud-computing/index.html> (accessed 30.01.17).

O'Driscoll, A., Daugelaite, J., Sleator, R.D., 2013. 'Big data', Hadoop and cloud computing in genomics. J. Biomed. Inform. 46 (5), 774–781.

Oh, S., Cha, J., Ji, M., Kang, H., Kim, S., Heo, E., et al., 2015. Architecture design of healthcare software-as-a-service platform for cloud-based clinical decision support service. Healthc. Inform. Res. 21 (2), 102–110.

Onsongo, G., Erdmann, J., Spears, M.D., Chilton, J., Beckman, K.B., Hauge, A., et al., 2014. Implementation of cloud based next generation sequencing data analysis in a clinical laboratory. BMC Res. Notes 7 (1), 1.

Pearson, S., 2013. Privacy, security and trust in cloud computing. Privacy and Security for Cloud Computing. Springer, London, pp. 3–42.

Ryan, M.D., 2013. Cloud computing security: the scientific challenge, and a survey of solutions. J. Syst. Softw. 86 (9), 2263–2268.

Sandhu, R., Gill, H.K., Sood, S.K., 2016. Smart monitoring and controlling of Pandemic Influenza A (H1N1) using Social Network Analysis and cloud computing. J. Comput. Sci. 12, 11–22.

Schweitzer, E.J., 2012. Reconciliation of the cloud computing model with US federal electronic health record regulations. J. Am. Med. Inform. Assoc. 19 (2), 161–165.

Seddon, J.J., Currie, W.L., 2013. Cloud computing and trans-border health data: unpacking US and EU healthcare regulation and compliance. Health Policy Technol. 2 (4), 229–241.

Silva, L.A.B., Costa, C., Oliveira, J.L., 2012. A PACS archive architecture supported on cloud services. Int. J. Comput. Assist. Radiol. Surg. 7 (3), 349–358.

Suciu, G., Suciu, V., Halunga, S., Fratu, O., 2015. Big data, internet of things and cloud convergence for e-health applications. New Contributions in Information Systems and Technologies. Springer International Publishing, pp. 151–160.

Taylor, R.C., 2010. An overview of the Hadoop/MapReduce/HBase framework and its current applications in bioinformatics. BMC Bioinform. 11 (Suppl. 12), S1.

Youseff, L., Butrico, M., Da Silva, D., 2008. Toward a unified ontology of cloud computing. In: 2008 Grid Computing Environments Workshop. IEEE, pp. 1–10.

Zhang, Q., Cheng, L., Boutaba, R., 2010. Cloud computing: state-of-the-art and research challenges. J. Internet Serv. Appl. 1 (1), 7–18.

Zhang, Z., Wu, C., Cheung, D.W., 2013. A survey on cloud interoperability: taxonomies, standards, and practice. ACM SIGMETRICS Perform. Eval. Rev. 40 (4), 13–22.

Chapter 18

Social and Consumer Informatics

Felix Greaves[1] and Ronen Rozenblum[2]

[1]Imperial College London, London, United Kingdom, [2]Harvard Medical School, Boston, MA, United States

INTRODUCTION

Beyond the increasingly well-developed information systems in hospitals and ambulatory care settings (see Chapters 2 and 3: Inpatient Clinical Information Systems; Outpatient Clinical Information Systems), other forms of technology are emerging that allow healthcare consumers in general and patients in particular to be engaged with managing their health and interacting with the healthcare delivery system in new and innovative ways. These include specific tools for patient engagement, as well as new social media and consumer devices such as the phones in people's pockets. This chapter considers some of these emerging trends.

THE IMPORTANCE OF PATIENT ENGAGEMENT IN CARE DELIVERY AND HEALTH MANAGEMENT

As providers continue to experiment with new healthcare delivery models, from Accountable Care Organizations to Patient Centered Medical Homes, the concepts of patient-centered care and patient engagement have become increasingly important in recent years. This interest reinforces the increasingly important roles that patients and families are assuming as more active and empowered consumers of healthcare services (Rozenblum et al., 2015b). One of the most prominent spotlights on patient-centered care appeared in the 2002 Institute of Medicine (now the National Academy of Medicine) report, "Crossing the Quality Chasm: A New Health System for the 21st Century," which highlighted the importance of incorporating patients' needs and perspectives into care delivery (Institute of Medicine, 2001; Berwick, 2002). Triggered in part by this report, patient-centered care and patient engagement have become a central focus in the national healthcare discussion

Key Advances in Clinical Informatics. DOI: http://dx.doi.org/10.1016/B978-0-12-809523-2.00018-2

within the United States and elsewhere (Institute of Medicine, 2001; Jha et al., 2008; US Government, 2009). As a result, meaningful collaboration with patients and families and their active participation in the care process and decision making are now considered key elements of healthcare quality and delivery.

Although most models of patient-centered care place an emphasis on the importance of the patient's and family's engagement in their care, there is a lack of consistency in how the concepts of patient and family engagement are defined (Gallivan et al., 2012). The term "patient engagement" is sometimes used interchangeably with "patient empowerment," "patient partnership," and "patient activation." While these terms are indeed related, they are not synonymous. Some definitions of patient engagement focus on individuals' behavior relative to their health care, while others focus on the relationship between patients and healthcare providers (Carman et al., 2013). Recently, a group of researchers from four institutions in the United States defined patient and family engagement as "an active partnership between health professionals and patients and families working at every level of the healthcare system to improve health and the quality, safety, and delivery of health care. Arenas for such engagement include, but are not limited to participation in direct care, communication of patient values and goals, and transformation of care processes to promote and protect individual respect and dignity" (Brown et al., 2015). This definition further elaborates that, "Patient and family engagement comprises five core concepts: collaboration; respect and dignity; activation and participation; information sharing; and decision making" (Brown et al., 2015). In this chapter, we have chosen to use the term "patient engagement" to denote a broader concept that includes patient empowerment, patient partnership, and patient activation.

Part of the impetus for supporting patient-centered care and patient engagement initiatives is the growing body of evidence that they often lead to greater patient satisfaction, improved clinical outcomes, health service efficiency, and improved health-related business metrics (Carman et al., 2013; Manary et al., 2012; Glickman et al., 2010; Isaac et al., 2010). In particular, studies have indicated that patient engagement and shared decision making leads to improvement in self-management and treatment adherence (Hibbard et al., 2007, 2013; Mosen et al., 2007; Greene and Hibbard, 2012; Remmers et al., 2009). Other benefits associated with patient engagement have been found in the form of more efficient health services utilization of diagnostic tests, referrals, emergency department visits, and hospital attendance (Greene and Hibbard, 2012; Remmers et al., 2009). Consistent with this idea, higher levels of patient activation have been shown to be correlated with lower predicted costs per head (Hibbard et al., 2013).

Over the last decade, many healthcare organizations have developed approaches to enhance patient engagement in care delivery. Yet, despite these efforts, difficulties in transforming their organizational culture from provider-focused to patient-centric have left many provider organizations short of achieving meaningful patient engagement and high patient experience scores (Jha et al., 2008; Hibbard and Cunningham, 2008; Bates and Wells, 2012; Rozenblum et al., 2015a). There are number of potential challenges that might explain why many healthcare organizations are not achieving this goal, such as change management, consistency in practice, resource management, healthcare professional engagement and buy-in, and staff education and training (Aboumatar et al., 2015). Consequently, healthcare organizations need to develop more effective approaches to engage patients as a strategic priority.

HEALTH INFORMATION TECHNOLOGIES THAT ENABLE AND OPTIMIZE PATIENT ENGAGEMENT

Health information technology (HIT) and consumer e-health tools have become central to promoting patient engagement and empowerment through better communication with providers (Rozenblum et al., 2015b; Grando et al., 2015). Examples of promising patient-facing technology include personal health records, patient portals, mobile health technologies, personal monitoring systems, secure e-mail messaging, Internet-based health information, education and consultation, and social media networking websites. Some tools give patients the opportunity to be more responsible for their care by providing them with the ability to access health information, choose healthcare providers, and manage their health care. Other tools allow patients to communicate directly with their care team, coordinate care across caregivers, and interact with other patients with similar health conditions, creating a broader and more connected healthcare network (see Table 18.1 for examples).

These tools have the potential to transform care into an active collaboration between providers and patients, with the goal of improving standards of care. Enhancing patient engagement has been shown to directly impact patient behavior that promotes positive health outcomes, patient satisfaction, care delivery efficiency, improved quality of care and patient safety as well as reduce costs (Rozenblum et al., 2015b; Wagner et al., 2012; Zhou et al., 2010; Giardina et al., 2014; Tang et al., 2006; Delbanco et al., 2012). Consistent with this notion, policy trends in the United States and elsewhere are promoting the use of HIT as a vital component to improving patient engagement and outcomes (Bitton et al., 2015). For example, the Office of the National Coordinator for Health Information Technology (ONC) in the United States has identified the adoption and meaningful use of HIT by healthcare providers and patients as a key factor in improving the nation's health system, with incentive programs designed to promote consumer access to health data (Ricciardi et al., 2013). A similar role, as Chief Clinical

TABLE 18.1 Patient Engagement Approaches

Channel of Patient Engagement	Enabling Technologies	Examples
Access to health information and patient education	• Electronic health records • Patient-facing online portals • Online health tutorials	NextMD, WebMD, MyHealtheVet, OpenNotes
Patient health monitoring	• Wearable monitor/trackers • Mobile health apps • Telehealth platforms	Omada, Fitbit, Apple Watch, Active blood glucose monitoring, AmericanWell
Communication with providers	• Secure messaging • Video conferencing	TigerText, BlueJeans
Peer to peer counseling	• Social media platforms	PatientsLikeMe
Patient experience feedback	• Online surveys and ratings	RateMDs, Yelp
Online patient scheduling	• Patient scheduling portal	ZocDoc

Information Officer for the National Health Service (NHS) has been created to drive HIT adoption in England.

Motivated by policy initiatives to accelerate adoption and meaningful use of HIT, healthcare organizations have begun to implement, use, and promote e-health tools (Bitton et al., 2015; Wells et al., 2014). However, the limited adoption data available suggest that ongoing patient usage rates of HIT modalities remain low (Bates and Wells, 2012; Ahern et al., 2011). Some of the challenges related to patient adoption of HIT could be related to lack of patient awareness, limited health literacy, lower socioeconomic status, older age, inadequate computer skills, and unmet technical support needs. Each of these factors have been identified and demonstrated in research literature to negatively affect patients' use of HIT (Ahern et al., 2011). Nevertheless, HIT tools that enable patient engagement are likely to continue to grow in importance as their potential is further understood and harnessed by policymakers, providers, and patients alike.

The following sections outline the major IT tools to improve patient engagement now being employed by early adopting providers. While this inventory of approaches and tools is not collectively exhaustive, it does represent the principal and most promising strategies to engage patients and their families in the delivery of healthcare services and the ongoing management of patient health.

PATIENT PORTALS/PERSONAL HEALTH RECORDS

Patient portals are Web-based platforms that allow patients to securely and remotely access their personal health information, bolstering their involvement in the ongoing care process and increasing their sense of ownership in that care (Deering and Baur, 2015; HealthIT.gov., 2016). Many of these portals also allow patients to communicate securely with members of their care team, request refills for prescriptions, schedule clinic appointments, access educational material, and pay medical bills. And although, various types of patient portals are being adopted by healthcare systems (Deering and Baur, 2015; Frost and Sullivan, 2013), the exact impact they have on care delivery remains unclear (Ricciardi et al., 2013; Wells et al., 2014, 2015). More specifically, even though positive associations between patient portals and improved health outcomes have demonstrated promising early results in some studies, there is still a need for additional research relating to their specific impact on health outcomes, utilization, and cost (Goldzweig et al., 2013; Rozenblum et al., 2013; Otte-Trojel et al., 2014; Ammenwerth et al., 2012).

At a policy level, no widely recognized standards exist to govern patient portal access on behalf of others and current meaningful use policy has not yet addressed the critical roles caregivers play in the healthcare system (Bitton et al., 2015; Sarkar and Bates, 2014). Significant legal and ethical issues arise around the best and safest ways to allow family and caregivers to access patient portals on behalf of their loved ones (Brown et al., 2015). As a result, providers need to carefully consider how to optimize the portal they are implementing as well as to clearly define their clinical goals and understand their patients' needs and preferences (Deering and Baur, 2015).

To date, adoption and use of patient portals has largely been limited to the ambulatory setting (Wells et al., 2014, 2015). Increasingly, patients and caregivers are interested in accessing health information and communicating with providers during a hospital stay (Collins et al., 2016; Dalal et al., 2016; Dykes et al., 2013). Preliminary research suggests that these portals should provide a single integrated experience across the inpatient and ambulatory settings. Core functionality includes tools that facilitate communication, personalize the patient, and deliver education to advance safe, coordinated, and dignified patient-centered care (Collins et al., 2016). Further qualitative and quantitative research should evaluate existing portals (e.g., various features and functionalities) and measure the impact on outcomes.

Case 1—OpenNotes

OpenNotes is a national initiative in the United States to give patients access to the notes that doctors, nurses, and other clinicians write after a clinical encounter (OpenNotes, 2016). Through OpenNotes, patients can read their provider-generated notes at home and can readily share those notes with people of their choice.

The current evidence suggests that inviting patients to read their clinicians' notes makes care more efficient, improves communication between patients and providers, increases patients' understanding of their medical issues, and helps patients become more actively involved with their health and health care (Delbanco et al., 2012; Weinert, 2015; Esch et al., 2016). Specifically, an OpenNotes study, conducted in 2010 at three major healthcare systems showed that patients were enthusiastic about being able to read the notes, and that a large majority reported increased medication adherence, while very few participants felt confused, worried, or offended by what they read (Delbanco et al., 2012). Providers experienced very low to moderate impact on their workflow, and all chose to continue to post their notes after the study ended (Delbanco et al., 2012). Since this study, similar findings have been replicated in various settings, leading many major healthcare organizations, like the Veterans Health Administration to adopt note sharing practices with patients (Weinert, 2015; Nazi et al., 2015; Esch et al., 2016; Delbanco et al., 2010; Walker et al., 2015).

TELEHEALTH

Recent advances in telecommunications and healthcare connectivity have converged to create a new telehealth service channel, and leveraging the same technologies that have revolutionized commerce in other industries. Centered on synchronous technologies that link the patient and provider in real time through video consults, telehealth is offering both patients and providers new avenues to access and deliver healthcare services (Dorsey and Topol, 2016; Spivack, 2005). In the wake of substantial and ongoing reform of the healthcare industry across the globe, telehealth now provides a potentially pivotal solution to help meet future patient demands and alleviate acute provider shortages. By efficiently matching patient demand with provider supply in a convenient and cost-effective online network, telehealth promises to alleviate not only issues of accessibility, but also spiraling costs.

Telehealth aims to meet the evolving needs of all three facets of the triple aim in health care—improving experience of care, population health, and reducing cost—and be an increasingly central avenue for providers to manage panels stressed by high-cost, chronic care patients.

While providers of innovative telehealth services, like American Well in the United States and Babylon in the United Kingdom, are assuming expanding roles in the healthcare delivery space, greater awareness of its appropriate role and efficacy will certainly be needed before telehealth is adopted as a mainstream healthcare delivery channel (Dorsey and Topol, 2016; Schoenberg, 2015; Wild et al., 2016). Moreover, the traditional expectation of in-person care delivery must evolve to accommodate this new platform that promises more efficient access of valuable health resources. However, the potential it holds to transform how health care is accessed, provided, and paid for may well be a key driver of the growth of consumer-driven healthcare.

REAL-TIME PATIENT EXPERIENCE FEEDBACK

Over the last decade, healthcare organizations have embarked on several strategies to improve patient experience. Collecting and reporting patient feedback about their experiences has been one of their main priorities. However, despite acknowledging the importance of improving patient's experience, many healthcare organizations have not been able to lift their reported patient experience scores (Jha et al., 2008; Rozenblum et al., 2012).

Furthermore, there is little evidence to show that the collection of patient experience data using traditional paper-based surveys results in significant improvements in care (Asprey et al., 2013; Davies et al., 2008). Organizations and providers often have difficulty responding to current patient survey methods and designing effective interventions to improve patient experience (Davies and Cleary, 2005; Wensing et al., 2003). These difficulties stem from infrequent administration of surveys, low response rates, recall bias, and selection bias. In addition, the logistics of collecting patient experience information is resource-intensive, requiring staff time to distribute surveys and collect and analyze the data. Therefore, innovative approaches are needed to enhance data collection methods of patient experience in order to enable effective improvement of care.

Growing interest now centers on developing HITs for systematic collection, analysis, and reporting of patient feedback about their experiences in real time or near real time (Wofford et al., 2015; Dirocco and Day, 2010). Examples of potential patient-facing technology to collect patient experience data include electronic kiosk surveys, mobile health technologies, patient portals, secure e-mail messaging, and other Internet-based resources. Preliminary studies have found that real-time feedback has the potential to enable healthcare organizations to respond promptly to patients' feedback and make timely improvements to services (Wofford et al., 2015; Dirocco and Day, 2010; NHS Practice Management Network, 2010). In addition, preliminary evidence shows that electronic survey methods receive higher acceptance and have less data error than paper-based surveys (Lee et al., 2007; Velikova et al., 1999). While associations between HITs and improved health outcomes are promising, further study is needed to evaluate these innovative tools to enhance patient experience data collection, dissemination, and improvement processes.

Another important application of HIT that is relatively new in the healthcare services world is patient-reported outcomes (PROs). PROs are defined as "any report of the status of a patient's health condition that comes directly from the patient, without interpretation of the patient's response by a clinician or anyone else" (US Department of Health and Human Services FDA Center for Drug Evaluation and Research et al., 2006). This includes collection of patient-generated data onsite through tablets/kiosks and offsite through patient portals, Internet, and smartphone applications. PROs and PRO measures (or PROMs) that describe the use of PRO aggregated data as performance measures allow both patients and providers to use the data both on an individual patient level as well as on a population level. While successful small-scale implementations have occurred, larger scale ones are still few and there are still significant challenges that need to be overcome, such as attaining high response rates for data reporting from patients and integrating PROs into the provider—patient encounter (Zimlichman, 2015; Hvitfeldt et al., 2009).

SOCIAL MEDIA USAGE FOR DISEASE DETECTION AND PATIENT ENGAGEMENT

Using social media has become increasingly normal behavior for many people. In 2015 more than 65% of US adults had used a social networking site, increasingly rapidly from less than 10% in 2005 (Fig. 18.1).

Social media covers a variety of ways for people to create and share information online and includes social networking (such as Facebook), micro-blogging (such as Twitter), messenger services (such as Whatsapp), and community fora (such as Patientslikeme or Healthunlocked).

People often use social media to share their real-life experiences and opinions. Data from social media have been used to try to answer a number of real-world questions. For example trying to predict the results of an election from the views people express online (Tumasjan et al., 2010) or trying to predict the price of the stock market on the basis of social media sentiment (Bollen et al., 2011).

There are many potential uses for social media to improve and better understand health. For example, digital disease detection approaches allow epidemiologists to map disease outbreaks, spread, and control from

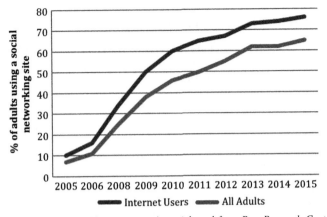

FIGURE 18.1   Social networking use over time. *Adapted from Pew Research Center, 2016b. Social media use over time. http://www.pewinternet.org/data-trend/social-media/social-media-use-all-users/.*

information on the Internet (Brownstein et al., 2009; Eysenbach, 2009). For example, search engine data (Dugas et al., 2012), Twitter data (Chew and Eysenbach, 2010), and Wikipedia usage data (McIver and Brownstein, 2014) have all been used to attempt to predict influenza activity. Twitter has been used to monitor outbreaks of rare diseases such as cholera (Chunara et al., 2012), dengue (Gomide et al., 2011), and *E. coli* (Diaz-Aviles and Stewart, 2012).

Other use cases include looking at social media conversations to understand attitudes toward topics such as vaccination (Love et al., 2013) and healthcare reform (King et al., 2013), or using social media conversations about drug side effects for safety surveillance of drugs.

In addition, social media has been widely adopted by the healthcare industry, with more than 90% of hospitals having a Facebook presence and more than 50% on Twitter in 2014 in the United States (Griffis et al., 2014). More than 99% of hospitals are rated on Yelp. Large, urban and teaching hospitals are all more likely to have a social media presence.

Social media provides people with opportunities to share their own personal experiences about care and for other patients to learn about providers and select where to receive care. For care provider organizations, it provides an opportunity for patient engagement, to understand their services from a patient perspective and a tool for learning and improvement. This has been described variously as the "digital patient experience economy" (Lupton, 2013a) and the "cloud of patient experience" (Greaves et al., 2012b). Use of physician and hospital rating websites is becoming more common—for example, 26% of patients had used them in Germany (Terlutter et al., 2014) and up to 50% in the United States (Rock Health, 2015). Examples of the different sources of data from social media are described in Table 18.2.

TABLE 18.2 Potential Sources of Information From Social Media

Type of Source	Examples	Information That Could Be Used	Advantages	Disadvantages
Rating and feedback websites	RateMDs	Ratings and free text descriptions of healthcare providers and individual clinicians	Comments usually directly relate to care experience	Comparatively low usage, possibility of deliberate gaming
Patient networks, discussion fora and blogs	Patientslikeme	Patients' and carers' shared descriptions of their care and experiences	Authentic voice of the patient, often well used in specific patient communities	May be a selection bias toward particular demographics (with higher socioeconomic status) or interest groups
Micro-blogs	Twitter	Tweets (short messages) directed toward hospitals or care providers	High volume of traffic, often tagged with service they relate to	Short, unstructured messages may contain minimal information about care quality
Social networks	Facebook	Comments left on hospital or friends' pages about care or specific signals of appreciation (e.g., likes)	High membership and usage by the public	Public rarely talks about healthcare on these platforms
				Content may be from employees rather than patients

Source: Adapted from Greaves, F., Ramirez-Cano, D., et al., 2012b. Harnessing the cloud of patient experience: using social media to detect poor quality healthcare. BMJ Qual. Saf. 22 (3), 251–252. http://www.ncbi.nlm.nih.gov/pubmed/23349387 (accessed 09.04.13).

Studies have demonstrated that both ratings and comments left on hospital review websites are correlated with quality (compared with objective measures such as mortality rates and robust patient surveys) in hospitals

(Bardach et al., 2013) and primary care (Greaves et al., 2012c). Even simple markers such as hospitals having more Facebook likes are associated with lower mortality rates and better patient experience scores (Timian et al., 2013). In addition, new technologies such as natural language processing and machine learning (see Chapter 19: Machine Learning in Healthcare) are allowing free-text patient comments to be collected, processed, and used to generate signals about patient care. Using these new approaches, it has been possible to monitor what patients are saying about the care they received on social media such as Twitter, and use this to identify specific comments about particular topics—for example, those about care quality and patient safety (Hawkins et al., 2015).

Case 2—NHS Choices

NHS Choices—the main website of the NHS in England (NHS Choices, 2011)—allows patients to rate and review their experience of care from all NHS providers, as part of a national effort to improve transparency and promote choice within the health system. More than 800,000 ratings have been left on the website, and more than 170,000 reviews of care (Greaves et al., 2014). Public ratings of hospital care were found to correlate with risk-adjusted mortality rates and representative patient experience surveys (Greaves et al., 2012a). Patient comments left online were incorporated into the national intelligent monitoring framework of the national health system quality regulator in 2013 (Care Quality Commission, 2013).

SMARTPHONES, WEARABLES, APPS, AND SELF-MONITORING

Most people now have a cellular or mobile phone (Fig. 18.2). Around 92% of US adults owned a cell phone in 2015, and 68% owned a smartphone (up from 35% in 2011) (Pew Research Center, 2016a). In addition to rising levels of ownership, they have also become substantially more powerful and capable—equipped with processers able to run sophisticated software, and sensors capable of measuring activity, location, and more.

These smartphones allow the use of specific pieces of software—known as applications, or apps for short. There are large numbers of health apps on the market, with industry estimates as high as 165,000 available in 2015, and doubling in the last 2 years (IMS Institute for Healthcare Informatics, 2015). Of these, the majority are related to fitness, lifestyle, and diet, but also with many for the prevention, monitoring, and management of specific medical conditions. Health app usage is increasing, with commercial estimates suggesting 33% of people having used health apps across a range of high-income countries (Accenture Consulting, 2016).

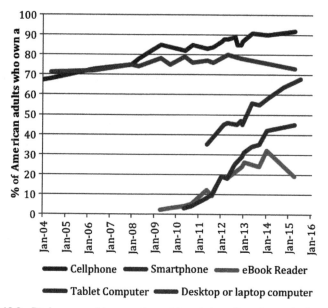

FIGURE 18.2 Device ownership over time. *Adapted from Pew Research Center, 2016a. Device ownership over time. http://www.pewinternet.org/data-trend/mobile/device-ownership/.*

Despite the large numbers of apps on the market, and high levels of downloads, there are issues around engagement with health apps, with studies finding comparatively few that are directly engaging to patients (Singh et al., 2016) and rapid attrition in usage rates (Yardley et al., 2016). There are also regulatory issues around measurement of safety and quality. The effectiveness of apps and digital interventions is often unclear, particularly when compared with traditional and face-to-face delivery of care. Regulatory mechanisms for these products are variable in different jurisdictions, and evaluator frameworks for this type of intervention are still in the developmental stages (Murray et al., 2016).

There is also an increasing consumer market in wearable technology, or "wearables." This covers a variety of specific devices that are worn on the body or on clothing, including fitness trackers, step counters, heart rate monitors, smart watches, and a range of other specific measurement tools. Market estimates in 2016 suggest between 12% (Rock Health, 2015) and 21% (Accenture Consulting, 2016) of people are using some form of wearable technology in the United States.

As a result of the availability and use of phones, apps, and wearables, large numbers of people are using these devices to measure and record their activity (a 2015 consumer survey found 17% of people were actively tracking one or more health-related factor) (Rock Health, 2015). This includes data on regular activity, such as walking or other forms of exercise, together with more topic specific data, including chronic diseases status. This act measuring personal activity has been described as the "quantified self" (Lupton, 2013b).

Emerging from these new devices, apps, and social media are whole new streams of data that have the potential to be important for patients and clinicians in providing care. Some have suggested that the digital footprint left by these various devices—which has been characterized as a "digital phenotype"—provide unique opportunities for data collection, investigation, and intervention (Jain et al., 2015). Although complex, if analyzed appropriately these data provide a rich source of data about people's habits and behaviors, and also capture the majority of people's normal lives, rather than the less frequent events where an individual has formal contact with the health system. They may provide opportunities for collecting PROs, and downstream effectiveness data for the delivery of value-based healthcare.

For these data to be most useful, they may need to be linked to other data sources about people's health. Linking between patients own data on their devices and the formal healthcare record held on EHRs remains difficult (Mandl and Kohane, 2016). There are substantial gaps in interoperability between different hardware, software, and labeling systems (Weber et al., 2014) and data are often held in separate silos between institutions and individuals. Some attempts to create interoperability platforms are now emerging, including the Substitutable Medical Applications Reusable Technologies Health IT apps interface (see Chapter 16: An Apps-Based Information Economy in Healthcare). However, the task remains complex, due to the diversity of systems, and variability of structured and unstructured data, and lack of common identifiers (Fig. 18.3).

With all of the personal data generated from people's devices and social media activity, there are substantial complex and unresolved issues around data ownership, data sharing, and ethics. As data becomes more linked, it becomes more useful, but it also becomes increasingly easy to identify individuals (Weber et al., 2014). Data on the extent to which the public are willing to share their health data, and for what purposes, shows that willingness is dynamic and context-dependent, with high levels of variability between individuals (Weitzman et al., 2012). Commercial data suggest that 80% of people are happy for their data to be shared to receive better care from their doctor, and 60% to help with medical research (Rock Health, 2015), but results often vary with the sample population and methodology. Approaches to mitigate these risks include techniques such as dynamic consent, where individuals are able to specify and change their preferences for data sharing in line with particular purposes (Williams et al., 2015).

While these new personal and social data provide a novel way to understand patient's behaviors and listen to their experiences, there are also potential risks in their interpretation and representativeness. Although use of both personal devices and social media is becoming increasingly normal, there are

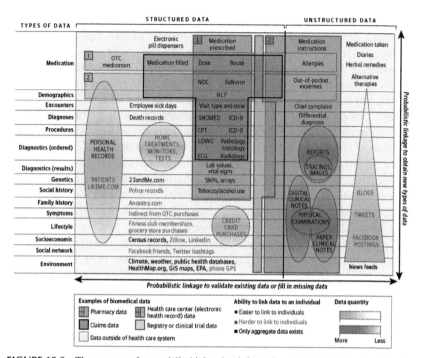

FIGURE 18.3 The tapestry of potentially high-value information sources that may be linked to an individual for use in health care. *From Weber, G., Mandl, K., Kohane, I., 2014. Finding the missing link for big biomedical data. JAMA, 311 (24), 2479–2480.*

still substantial and nonrandom differences in usage, with a lower usage patterns in older age groups, those in certain ethnic groups, and often in those with lower socioeconomic status (Mislove et al., 2011). For example, Facebook and Twitter usage are both higher in groups with higher educational attainment and higher household income (Pew Research Center, 2015) in the United States and UK data shows similar patterns (Fig. 18.4). Smartphone ownership is markedly lower in those over 65 years (Fig. 18.5). Users of these technologies and data streams will need to be aware of this "digital divide" when interpreting data, or risk reinforcing the potential for inequalities that might be generated.

FUTURE OPPORTUNITIES

Beyond the patient-facing technologies described, there are also emerging forms of technology that are likely to interact with health service provision in the near future, often driven by consumer demand and the creativity of new technology companies. For example, rideshare apps are now utilized to transport patients to and from their hospital appointments

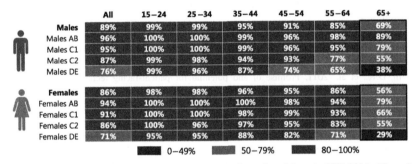

	All	**15–24**	**25–34**	**35–44**	**45–54**	**55–64**	**65+**
Males	20%	37%	27%	28%	17%	10%	5%
Males AB	29%	53%	53%	45%	27%	17%	6%
Males C1	23%	41%	20%	35%	19%	12%	8%
Males C2	17%	34%	19%	16%	19%	4%	4%
Males DE	9%	27%	8%	11%	1%	4%	1%
Females	16%	35%	21%	19%	14%	10%	2%
Females AB	22%	39%	30%	29%	25%	18%	3%
Females C1	17%	38%	20%	20%	17%	7%	4%
Females C2	14%	40%	20%	11%	3%	10%	3%
Females DE	10%	28%	14%	13%	3%	3%	0%

0–24%   25–49%   50–100%

FIGURE 18.4   Percentage accessing Twitter, by gender and social grade (UK, 2016, AB grade is highest). *From Ipsos MORI, 2016. Tech Tracker. https://www.ipsos-mori.com/researchpublications/publications/1866/Just-over-half-of-females-65-have-accessed-the-internet-over-the-last-three-months.aspx.*

	All	**15–24**	**25–34**	**35–44**	**45–54**	**55–64**	**65+**
Males	89%	99%	99%	95%	91%	85%	69%
Males AB	96%	100%	100%	99%	96%	98%	89%
Males C1	95%	100%	100%	99%	96%	95%	79%
Males C2	87%	99%	98%	94%	93%	77%	55%
Males DE	76%	99%	96%	87%	74%	65%	38%
Females	86%	98%	98%	96%	95%	86%	56%
Females AB	94%	100%	100%	100%	98%	94%	79%
Females C1	91%	100%	100%	98%	99%	93%	66%
Females C2	86%	100%	96%	97%	95%	83%	55%
Females DE	71%	95%	95%	88%	82%	71%	29%

0–49%   50–79%   80–100%

FIGURE 18.5   Percentage own a smartphone, by gender and social grade (UK, 2016, AB grade is highest). *From Ipsos MORI, 2016. Tech Tracker. https://www.ipsos-mori.com/researchpublications/publications/1866/Just-over-half-of-females-65-have-accessed-the-internet-over-the-last-three-months.aspx.*

(Powers et al., 2016), and drones are being used to deliver drugs and other therapeutics to patients in a timely manner (Pulver et al., 2016). New artificial intelligence techniques may provide options for personalization and optimization of treatment across a network of increasingly connected devices (Deo, 2015). Smart clothes and accessories (sometimes known as wearables) may allow continuous monitoring of activity and physiology. Together, these and other novel technologies represent new opportunities in patient engagement, for example, personalized real-time treatment and preventative advice on the basis of an individual's own behavior patterns.

CONCLUSIONS

HIT has a central, but still emerging role in engaging patients. Patient-facing technologies such as patient portals, telehealth, and real-time feedback have the potential to substantially change the way people interact with their care, leading to better informed and engaged patients, and ultimately, improved care outcomes. The ubiquity of sophisticated smartphones, interactive social media, and continuous lifestyle and activity tracking all provide unique opportunities to leverage new analytic techniques such as machine learning to make sense of the growing data tsunami for the benefit of individuals, professionals, and the wider health system. However, there are still substantial challenges around effectiveness, access, interoperability, and data sharing, before such opportunities can be fully realized. Our ability to overcome such challenges expediently will inform the rate of adoption for many of these exciting technologies in the health sector.

REFERENCES

Aboumatar, H.J., et al., 2015. Promising practices for achieving patient-centered hospital care. Med. Care 53 (9), 758−767, http://www.ncbi.nlm.nih.gov/pubmed/26147867 (accessed 12.02.17).

Accenture Consulting, 2016. Accenture 2016 Consumer Survey on Patient Engagement. https://www.accenture.com/t20160629T045304__w__/us-en/_acnmedia/PDF-15/Accenture-Patients-Want-A-Heavy-Dose-of-Digital-Research-Global-Report.pdf#zoom = 50.

Ahern, D.K., et al., 2011. Promise of and potential for patient-facing technologies to enable meaningful use. Am. J. Prev. Med. 40 (5), S162−S172.

Ammenwerth, E., Schnell-Inderst, P., Hoerbst, A., 2012. The impact of electronic patient portals on patient care: a systematic review of controlled trials. J. Med. Internet Res. 14 (6), e162.

Asprey, A., et al., 2013. Challenges to the credibility of patient feedback in primary healthcare settings: a qualitative study. Br. J. Gen. Pract. 63 (608), e200−e208.

Bardach, N., et al., 2013. The relationship between commercial website ratings and traditional hospital performance measures in the USA. BMJ Qual. Saf. 22 (3), 194−202, http://qualitysafety.bmj.com/content/22/3/194.abstract (accessed 06.03.13).

Bates, D.W., Wells, S., 2012. Personal health records and health care utilization. JAMA 308 (19), 2034−2036.

Berwick, D.M., 2002. A user's manual for the IOM's "Quality Chasm" report. Health Aff. 21 (3), 80−90.

Bitton, A., Poku, M., Bates, D.W., 2015. Policy context and considerations for patient engagement with health information technology. In: Grando, M., Rozenblum, R., Bates, D. (Eds.), Information Technology for Patient Empowerment in Healthcare, 2015. Walter de Gruyter, Inc., Berlin, pp. 3−22.

Bollen, J., Mao, H., Zeng, X., 2011. Twitter mood predicts the stock market. J. Comput. Sci. 2 (1), 1−8, http://www.sciencedirect.com/science/article/pii/S187775031100007X.

Brown, S.M., et al., 2015. Defining patient and family engagement in the intensive care unit. Am. J. Respir. Crit. Care Med. 191 (3), 358−360.

Brownstein, J.S., Freifeld, C.C., Madoff, L.C., 2009. Digital disease detection—harnessing the web for public health surveillance. N. Engl. J. Med. 360 (21), 2153−2157, http://www.nejm.org/doi/abs/10.1056/NEJMp0900702 (accessed 30.10.16).

Care Quality Commission, 2013. Intelligent Monitoring: NHS Acute Hospitals Indicators and Methodology. London. http://www.cqc.org.uk/sites/default/files/media/documents/20131119_intelligent_monitoring_indicators_and_methodology_v12_for_publication.pdf.

Carman, K.L., et al., 2013. Patient and family engagement: a framework for understanding the elements and developing interventions and policies. Health Aff. 32 (2), 223–231.

Chew, C., Eysenbach, G., 2010. Pandemics in the age of Twitter: content analysis of Tweets during the 2009 H1N1 outbreak. PLoS One 5 (11), e14118, http://www.pubmedcentral.nih.gov/articlerender.fcgi?artid = 2993925&tool = pmcentrez&rendertype = abstract (accessed 07.08.13).

Chunara, R., Andrews, J.R., Brownstein, J.S., 2012. Social and news media enable estimation of epidemiological patterns early in the 2010 Haitian cholera outbreak. Am. J. Trop. Med. Hyg. 86 (1), 39–45, http://www.ncbi.nlm.nih.gov/pubmed/22232449 (accessed 30.10.16).

Collins, S.A., et al., 2016. Acute care patient portals: a qualitative study of stakeholder perspectives on current practices. J. Am. Med. Inform. Assoc.http://www.ncbi.nlm.nih.gov/pubmed/27357830 (accessed 01.11.16).

Dalal, A.K., et al., 2016. A web-based, patient-centered toolkit to engage patients and caregivers in the acute care setting: a preliminary evaluation. J. Am. Med. Inform. Assoc. 23 (1), 80–87.

Davies, E., Cleary, P.D., 2005. Hearing the patient's voice? Factors affecting the use of patient survey data in quality improvement. Qual. Saf. Health Care 14 (6), 428–432, http://quality-safety.bmj.com/content/14/6/428.abstract.

Davies, E., et al., 2008. Evaluating the use of a modified CAHPS® survey to support improvements in patient-centred care: lessons from a quality improvement collaborative. Health Expect. 11 (2), 160–176.

Deering, M. & Baur, C., 2015. Patient portals can enable provider-patient collaboration and person-centered care. In: Grando, M., Rozenblum, R., Bates, D., (Eds.), Information Technology for Patient Empowerment in Healthcare. Berlin, pp. 93–111.

Delbanco, T., et al., 2010. Open notes: doctors and patients signing on. Ann. Intern. Med. 153 (2), 121–125.

Delbanco, T., et al., 2012. Inviting patients to read their doctors' notes: a quasi-experimental study and a look ahead. Ann. Intern. Med. 157 (7), 461–470.

Deo, R.C., 2015. Machine learning in medicine. Circulation 132 (20), 1920–1930, http://circ.ahajournals.org/content/132/20/1920.abstract.

Diaz-Aviles, E. & Stewart, A., 2012. Tracking Twitter for epidemic intelligence: case study: EHEC/HUS outbreak in Germany, 2011. In: Proceedings of the 4th Annual ACM Web Science Conference. WebSci '12. ACM, New York, NY, pp. 82–85. http://doi.acm.org/10.1145/2380718.2380730.

Dirocco, D.N., Day, S.C., 2010. Obtaining patient feedback at point of service using electronic kiosks. Am. J. Manag. Care 17 (7), e270–e276.

Dorsey, E.R. & Topol, E.J., 2016. State of telehealth E. W. Campion, ed. N. Engl. J. Med. 375 (2), 154–161. http://www.nejm.org/doi/10.1056/NEJMra1601705 (accessed 01.11.16).

Dugas, A.F., et al., 2012. Google flu trends: correlation with emergency department influenza rates and crowding metrics. Clin. Infect. Dis.http://cid.oxfordjournals.org/content/early/2012/01/02/cid.cir883.abstract.

Dykes, P.C., et al., 2013. Building and testing a patient-centric electronic bedside communication center. J. Gerontol. Nurs. 39 (1), 15–19.

Esch, T., et al., 2016. Engaging patients through open notes: an evaluation using mixed methods. BMJ Open 6 (1), e010034.

Eysenbach, G., 2009. Infodemiology and infoveillance: framework for an emerging set of public health informatics methods to analyze search, communication and publication behavior on the internet. J. Med. Internet Res. 11 (1), e11, http://www.jmir.org/2009/1/e11.

Frost and Sullivan, 2013. Market Disruption Imminent as Hospitals and Physicians Aggressively Adopt Patient Portal Technology. http://www.frost.com/prod/servlet/press-release.pag?docid = 285477570 (accessed 30.10.16).

Gallivan, J., et al., 2012. The many faces of patient engagement. J. Particip. Med. 4, e32.

Giardina, T.D., et al., 2014. Patient access to medical records and healthcare outcomes: a systematic review. J. Am. Med. Inform. Assoc. 21 (4), 737−741.

Glickman, S.W., et al., 2010. Patient satisfaction and its relationship with clinical quality and inpatient mortality in acute myocardial infarction. Circ. Cardiovasc. Qual. Outcomes 3 (2), 188−195, http://www.ncbi.nlm.nih.gov/pubmed/20179265 (accessed 29.01.14).

Goldzweig, C.L., et al., 2013. Electronic patient portals: evidence on health outcomes, satisfaction, efficiency, and attitudes: a systematic review. Ann. Intern. Med. 159 (10), 677−687.

Gomide, J. et al., 2011. Dengue surveillance based on a computational model of spatio-temporal locality of Twitter. Proceedings of the ACM WebSci'11. 14−17 June 2011, Koblenz, Germany, pp. 1−8. http://journal.webscience.org/429/ (accessed 30.10.16).

Grando, M., Rozenblum, R., Bates, D. (Eds.), 2015. Information Technology for Patient Empowerment in Healthcare. Walter de Gruyter, Inc., Berlin.

Greaves, F., Pape, U.J., King, D., et al., 2012a. Associations between web-based patient ratings and objective measures of hospital quality. Arch. Intern. Med. 172 (5), 435−436, http://www.ncbi.nlm.nih.gov/pubmed/22331980 (accessed 04.03.13).

Greaves, F., Ramirez-Cano, D., et al., 2012b. Harnessing the cloud of patient experience: using social media to detect poor quality healthcare. BMJ Qual. Saf. 22 (3), 251−252, http://www.ncbi.nlm.nih.gov/pubmed/23349387 (accessed 09.04.13).

Greaves, F., Pape, U.J., Lee, H., et al., 2012c. Patients' ratings of family physician practices on the internet: usage and associations with conventional measures of quality in the English National Health Service. J. Med. Internet Res. 14 (5), e146, http://www.jmir.org/2012/5/e146/ (accessed 18.10.12).

Greaves, F., Millett, C., Nuki, P., 2014. England's Experience incorporating "anecdotal" reports from consumers into their national reporting system: lessons for the United States of what to do or not to do? Med. Care Res. Rev. 71 (5 Suppl), 65S−80S, http://mcr.sagepub.com/content/71/5_suppl/65S.abstract.

Greene, J., Hibbard, J.H., 2012. Why does patient activation matter? An examination of the relationships between patient activation and health-related outcomes. J. Gen. Intern. Med. 27 (5), 520−526, http://dx.doi.org/10.1007/s11606-011-1931-2.

Griffis, H.M., et al., 2014. Use of social media across US hospitals: descriptive analysis of adoption and utilization. J. Med. Internet Res. 16 (11), e264, http://www.pubmedcentral.nih.gov/articlerender.fcgi?artid=4260061&tool=pmcentrez&rendertype=abstract (accessed 13.02.15).

Hawkins, J.B., et al., 2015. Measuring patient-perceived quality of care in US hospitals using Twitter. BMJ Qual. Saf.http://qualitysafety.bmj.com/content/early/2015/10/07/bmjqs-2015-004309.abstract.

HealthIT.gov., 2016. What is a patient portal? http://www.healthit.gov/providers-professionals/faqs/what-patient-portal (accessed 30.10.16).

Hibbard, J.H., Cunningham, P.J., 2008. How engaged are consumers in their health and health care, and why does it matter. Res. Brief 8, 1−9.

Hibbard, J.H., et al., 2007. Do increases in patient activation result in improved self-management behaviors?. Health Serv. Res. 42 (4), 1443−1463.

Hibbard, J.H., Greene, J., Overton, V., 2013. Patients with lower activation associated with higher costs; delivery systems should know their patients' "scores". Health Aff. 32 (2), 216−222.

Hvitfeldt, H., et al., 2009. Feed forward systems for patient participation and provider support. Qual. Manag. Health Care 18 (4), 247–256, http://www.ncbi.nlm.nih.gov/pubmed/19851232 (accessed 12.02.17).

IMS Institute for Healthcare Informatics, 2015. Patient Adoption of mHealth (September). www.theimsinstitute.org.

Institute of Medicine, 2001. Crossing the Quality Chasm: A New Health System for the 21st Century. National Academies Press, Washington, DC.

Ipsos MORI, 2016. Tech Tracker. https://www.ipsos-mori.com/researchpublications/publications/1866/Just-over-half-of-females-65-have-accessed-the-internet-over-the-last-three-months.aspx.

Isaac, T., et al., 2010. The relationship between patients' perception of care and measures of hospital quality and safety. Health Serv. Res. 45 (4), 1024–1040. Available from: http://dx.doi.org/10.1111/j.1475-6773.2010.01122.x.

Jain, S.H., et al., 2015. The digital phenotype. Nat. Biotechnol. 33 (5), 462–463.

Jha, A.K., et al., 2008. Patients' perception of hospital care in the United States. N. Engl. J. Med. 359 (18), 1921–1931.

King, D., et al., 2013. Twitter and the health reforms in the English National Health Service. Health Policy 110 (2–3), 291–297, http://www.ncbi.nlm.nih.gov/pubmed/23489388 (accessed 29.03.13).

Lee, S.J., Kavanaugh, A., Lenert, L., 2007. Electronic and computer-generated patient questionnaires in standard care. Best. Pract. Res. Clin. Rheumatol. 21 (4), 637–647.

Love, B., et al., 2013. Twitter as a source of vaccination information: content drivers and what they are saying. Am. J. Infect. Control 41 (6), 568–570. Available from: http://dx.doi.org/10.1016/j.ajic.2012.10.016 (accessed 03.06.13).

Lupton, D., 2013a. The commodification of patient opinion: the digital patient experience economy in the age of big data. Sociol. Health Illn. 36 (6), 856–869, http://ses.library.usyd.edu.au//bitstream/2123/9063/1/Commodification of Patient Opinion - Working Paper No. 3.pdf (accessed 02.06.13).

Lupton, D., 2013b. The digitally engaged patient: self-monitoring and self-care in the digital health era. Soc. Theory Health 11 (3), 256–270, http://www.palgrave-journals.com/sth/journal/v11/n3/abs/sth201310a.html (accessed 10.02.14).

Manary, M.P., et al., 2012. The patient experience and health outcomes. N. Engl. J. Med. 368 (3), 201–203. Available from: http://dx.doi.org/10.1056/NEJMp1211775.

Mandl, K.D., Kohane, I.S., 2016. Time for a patient-driven health information economy?. N. Engl. J. Med. 374 (3), 205–208. Available from: http://dx.doi.org/10.1056/NEJMp1512142.

McIver, D.J., Brownstein, J.S., 2014. Wikipedia usage estimates prevalence of influenza-like illness in the United States in near real-time. PLoS Comput. Biol. 10 (4), e1003581, http://www.ncbi.nlm.nih.gov/pubmed/24743682 (accessed 30.10.16).

Mislove, A., et al., 2011. Understanding the demographics of twitter users. Proceedings of the Fifth International Conference on Weblogs and Social Media. The AAAI Press, Barcelona, Spain.

Mosen, D.M., et al., 2007. Is patient activation associated with outcomes of care for adults with chronic conditions? J. Ambul. Care Manage. 30 (1), 21–29.

Murray, E., et al., 2016. Evaluating digital health interventions: key questions and approaches. Am. J. Prev. Med. 51 (5), 843–851. Available from: http://dx.doi.org/10.1016/j.amepre.2016.06.008.

Nazi, K.M., et al., 2015. VA OpenNotes: exploring the experiences of early patient adopters with access to clinical notes. J. Am. Med. Inform. Assoc. 22 (2), 380–389.

NHS Choices, 2011. NHS Choices. http://www.nhs.uk/.

NHS Practice Management Network, 2010. A best practice guide to using real-time patient feedback. http://www.practicemanagement.org.uk/uploads/best_practice_guide_to_using_real-time_patient_feedback,_final_version_august_2010.pdf (accessed 01.11.16).

OpenNotes, 2016. What is OpenNotes? http://www.opennotes.org/about-opennotes/.

Otte-Trojel, T., et al., 2014. How outcomes are achieved through patient portals: a realist review. J. Am. Med. Inform. Assoc. 21 (4), 751–757.

Pew Research Center, 2015. The demographics of social media users. http://www.pewinternet.org/2015/08/19/the-demographics-of-social-media-users/.

Pew Research Center, 2016a. Device ownership over time. http://www.pewinternet.org/data-trend/mobile/device-ownership/.

Pew Research Center, 2016b. Social media use over time. http://www.pewinternet.org/data-trend/social-media/social-media-use-all-users/.

Powers, B.W., Rinefort, S., Jain, S.H., 2016. Nonemergency medical transportation—delivering care in the era of Lyft and Uber. JAMA 316 (9), 921, http://jama.jamanetwork.com/article.aspx?doi=10.1001/jama.2016.9970 (accessed 28.10.16).

Pulver, A., Wei, R., Mann, C., 2016. Locating AED enabled medical drones to enhance cardiac arrest response times. Prehosp. Emerg. Care 20 (3), 378–389. Available from: http://dx.doi.org/10.3109/10903127.2015.1115932.

Remmers, C., et al., 2009. Is patient activation associated with future health outcomes and healthcare utilization among patients with diabetes? J. Ambul. Care Manage. 32 (4), 320–327.

Ricciardi, L., et al., 2013. A national action plan to support consumer engagement via e-health. Health Aff. 32 (2), 376–384.

Rock Health, 2015. Digital health consumer adoption: 2015. https://rockhealth.com/reports/digital-health-consumer-adoption-2015/.

Rozenblum, R., et al., 2012. The patient satisfaction chasm: the gap between hospital management and frontline clinicians. BMJ Qual. Saf. 22 (3), 242–250, http://www.ncbi.nlm.nih.gov/pubmed/23178858 (accessed 28.04.15).

Rozenblum, R., et al., 2013. The impact of medical informatics on patient satisfaction: a USA-based literature review. Int. J. Med. Inform. 82 (3), 141–158.

Rozenblum, R., et al., 2015a. Clinicians' perspectives on patient satisfaction in adult congenital heart disease clinics—a dimension of health care quality whose time has come. Congenit. Heart Dis. 10 (2), 128–136.

Rozenblum, R., et al., 2015b. Patient-centered healthcare, patient engagement and health information technology: the perfect storm. In: Grando, M., Rozenblum, R., Bates, D. (Eds.), Information Technology for Patient Empowerment in Healthcare. Walter de Gruyter, Inc., Berlin, pp. 3–22.

Sarkar, U., Bates, D.W., 2014. Care partners and online patient portals. JAMA 311 (4), 357–358.

Schoenberg, R., 2015. Telehealth: connecting patients with providers in the 21st century. In: Grando, M., Rozenblum, R., Bates, D. (Eds.), Information Technology for Patient Empowerment in Healthcare. Berlin, pp. 125–140.

Singh, K., et al., 2016. Developing a framework for evaluating the patient engagement, quality, and safety of mobile health applications. Issue Brief (Commonw. Fund) 5 (1863), 1–12.

Spivack, R., 2005. Innovation in Telehealth and a Role for the Government. Future of Intelligent and Extelligent Health Environment. IOS Press.

Tang, P.C., et al., 2006. Personal health records: definitions, benefits, and strategies for overcoming barriers to adoption. J. Am. Med. Inform. Assoc. 13 (2), 121–126.

Terlutter, R., Bidmon, S., Röttl, J., 2014. Who uses physician-rating websites? Differences in sociodemographic variables, psychographic variables, and health status of users and nonusers of physician-rating websites. J. Med. Internet Res. 16 (3), e97, http://www.ncbi.nlm.nih.gov/pubmed/24686918 (accessed 30.10.16).

Timian, A., et al., 2013. Do patients "like" good care? Measuring hospital quality via Facebook. Am. J. Med. Qual. 28 (5), 374–382, http://ajm.sagepub.com/content/early/2013/01/31/1062860612474839.full#ref-28 (accessed 23.05.13).

Tumasjan, A., et al., 2010. Predicting elections with Twitter: what 140 characters reveal about political sentiment. Proceedings of the 4th International AAAI Conference on Weblogs and Social Media. AAAI Press, Washington, DC.

US Department of Health and Human Services FDA Center for Drug Evaluation and Research, US Department of Health and Human Services FDA Center for Biologics Evaluation and Research & US Department of Health and Human Services FDA Center for Devices and Radiological Health, 2006. Guidance for industry: patient-reported outcome measures: use in medical product development to support labeling claims: draft guidance. Health Qual. Life Outcomes. 4 (1), 79, http://www.ncbi.nlm.nih.gov/pubmed/17034633 (accessed 12.02.17).

US Government, 2009. The Patient Protection and Affordable Care Act.

Velikova, G., et al., 1999. Automated collection of quality-of-life data: a comparison of paper and computer touch-screen questionnaires. J. Clin. Oncol. 17 (3), 998–1007, http://www.ncbi.nlm.nih.gov/pubmed/10071295 (accessed 01.11.16).

Wagner, P.J., et al., 2012. Personal health records and hypertension control: a randomized trial. J. Am. Med. Inform. Assoc. 19 (4), 626–634.

Walker, J., Meltsner, M., Delbanco, T., 2015. US experience with doctors and patients sharing clinical notes. BMJ 350, g7785.

Weber, G., Mandl, K., Kohane, I., 2014. Finding the missing link for big biomedical data. JAMA 311 (24), 2479–2480.

Weinert, C., 2015. Giving doctors' daily progress notes to hospitalized patients and families to improve patient experience. Am. J. Med. Qual., p.1062860615610424.

Weitzman, E.R., et al., 2012. Willingness to share personal health record data for care improvement and public health: a survey of experienced personal health record users. BMC Med. Inform. Decis. Mak. 12 (1), 39.

Wells, S., et al., 2014. Personal health records for patients with chronic disease: a major opportunity. Appl. Clin. Inform. 5 (2), 416–429.

Wells, S., et al., 2015. Organizational strategies for promoting patient and provider uptake of personal health records. J. Am. Med. Inform. Assoc. 22 (1), 213–222.

Wensing, M., Vingerhoets, E., Grol, R., 2003. Feedback based on patient evaluations: a tool for quality improvement? Patient Educ. Couns. 51 (2), 149–153.

Wild, S.H. et al., 2016. Supported telemonitoring and glycemic control in people with type 2 diabetes: the Telescot diabetes pragmatic multicenter randomized controlled trial. PLoS Med. 13 (7), e1002098. http://dx.plos.org/10.1371/journal.pmed.1002098 (accessed 12.02.17).

Williams, H., et al., 2015. Dynamic consent: a possible solution to improve patient confidence and trust in how electronic patient records are used in medical research. JMIR Med. Inform. 3 (1), e3, http://medinform.jmir.org/2015/1/e3/.

Wofford, J.L., et al., 2015. Real-time patient survey data during routine clinical activities for rapid-cycle quality improvement. JMIR Med. Inform. 3 (1).

Yardley, L., et al., 2016. Understanding and promoting effective engagement with digital behavior change interventions. Am. J. Prev. Med. 51 (5), 833–842, http://linkinghub.elsevier.com/retrieve/pii/S0749379716302434.

Zhou, Y.Y., et al., 2010. Improved quality at Kaiser Permanente through e-mail between physicians and patients. Health Aff. 29 (7), 1370–1375.

Zimlichman, E., 2015. Using patient-reported outcomes to drive patient-centered care. In: Grando, M., Rozenblum, R., Bates, D. (Eds.), Information Technology for Patient Empowerment in Healthcare. Walter de Gruyter, Inc., Berlin, pp. 241–256.

Chapter 19

Machine Learning in Healthcare

Alison Callahan and Nigam H. Shah
Stanford University, Stanford, CA, United States

INTRODUCTION

Studying patient populations to identify causes, risk factors, effective treatments, and subtypes of disease has long been the purview of epidemiology (Dicker et al., 2006). Epidemiological methods such as randomized controlled trials and case-control studies are the cornerstones of evidence-based medicine. However, such methods are time-consuming and expensive, may not be free of the biases they are designed to combat (Delgado-Rodríguez and Llorca, 2004), and their results may not be applicable to real-world patient populations (Weng et al., 2014). Such studies are also difficult to execute over large patient populations because of restrictive inclusion/exclusion criteria and the operational challenges of running large prospective studies that follow patients over long periods of time. Internationally, the adoption of electronic health records (EHRs) is increasing due to strategies and agencies that incentivize their use such as the Health Information for Economic and Clinical Health Act in the United States (Maxson et al., 2010), the National Agency for Health IT in Denmark, and the National eHealth Authority in India (Mossialos et al., 2016). As a result, methods that leverage EHRs to answer questions tackled by epidemiologists (Casey et al., 2016) and to increase precision in healthcare delivery are now commonplace (Parikh et al., 2016).

Data analysis approaches broadly fall into the following categories: descriptive, exploratory, inferential, predictive, and causal (Leek and Peng, 2015). A descriptive analysis reports summaries of data without interpretation and an exploratory analysis identifies associations between variables in a dataset. An inferential study quantifies the degree to which an observed association in a population will hold outside the dataset from which it was derived, and a predictive study attempts to quantify the probability of an outcome at the level of an individual. Finally, a causal analysis determines how changes in one variable affect another. It is crucial to define the type of

Key Advances in Clinical Informatics. DOI: http://dx.doi.org/10.1016/B978-0-12-809523-2.00019-4

question being asked in a given study to determine the type of data analysis that is appropriate to use in answering the question.

Inferential analyses are widely used in clinical research. The primary goal of such studies is to quantify how a change in one or more variables affects an observed outcome at a population level (Leek and Peng, 2015). Such analysis (referred to as the culture of data modeling by Breiman, 2001) assumes the existence of a model that produces observed outcomes from input variables and evaluates a model's quality by measuring its goodness of fit to existing data. A model that has a good fit to existing data is considered adequately to define the process by which data are generated.

Predictive analyses, on the other hand, aim to predict outcomes for individuals (Leek and Peng, 2015) by constructing a statistical model from observed data and using this model to generate a prediction for an individual based on their unique features. Predictive modeling is a type of algorithmic modeling (Breiman, 2001), which considers the process by which data are generated to be unknown (and perhaps unknowable). Such modeling approaches measure performance by metrics such as precision, recall, and calibration (Table 19.1), which quantify different notions of the frequency with which a model's predictions are correct.

Machine learning is the process of learning a good enough statistical model using observed data to predict outcomes or categorize observations in future data. Specifically, supervised machine learning methods train a model using observations on samples where the categories or predicted value of the outcome of interest are already known (a gold standard). The resulting model—which is often a penalized regression of some form—is typically applied to new samples to categorize or predict values of the outcome for previously unseen observations, and its performance evaluated by comparing predicted values to actual values for a set of test samples. Therefore, machine learning "lives" in the world of algorithmic modeling and should be evaluated as such. Regression models developed using machine learning methods cannot and should not be evaluated using criteria from the world of data modeling (Breiman, 2001). To do so would produce inaccurate assessments of a model's performance for its intended task, potentially misleading users into incorrect interpretation of the model's output.

EHRs provide access to a large number and variety of variables that enable high-quality classification and prediction (Kennedy et al., 2013), while machine learning offers the methods to handle the large volumes of high-dimensional data that are typical in a healthcare setting. As a result, the application of machine learning to EHR data analysis is at the forefront of modern clinical informatics (Goldstein et al., 2016), fueling advances in both the science and practice of medicine. In the following sections, we present an overview of how machine learning has been applied in clinical settings and summarize the advantages it offers over traditional analysis methods and caveats when using machine learning in real-world settings. We describe the

TABLE 19.1 Common Performance Measures Used to Evaluate Machine Learning Models

Performance Measure	Definition	Formula or Plot
Accuracy	The number of correct classifications made by a model (true positives and true negatives) divided by the total number of predictions made	$A = \frac{TP + TN}{TP + FP + TN + FN}$
Calibration	A measure of how closely predicted probabilities for an outcome match the observed outcome in test data, e.g., the Brier score	$\text{Brier score} = \frac{1}{N}\sum_{i=1}^{N}(p_i - o_i)$
Discrimination	A measure of how well a model discriminates between randomly selected true positive cases and true negative cases, usually measured as the area under the receiver operator curve (AUC)	
Negative predictive value	The total number of correct negative classifications made (true negatives) divided by the total number of negative classifications made (true negatives and false negatives)	$NPV = \frac{TN}{TN + FN}$
Precision (also called positive predictive value)	The total number of correct positive classifications made (true positives) divided by the total number of positive classifications made (true positives and false positives)	$P \text{ or } PPV = \frac{TP}{TP + FP}$
Recall (also called sensitivity or the true positive rate)	The total number of correct positive classifications made (true positives) divided by the number of positive class members in the data (true positives and false negatives)	$R = \frac{TP}{TP + FN}$

(Continued)

TABLE 19.1 (Continued)

Performance Measure	Definition	Formula or Plot
Specificity (also called the true negative rate)	The total number of correct negative classifications made (true negatives) divided by the number of negative class members in the data (true negatives and false positives)	$S = \frac{TN}{TN + FP}$

For a binary classifier that classifies data points to be a member of either a positive class or a negative class, the following definitions are used in the measure formulas: True positive (TP)— a correct classification that a data point is a member of the positive class. True negative (TN)—a correct classification that a data point is a member of the negative class. False positive (FP)—an incorrect classification that a data point is a member of the positive class. False negative (FN)—an incorrect classification that a data point is a member of the negative class. For a classifier that predicts the probability of an outcome, p_i is the predicted probability of the outcome for data point i, and o_i is the whether the outcome was observed for that data point or not.

methodological and operational challenges of using machine learning in research and practice. Lastly, we offer our perspective on opportunities for machine learning in medicine and applications that have the highest potential for impacting health and healthcare delivery.

FROM SCORING SYSTEMS TO STATISTICAL MODELS

In medicine, the use of scoring systems to categorize patients into different risk strata is quite common. For example, the Charlson Comorbidity Index (CCI) (Charlson et al., 1987) quantifies an individual's burden of disease and corresponding 1-year mortality risk. The widely used APACHE score (Haddad et al., 2008) quantifies the severity of disease for patients admitted to intensive care units. The Apgar score summarizes physiological measures as an indicator of infant morbidity (Casey et al., 2001). A recent entrant to risk scoring systems is the Rothman Index (Finlay et al., 2013), which is a score that quantifies risk of death based on vital signs, nursing assessments, and laboratory results. Disease-specific scoring systems have also been developed, such as the Dutch Lipid Clinic Network (DLCN) criteria for diagnosing "unlikely," "possible," "probable," or "definite" familial hypercholesterolemia (FH) (European Association for Cardiovascular Prevention and Rehabilitation et al., 2011).

Scoring systems have the advantage of being easily interpretable and are usually straightforward to administer, but may not perform well at accurately reflecting the risk for individuals. For example, the DLCN criteria for FH have specificity ranging from 5.9% to 89.4% and sensitivity ranging from 41.5% to 99.3% for predicting the results of a genetic test for the disease

(Damgaard et al., 2005). However, due to the association between age and several features of "definite" FH, the scores are more likely to misclassify younger individuals (Besseling et al., 2016). Similarly, the APACHE scores were found to have excellent discrimination, but poor calibration (Haddad et al., 2008) meaning that while the APACHE score could separate the high-risk group from the low-risk group, the outcome probabilities for specific individuals were inaccurate. This has implications for deciding how to use such scores in practice. For example, the excellent discrimination of the APACHE score makes it highly useful to decide whether to transfer a patient out of an ICU or not. However, it is inappropriate to use the score to quantify that an individual patient will need to be resuscitated with a 90% probability.

Given such limitations of traditional scoring systems, approaches that use machine learning to build well-calibrated statistical models that correctly classify individuals and provide accurate estimates of risk are growing in popularity. We have, for example, recently developed a statistical model—a penalized logistic regression model—to classify patients as "unlikely," "possible," "probable," or "definite" FH based on the same criteria as the DLCN, but learning the variable weights from EHR data. Our model augments the DLCN criteria with variables that quantify health system utilization and summary statistics of cholesterol levels and achieves an AUC of 95% with 82% precision and 86% recall. Other approaches have yielded comparable results for predicting FH (Weng et al., 2015; Besseling et al., 2016) and have significant potential cost savings by reducing unnecessary genetic testing and simultaneously ensuring that true cases are not missed.

Using a similar approach, researchers in Canada updated and reevaluated the CCI with more recent administrative data and larger more diverse patient populations, and without relying on chart review (Quan et al., 2011). Rather than replacing the CCI altogether with a statistical model, they used statistical models to assign new weights to comorbidities used in the CCI. Doing so resulted in a decrease in the number of conditions included in the Index from 17 to 12. This simpler index achieved comparable performance to the CCI, based on evaluation with data from six economically developed countries. Other risk models learned using EHR data include PhysiScore (Saria et al., 2010), a system that predicts preterm infant morbidity using similar physiological features as the Apgar score, and TrewScore (Henry et al., 2015), an early warning system for sepsis. Both achieved improved performance compared to traditional point-based scoring systems. Numerous other examples exist for models that predict sepsis (Gultepe et al., 2014), delayed wound healing (Jung et al., 2016), risk of being depressed (Huang et al., 2014), and cardiovascular adverse events (Ross et al., 2016).

These studies demonstrate that using machine learning approaches to do the "heavy lifting" of deriving appropriate variable weights (and in many cases, performing automated variable selection) can produce a model that outperforms the corresponding expert-derived scoring system and can update existing scoring models as data change over time (Jung and Shah, 2015). Proper use of machine learning methods that can handle large numbers of sparsely measured variables and that can identify variables with the most significant effect on model performance are crucial to the success of models built from heterogeneous and often messy EHR data.

CHALLENGES IN BUILDING AND DEPLOYING EHR-BASED MODELS

The challenges inherent in retrospectively analyzing EHR data collected for administrative and clinical purposes to do research have been well characterized (Hersh et al., 2013). They include biases due to using data from individuals generated only when they visit their healthcare provider (and are therefore likely ill), incomplete records (and the related issue of missing data violating the missing-at-random assumption), loss to follow-up, the modification of records for billing purposes, and potential errors introduced to a patient's record as they move through a healthcare system such as inaccurate coding or data entry (Table 19.2). Statistical methods commonly employed to address these difficulties include imputation to fill in missing data and propensity score based matching, weighting or stratification to select patient samples as controls (Schneeweiss et al., 2009; Brookhart et al., 2010; Toh et al., 2011; Patorno et al., 2013). These challenges and biases that affect retrospective studies also impact the use of machine learning to build models based on EHR data. The effectiveness of methods that minimize biases in retrospective studies when applied to high-dimensional EHR data is an area of active research.

A recent systematic review of published research on developing predictive models using EHR data (Goldstein et al., 2016) found that most predictive models developed for healthcare applications did not take advantage of the wealth of EHR data available. For example, only about half of the 60 studies that predicted a clinical outcome used information beyond diagnosis codes to define the outcome, and only eight studies used longitudinal measurements. The task of accurately ascertaining whether a specific outcome occurred (or not) is called electronic phenotyping (Richesson et al., 2016) and is the most important step in any statistical modeling exercise done with EHR data. A related study (Wei et al., 2016) found that in 4 out of 10 medical conditions examined, the medical records of less than 10% of sampled patients known to have the condition contained the International Classification of Diseases diagnosis codes necessary to identify the condition. For the remainder, data derived from clinical notes or medication records (just two of many possible EHR-derived sources) were required to determine the diagnosis. In light of the importance of accurate phenotype determination—both for the outcome as well as for the presence of

TABLE 19.2 Types of Errors in EHR and Claims Data, and Possible Causes

Type of Error	Definition	Possible Causes
Missing data	A patient's record is incomplete and has gaps in time when they received healthcare but it was not recorded	— Patient does not seek care because they lack healthcare insurance coverage — Loss to follow-up—a patient leaves their insurer or stops seeing their healthcare provider — Lag time or mistake in insurance claim processing — Error when populating insurance claims database — Incomplete record linkage—records belonging to a single patient are not linked
Missing data not missing at random	The reason data is missing from a patient's record is related to unobserved environmental or patient-specific factors	— Patient does not seek healthcare because they are very sick or healthy — Healthcare provider misdiagnoses a patient, makes an error when writing a note for a patient visit, or does not include all information in their note — Hospital/clinic staff miscodes diagnoses and/or medical procedures, or does not file insurance claim
Outcome or exposure misclassification	Incorrect labeling of patients who have a disease or do not, or who had an exposure (e.g., a medical treatment, a risk factor for a disease) or did not	

comorbidities—several research efforts are focused on employing machine learning methods for the purpose of phenotyping itself (Agarwal et al., 2016; Halpern et al., 2016).

Another significant issue faced by researchers using machine learning to derive insights from EHR data is that of external validity or generalizability (Iezzoni, 1999)—the performance of learned models at sites other than the that which sourced the data used for training. One prevailing opinion is that models not validated on data from an external site (or sites) are not useful, and that models lacking external validity have failed in their task. Such thinking affects model design decisions and evaluation; for example, a model with fewer variables is often considered more generalizable, and therefore more useable, than one with more variables. We posit that one of the benefits of machine learning is that a model can be trained with data from any local site and evaluated using data from that site itself. Therefore, if a model achieves sufficient performance at a local site then it has achieved its purpose. So the discussion should be about sharing the model building workflow to retrain a model at a new site, rather than about "external validity." Indeed, learning a model using site-specific data and updating its parameters as a dataset changes over time will produce a model that is best suited for prediction at that site and with that data (Lee and Maslove, 2015). The success of such an approach across multiple sites (Peck et al., 2013) is evidence that the model building approach is valid.

The success of building and deploying a statistical model at a new site rests on the availability and consistent representation of data across sites. The Observational Health Data Sciences and Informatics (OHDSI) network is an international collaboration of observational health researchers that has developed a Common Data Model (CDM) for EHR data currently used to structure hundreds of millions of patient records from more than 50 sites in 12 countries. A query to a CDM-compliant database to construct a cohort at one site can be deployed at any other OHDSI site, without modification. Such portability has already facilitated large-scale research into treatment pathways (Hripcsak et al., 2016) and presents the opportunity to develop statistical models that are portable across sites. Researchers can share the workflow used to construct and train a model, such that it can be retrained using data from a new site and deployed locally.

In addition to technical challenges around data models, external validity, and the need to use multiple data types, operational challenges also exist for the construction and deployment of a statistical model in a clinical setting. These challenges include data access, availability of personnel with the necessary skills to develop a model and interpret its output, and, perhaps most essential, a system for integrating the model into the healthcare practitioner's workflow it is designed to inform. For effective integration into the clinical workflow it is essential that the model's output be actionable, that is, that there are concrete interventions that can be executed in response to the model output. For example, if a model predicts a high risk of delayed wound healing, a physician can opt to begin hyperbaric oxygen therapy early (Jung et al., 2016). The availability of an effective intervention is a key consideration in assessing the utility of a predictive model built using EHR data.

The success of models deployed for practice management therefore depends on more than just model performance—additional factors such as the cost of deploying the model and the effectiveness of any action triggered by the model are also necessary to consider. For example, factors to consider include the size of the patient population where intervention is possible, the prevalence of the outcome being predicted, the expected number of patients with the predicted outcome, the cost savings per patient exposed to the intervention, the estimated cost of implementing the intervention, the total target savings, and the number of patients subjected to the intervention required to meet that target. If deploying a model is expensive, or the affected population too small to result in cost savings for the practice in question, the model will not be "useful" even if its performance as measured in terms of precision, recall and calibration is exceptional.

OPPORTUNITIES FOR MACHINE LEARNING IN HEALTHCARE

Despite the methodological challenges of working with EHR data (see Chapter 4: Electronic Clinical Documentation) and that researchers have yet to take full advantage of the universe of EHR-derived variables available for predictive modeling, there are many exciting opportunities for machine learning to improve health and healthcare delivery. Models that stratify patients into different risk categories to inform practice management have enormous potential impact on healthcare value (see Chapter 8: Health Information Technology and Value) (Parikh et al., 2016), and methods that can predict outcomes for individual patients bring clinical practice one step closer to precision medicine (see Chapter 11: Bioinformatics and Precision Medicine) (Pencina and Peterson, 2016). Identifying high-cost and high-risk patients (Bates et al., 2014) in time to attempt targeted intervention will become increasingly necessary as healthcare providers take on the financial risk of treating their patients.

Machine learning approaches have already been used to characterize and predict a variety of health risks. Recent work in our group using penalized logistic regression to identify patients with undiagnosed peripheral artery disease and predict their mortality risk found that such an approach outperforms a simpler stepwise logistic regression in terms of accuracy, calibration, and net reclassification (Ross et al., 2016). Such predictive models have been implemented in medical practice, resulting in more efficient and better quality care. For example, a predictive model to stratify newborns' risk of sepsis decreased antibiotic prescribing by 33%−60% when deployed by Kaiser Permanente (Escobar et al., 2014). Recent work to learn reference ranges for pediatric heart and respiratory rates from EHR data resulted in a decrease in heart rate alarms in a pediatric acute care unit following implementation of the learned ranges in unit monitors (Goel et al., 2015), potentially reducing alarm fatigue.

Machine learning has also been applied to hospital and practice management, to streamline operations and improve patient outcomes. For example, models have been developed to predict demand for emergency department beds (Peck et al., 2012) and elective surgery case volume (Tiwari et al., 2014), to inform hospital staffing decisions. The Veterans Health Administration's Clinical Data Warehouse, which houses clinical data for more than 20 million patients, was used to develop models for risk of hospitalization and death of individual patients with AUC values ranging from 0.81 to 0.87 (Fihn et al., 2014). The risk scores calculated by these models are presented to Patient-Aligned Care Teams and are accessed by more than 1200 healthcare providers per month as part of their routine practice. In another successful example, a 30-day readmission risk score for heart failure patients at the Parkland Health and Hospital System, called e-Score, was implemented in a prospective study to assess the effectiveness of a targeted intervention (involving intensive follow-up care) in patients with a high predicted risk of readmission (Amarasingham et al., 2013). The study found that the intervention, triggered by the predicted risk score, resulted in significantly reduced 30-day readmission compared to a control group. These examples demonstrate that predictive models can be deployed in healthcare settings and used effectively to improve resource allocation and patient care.

Finally, machine learning is not a panacea, and not all things that can be predicted will be actionable. For example, we may be able to accurately predict progression from stage 3 to stage 4 chronic renal failure. In the absence of effective treatment options—besides dialysis and kidney transplant—the prediction does not do much to improve the management of the patient. As another extreme example, imagine a model that predicts that treatment of a particular cancer (say lung cancer) will be ineffective based on the genotypic profile of the tumor. Would we withhold treatment based on such a prediction? Such extreme situations aside, there is ample middle ground where the use of predictive modeling in healthcare will enable proactive treatment, more efficient use of resources, and deliver better care at lower cost. Multiple other industries—such as finance, retail, aviation, web commerce, and election campaigning—have made the transition to incorporate machine learning, and the resulting predictive models, into their respective workflows. It is time for healthcare to make that transition.

CONCLUSIONS

Machine learning techniques applied to EHR data can generate actionable insights, from improving upon patient risk score systems, to predicting the onset of disease, to streamlining hospital operations. Statistical models that leverage the variety and richness of EHR-derived data (as opposed to using a small set of expert-selected and/or traditionally used features) are still relatively rare and offer an exciting avenue for further research. New types of data, such as from wearables, bring their own opportunities and challenges. Challenges in effectively using machine learning methods include the availability of personnel with the skills to build, evaluate, and apply learned models, as well as the assessing the real-world cost—benefit trade-off of embedding a model in a healthcare workflow. As machine learning using EHR data touches a large number of healthcare problems, it will motivate applied research in statistics and computer science. Best practices informed by the success (and failure) of machine learning models will move clinical informatics forward as a field and continue to transform medicine and the delivery of clinical care.

REFERENCES

Agarwal, V., Vibhu, A., Tanya, P., Banda, J.M., Veena, G., Leung, T.I., et al., 2016. Learning statistical models of phenotypes using noisy labeled training data. J. Am. Med. Inform. Assoc. 23, 1166–1173.

Amarasingham, R., Patel, P.C., Toto, K., Nelson, L.L., Swanson, T.S., Moore, B.J., et al., 2013. Allocating scarce resources in real-time to reduce heart failure readmissions: a prospective, controlled study. BMJ Qual. Saf. 22, 998–1005.

Bates, D.W., Saria, S., Ohno-Machado, L., Shah, A., Escobar, G., 2014. Big data in health care: using analytics to identify and manage high-risk and high-cost patients. Health Aff. 33, 1123–1131.

Besseling, J., Reitsma, J.B., Gaudet, D., Brisson, D., Kastelein, J.J.P., Hovingh, G.K., et al., 2016. Selection of individuals for genetic testing for familial hypercholesterolaemia: development and external validation of a prediction model for the presence of a mutation causing familial hypercholesterolaemia. Eur. Heart J. 38 (8), 565–573.

Breiman, L., 2001. Statistical modeling: the two cultures (with comments and a rejoinder by the author). Stat. Sci. 16, 199–231.

Brookhart, M.A., Stürmer, T., Glynn, R.J., Rassen, J., Schneeweiss, S., 2010. Confounding control in healthcare database research: challenges and potential approaches. Med. Care 48, S114–S120.

Casey, B.M., McIntire, D.D., Leveno, K.J., 2001. The continuing value of the Apgar score for the assessment of newborn infants. N. Engl. J. Med. 344, 467–471.

Casey, J.A., Schwartz, B.S., Stewart, W.F., Adler, N.E., 2016. Using electronic health records for population health research: a review of methods and applications. Annu. Rev. Public Health 37, 61–81.

Charlson, M.E., Pompei, P., Ales, K.L., MacKenzie, C.R., 1987. A new method of classifying prognostic comorbidity in longitudinal studies: development and validation. J. Chronic Dis. 40, 373–383.

Damgaard, D., Larsen, M.L., Nissen, P.H., Jensen, J.M., Jensen, H.K., Soerensen, V.R., et al., 2005. The relationship of molecular genetic to clinical diagnosis of familial hypercholesterolemia in a Danish population. Atherosclerosis 180, 155–160.

Delgado-Rodríguez, M., Llorca, J., 2004. Bias. J. Epidemiol. Community Health 58, 635–641.

Dicker, R., Coronado, F., Koo, D., Parrish, R.G., 2006. Principles of Epidemiology in Public Health Practice. US Department of Health and Human Services, Atlanta, GA.

Escobar, G.J., Puopolo, K.M., Wi, S., Turk, B.J., Kuzniewicz, M.W., Walsh, E.M., et al., 2014. Stratification of risk of early-onset sepsis in newborns ≥ 34 weeks' gestation. Pediatrics 133, 30–36.

European Association for Cardiovascular Prevention and Rehabilitation, Reiner, Z., Catapano, A. L., De Backer, G., Graham, I., Taskinen, M.-R., et al., 2011. ESC/EAS Guidelines for the management of dyslipidaemias: the Task Force for the management of dyslipidaemias of the European Society of Cardiology (ESC) and the European Atherosclerosis Society (EAS). Eur. Heart J. 32, 1769–1818.

Fihn, S.D., Francis, J., Clancy, C., Nielson, C., Nelson, K., Rumsfeld, J., et al., 2014. Insights from advanced analytics at the Veterans Health Administration. Health Aff. 33, 1203–1211.

Finlay, G.D., Duncan Finlay, G., Rothman, M.J., Smith, R.A., 2013. Measuring the modified early warning score and the Rothman Index: advantages of utilizing the electronic medical record in an early warning system. J. Hosp. Med. 9, 116–119.

Goel, V., Poole, S., Kipps, A., Palma, J., Platchek, T., Pageler, N., et al., 2015. Implementation of data driven heart rate and respiratory rate parameters on a pediatric acute care unit. Stud. Health Technol. Inform. 216, 918.

Goldstein, B.A., Navar, A.M., Pencina, M.J., Ioannidis, J.P., 2016. Opportunities and challenges in developing risk prediction models with electronic health records data: a systematic review. J. Am. Med. Inform. Assoc. 27 (1), 198–208.

Gultepe, E., Green, J.P., Nguyen, H., Adams, J., Albertson, T., Tagkopoulos, I., 2014. From vital signs to clinical outcomes for patients with sepsis: a machine learning basis for a clinical decision support system. J. Am. Med. Inform. Assoc. 21, 315–325.

Haddad, Z., Falissard, B.F., Chokri, K.C., Kamel, B.K., Nader, B.N., Nagi, S.N., et al., 2008. Disparity in outcome prediction between APACHE II, APACHE III and APACHE IV. Crit. Care 12, P501.

Halpern, Y., Yoni, H., Steven, H., Youngduck, C., David, S., 2016. Electronic medical record phenotyping using the anchor and learn framework. J. Am. Med. Inform. Assoc. 23, 731–740.

Henry, K.E., Hager, D.N., Pronovost, P.J., Saria, S., 2015. A targeted real-time early warning score (TREWScore) for septic shock. Sci. Transl. Med. 7, 299ra122.

Hersh, W.R., Weiner, M.G., Embi, P.J., Logan, J.R., Payne, P.R.O., Bernstam, E.V., et al., 2013. Caveats for the use of operational electronic health record data in comparative effectiveness research. Med. Care 51, S30–S37.

Hripcsak, G., Ryan, P.B., Duke, J.D., Shah, N.H., Park, R.W., Huser, V., et al., 2016. Characterizing treatment pathways at scale using the OHDSI network. Proc. Natl. Acad. Sci. U. S. A. 113, 7329–7336.

Huang, S.H., Paea, L., Iyer, S.V., Ming, T.-S., David, C., Shah, N.H., 2014. Toward personalizing treatment for depression: predicting diagnosis and severity. J. Am. Med. Inform. Assoc. 21, 1069–1075.

Iezzoni, L.I., 1999. Statistically derived predictive models. Caveat emptor. J. Gen. Intern. Med. 14, 388–389.

Jung, K., Shah, N.H., 2015. Implications of non-stationarity on predictive modeling using EHRs. J. Biomed. Inform. 58, 168–174.

Jung, K., Covington, S., Sen, C.K., Januszyk, M., Kirsner, R.S., Gurtner, G.C., et al., 2016. Rapid identification of slow healing wounds. Wound Repair Regen. 24, 181–188.

Kennedy, E.H., Wiitala, W.L., Hayward, R.A., Sussman, J.B., 2013. Improved cardiovascular risk prediction using nonparametric regression and electronic health record data. Med. Care 51, 251–258.

Lee, J., Maslove, D.M., 2015. Customization of a severity of illness score using local electronic medical record data. J. Intensive Care Med. 32 (1), 38–47.

Leek, J.T., Peng, R.D., 2015. Statistics. What is the question? Science 347, 1314–1315.

Maxson, E., Jain, S., Kendall, M., Mostashari, F., Blumenthal, D., 2010. The regional extension center program: helping physicians meaningfully use health information technology. Ann. Intern. Med. 153, 666–670.

Mossialos, E., Wenzl, M., Osborn, R., Sarnak, D., 2016. 2015 International Profiles of Health Care Systems. The Commonwealth Fund.

Parikh, R.B., Kakad, M., Bates, D.W., 2016. Integrating predictive analytics into high-value care: the dawn of precision delivery. JAMA 315, 651−652.

Patorno, E., Grotta, A., Bellocco, R., Schneeweiss, S., 2013. Propensity score methodology for confounding control in health care utilization databases. Epidemiol. Biostat. Public Health 10.

Peck, J.S., Benneyan, J.C., Nightingale, D.J., Gaehde, S.A., 2012. Predicting emergency department inpatient admissions to improve same-day patient flow. Acad. Emerg. Med. 19, E1045−E1054.

Peck, J.S., Gaehde, S.A., Nightingale, D.J., Gelman, D.Y., Huckins, D.S., Lemons, M.F., et al., 2013. Generalizability of a simple approach for predicting hospital admission from an emergency department. Acad. Emerg. Med. 20, 1156−1163.

Pencina, M.J., Peterson, E.D., 2016. Moving from clinical trials to precision medicine: the role for predictive modeling. JAMA 315, 1713−1714.

Quan, H., Li, B., Couris, C.M., Fushimi, K., Graham, P., Hider, P., et al., 2011. Updating and validating the Charlson comorbidity index and score for risk adjustment in hospital discharge abstracts using data from 6 countries. Am. J. Epidemiol. 173, 676−682.

Richesson, R.L., Jimeng, S., Jyotishman, P., Abel, K., Denny, J.C., 2016. A survey of clinical phenotyping in selected national networks: demonstrating the need for high-throughput, portable, and computational methods. Artif. Intell. Med. 71, 57−61.

Ross, E.G., Shah, N.H., Dalman, R.L., Nead, K., Cooke, J., Leeper, N.J., 2016. The use of machine learning for the identification of peripheral artery disease and future mortality risk. J. Vasc. Surg. 64, 1515−1522.e3.

Saria, S., Rajani, A.K., Gould, J., Koller, D., Penn, A.A., 2010. Integration of early physiological responses predicts later illness severity in preterm infants. Sci. Transl. Med. 2, 48ra65.

Schneeweiss, S., Rassen, J.A., Glynn, R.J., Avorn, J., Mogun, H., Brookhart, M.A., 2009. High-dimensional propensity score adjustment in studies of treatment effects using health care claims data. Epidemiology 20, 512−522.

Tiwari, V., Furman, W.R., Sandberg, W.S., 2014. Predicting case volume from the accumulating elective operating room schedule facilitates staffing improvements. Anesthesiology 121, 171−183.

Toh, S., García Rodríguez, L.A., Hernán, M.A., 2011. Confounding adjustment via a semi-automated high-dimensional propensity score algorithm: an application to electronic medical records. Pharmacoepidemiol. Drug Saf. 20, 849−857.

Wei, W.-Q., Teixeira, P.L., Mo, H., Cronin, R.M., Warner, J.L., Denny, J.C., 2016. Combining billing codes, clinical notes, and medications from electronic health records provides superior phenotyping performance. J. Am. Med. Inform. Assoc. 23, e20−e27.

Weng, C., Li, Y., Ryan, P., Zhang, Y., Liu, F., Gao, J., et al., 2014. A distribution-based method for assessing the differences between clinical trial target populations and patient populations in electronic health records. Appl. Clin. Inform. 5, 463−479.

Weng, S.F., Kai, J., Andrew Neil, H., Humphries, S.E., Qureshi, N., 2015. Improving identification of familial hypercholesterolaemia in primary care: derivation and validation of the familial hypercholesterolaemia case ascertainment tool (FAMCAT). Atherosclerosis 238, 336−343.

Chapter 20

The Future of Medical Informatics

David W. Bates[1], Kathrin M. Cresswell[2],
Adam Wright[1] and Aziz Sheikh[1,2]

[1]*Brigham and Women's Hospital/Harvard Medical School, Boston, MA, United States,*
[2]*The University of Edinburgh, Edinburgh, United Kingdom*

INTRODUCTION

In this book, we have described the main developments in medical informatics and health information technology (HIT), with the aim of giving readers a broad picture of what HIT is in place today and how it is being used to change care. While we have described many areas, inevitably we have had to leave some things out, for example, not covering the role of robotics. It is clear from the chapters in this book that HIT is already proving fundamentally important to improving healthcare and health and given the concerted attention now being focused on healthcare and health by policymakers, health systems, and the business community, we believe its influence will markedly increase in the years ahead. In this last chapter, we speculate how that may happen and discuss some specific areas which are likely to be important with respect to these changes.

Healthcare is already experiencing a "sea change," in that most healthcare in developed countries is now being delivered using electronic health records (EHRs). While these records have many advantages over paper, so far they are delivering only a fraction of their potential benefit. In the coming years, they will become easier to use, do a better job of anticipating provider and patient needs, be used much more often by patients themselves, and offer many more opportunities to extract data and measure how care is being delivered and patients are doing.

The picture is different in developing countries, in that this transition has yet to take place, but also in these settings electronic tools and records offer powerful advantages compared to paper. There are many examples of use of open-source records to care for patients that show that care can be improved (Seebregts et al., 2009). And just collecting a modicum of data in a hospital

Key Advances in Clinical Informatics. DOI: http://dx.doi.org/10.1016/B978-0-12-809523-2.00020-0

for example about the number of admissions and what some simple outcomes are can be potentially transformative. But some tools such as mobile technologies have already spread like wildfire in developing countries, and the potential exists for these nations to leapfrog developed countries and avoid some of the issues that developed nations have struggled with in the digital health space, such as investing large sums of money in EHRs with poor usability, or setting up data exchange that works routinely.

THE FUTURE OF HEALTHCARE AND HIT

We always overestimate the change that will occur in the next two years and underestimate the change that will occur in the next ten. Don't let yourself be lulled into inaction.

Bill Gates

It's tough to make predictions, especially about the future.

Yogi Berra

Taking the words of Bill Gates and Yogi Berra into account, we will make a few predictions about how HIT is likely to evolve in the coming years. To imagine how transformational change may be, it is useful to think back to what life was like before the personal computer, Internet, or digital phones. All these innovations have permeated our lives to the degree that it is hard to think about doing without them.

Patient Engagement

First, patients are likely to become much more involved with their care, and much more care will be delivered outside the walls of hospitals. Hospitals are fundamentally expensive and dangerous places, and it will be possible to monitor patients at home in ways that are not possible today, with wearable devices that will track not just pulse, respiratory rate, and blood pressure, but dozens of parameters simultaneously. Patients will be accessing their information through their personal health records, and they will record many more things than they do today; much of this will be done by ambient, connected devices.

A number of specific technologies are likely to be especially transformative in healthcare. These include social media, mobile technologies, big data, the cloud, telemedicine, and the Internet of Things (IoT). Each of these and potential developments that relate to them will be described next.

Social Media

Social media are already exceptionally important in terms of how people interact with the world, and they have broken down many barriers, for example, enabling Arab Spring, which resulting in major changes in multiple countries. People love communicating with each other, and health is one of the topics they are most interested in. One of the main ways that social media are likely to be important in health is through development of online communities, in which patients and families dealing especially with rare diseases can communicate, share stories, and support each other (Nambisan et al., 2016). This has already been found to be extremely important, for example, in the groups of PatientsLikeMe, but so far only a fraction of conditions are covered (Wicks et al., 2010). This could go further with patients volunteering their data in various ways, including genetic information or about their microbiome. uBiome is a new citizen science effort, which includes a sequencing-based clinical screening test which is intended to help people understand their gut health (uBiome, 2017).

But traditional healthcare has been relatively slow to embrace social media as Greaves and Rozenblum describe in their chapter, and in many ways has lagged behind the rest of the world (Wicks et al., 2010). In the future, we expect that patients will routinely use social media to help select physicians, to decide which clinics or hospitals to go to, and to share their care experiences much as they do with sites such as TripAdvisor. This will undoubtedly create a lot of grumbling from providers at first, but they will need to get past it. In developing countries, there appears to be even more variability than in developed countries among facilities, and this is likely to be a powerful influence.

Imagine a woman living in Nairobi, Kenya, who is about to give birth to her first child. She has a huge number of clinics to choose from, but little information to go on about which ones treat patients as they wish, or have the best outcomes. In the near future, we suspect many such choices will be made using social media, and while this will be far from perfect, it will also be a major improvement compared to the status quo.

Mobile Technologies

Mobile technologies have already profoundly influenced our lives in many ways, for example, changing the way we decide what route to take, and making it simple to decide whether taking the subway or driving across a crowded city makes more sense. However, as Singh and Landman describe, the apps patients are using in healthcare so far—even though there a huge number of them—leave a lot to be desired.

We predict that there will soon be multiple apps to choose from which will be available for most chronic diseases—and for multiple diseases at the same time—which will make it much easier for patients with chronic conditions to manage their health. One of the things that emerged in an exercise sponsored by the Commonwealth Fund in which consumers were asked what they wanted in an app, was that they wanted a single app (Sage Health Advisor, 2017). They wanted it to offer advice and information, health tracking, a holistic picture of how they are doing, and a set of coordination and communication tools which would let them work with their care team—the tool proposed is called Sage. While this was an idealized version of what might be developed, a tool like this is within reach today, and we think that apps like this will soon become the norm.

Something like Sage would allow patients to do a variety of things they struggle with today, for example, to get vetted answers to important questions, like—"I have abdominal pain. What are the things I should be looking for that should make me call the doctor?" It could also make tracking much easier. There are already many tools for tracking a variety of parameters, but many do not work together and it is very hard to get information back to their provider. A tool like this could make it easy. Having the big picture of how a patient is doing, and describing a shared set of goals and objectives would also be a big advance. Finally, the communication function described earlier is pivotal. It should be easy for providers to, for example, give a patient an exercise prescription, for example, for both number of steps and a specific duration and then track how the patient is doing, but today it is not.

Mobile tools will also make a huge difference on the provider side. They will be able to access information routinely about the patients they are caring for on tablets or handhelds and manage many common issues from them. Especially in developing countries they have already made a huge difference—for example, enabling communication of tuberculosis test results, screening for retinal disorders, or ensuring that all health centers have an adequate supply of antimalarials at a given time.

A grandmother in Singapore wants to determine what she should do when she develops abdominal pain. She has been generally healthy, but tends to minimize her symptoms. In the future, she will likely to consult the equivalent of Sage, and find out that her symptoms are sufficiently worrisome that she should definitely contact the on-call provider (Sage Health Advisor, 2017). *This is the kind of thing that can enable detection of infrequent but serious conditions, like an infected gall bladder.*

Big Data

Access to truly big data, and the use of techniques like machine learning and artificial intelligence, is potentially transformative, as described by Callahan and Shah in their chapter. Until recently, it was very difficult to get large amounts of health data. The advent of EHRs and wearable technologies has changed that, and it can now be linked to other types of data such as demographic and social data. The net result is that it will soon be possible to address many questions in ways that were not previously possible.

These approaches are already widely used in other situations. For example, when Amazon is making recommendations to you about what things you may want to purchase, they are leveraging big data and machine learning to predict what items may be of most interest to you. These approaches have innumerable medical implications, but a few examples predict what might be the right next antihypertensive medication for a patient, based on their profile, what they have taken before, and what seemed to help patients who were most like them previously; sifting through multiple parameters in the intensive care unit (ICU) to predict who is developing sepsis, and using algorithms to determine what patients may be most likely to be readmitted early, and what interventions might be most helpful for them (Bates et al., 2014).

A 43-year-old patient is seeing his primary care provider in Australia. He has high cholesterol. In the past, the primary care provider would have just picked a cholesterol medication. But now, the patient's information can be synthesized with models that go far beyond those of Framingham and consider the trajectories of similar patients to predict their risk of heart attack, their genome has been sequenced which affects both the risk prediction and choice of medication, and the patient's family history and preferences can all be factored in. The net will be greater precision in the choice of medication which is much more likely to result in long-term benefit.

The Cloud

For decades, though we have been using computers, our data has resided on our local machines. That has many advantages—you know where it is, others cannot access it—but it has many downsides—being just one, what if it gets lost. Perhaps even more important, using just one computer means that there is a great deal of unused capacity at any given time. But as Mizani describes in his chapter, the cloud brings these and many other advantages.

Healthcare organizations have been reluctant to use the cloud. There is justifiably a great deal of concern about ensuring privacy with healthcare data in particular. But cloud architectures offer so many advantages that they will undoubtedly become widely used in healthcare in the near future, and cloud providers are increasingly understanding and embracing the technical and policy safeguards needed to store clinical data.

A primary provider in a small town in Idaho in the United States is using an EHR with his practice partner. They've liked it, but have struggled with dealing with the upgrades, and with managing issues when the system goes down. It has been hard for them to access the data when their patients are hospitalized in a small town 30 miles away. They then switch to a cloud-based EHR. This has many advantages for them; the upgrades are handled remotely. Fixes are added on an on-going basis, so they don't have to wait for the annual upgrade. Maintenance is handled by their IT provider, and data from other locations come in and out about their patients in a way that lessens their burden. Perhaps best, they no longer worry about losing their data if their system fails because it is backed up on a regular basis with multiple copies.

Telemedicine

Telemedicine will undoubtedly take off globally in the coming years, as is already in the United States. Many of the barriers to its use have related to payment and regulations; physicians understandably find it threatening if a provider in another much lower cost location can for example read a radiograph at much lower cost. But the regulations will catch up, and there are strong incentives to shift more care outside the walls of healthcare organizations.

There are already notable examples—for example, Mercy Virtual is a "hospital without beds" which runs a large ICU, delivers telestroke care, offers virtual hospitalists, and performs home monitoring (Mercy Virtual, 2017). Furthermore companies such as American Well, which is the largest US telehealth company, already offer on-demand patient visits for patients with acute needs (American Well, 2017). Traditional providers will undoubtedly start to incorporate these services routinely. These technologies will likely be woven into our more traditional medical institutions.

Jose, who lives in Uruguay, develops a bad cough, and shaking chills. It is the weekend and his primary care provider is not available. But he logs into a local service via his smartphone which connects him with a provider in Chile who assesses him and tells him there is a significant likelihood of pneumonia, and that he needs to be seen in his local emergency room.

The Internet of Things

One of the most profound changes that is likely to occur in the next 10 years in information technology is development of the IOT (Boulos and Al-Shorbaji, 2014). This refers to enabling everyday objects electronically, so that they can capture and send data. An example might be your refrigerator, which has a great deal of information about what food you are keeping in it.

The IOT may be especially helpful in helping older patients manage at home. Societies globally are aging, especially in developed countries. From the economic perspective, this is one of the most prominent problems that developed economies face today. The IOT will be absolutely pivotal in enabling patients to manage much more effectively in their homes than has been the case.

An elderly couple is living in England. The husband has mild cognitive deficiency—while not frankly demented, he has difficulty with his short-term memory. The wife is cognitively intact, but suffers from congestive heart failure and rheumatoid arthritis. They are adamant that they want to stay at home. Today, many small changes could break this fragile equilibrium. But the IoT could make it much easier to remain at home longer. Their home would be equipped with a fall detector, so that if either does fall, the emergency team will be notified. Both will likely use automated pill dispensing devices, so that they get reminders to take their medications, and their daughter can be notified if her mother skips her diuretic too frequently which she is wont to do. Their refrigerator will track how much food they have, and make it easy to order more if there are certain key things they are low on. It might even warn the wife if she happens to try to purchase something that contains too much salt for her, like a large jar of pickles.

CONCLUSIONS

Most powerful transformations will likely arise when changes in incentives including quality and safety goals, new practices, and new technologies all are aligned. Some of these projections may take longer or less time than we expect. But these and many other changes that we have not even considered are likely to occur in a reasonable time frame. And they will have profound implications for healthcare. It is predicted that self-driving cars will have a major impact—they will be safer. Because they can be used by many people, fewer will be needed, and we will need many fewer parking lots. The broader use of information technology will be similarly transformative. Future trends will likely be guided by attempts to develop solutions that address wider global challenges, many of them outside healthcare. But they will change healthcare nonetheless. We believe that healthcare will look very different in 10−20 years than it does today, and information technology will be a key force behind that change.

REFERENCES

American Well. https://www.americanwell.com/ (accessed 26.01.17).

Bates, D.W., Saria, S., Ohno-Machado, L., Shah, A., Escobar, G., 2014. Big data in health care: using analytics to identify and manage high-risk and high-cost patients. Health Aff. 33 (7), 1123−1131.

Boulos, M.N., Al-Shorbaji, N.M., 2014. On the Internet of Things, smart cities and the WHO Healthy Cities. Int. J. Health Geogr. 13 (1), 10.

Mercy Virtual. http://www.mercyvirtual.net/ (accessed 26.01.17).

Nambisan, P., Gustafson, D.H., Hawkins, R., Pingree, S., 2016. Social support and responsiveness in online patient communities: impact on service quality perceptions. Health Expect. 19 (1), 87−97.

Sage Health Advisor. http://www.sagehealthadvisor.com/ (accessed 26.01.17).

Seebregts, C.J., Mamlin, B.W., Biondich, P.G., Fraser, H.S., Wolfe, B.A., Jazayeri, D., et al., 2009. The OpenMRS implementers network. Int. J. Med. Inform. 78 (11), 711−720.

uBiome. https://ubiome.com/ (accessed 02.02.17).

Wicks, P., Massagli, M., Frost, J., Brownstein, C., Okun, S., Vaughan, T., et al., 2010. Sharing health data for better outcomes on PatientsLikeMe. J. Med. Internet Res. 12 (2), e19.

FURTHER READING

Greaves, F., Ramirez-Cano, D., Millett, C., Darzi, A., Donaldson, L., 2013. Harnessing the cloud of patient experience: using social media to detect poor quality healthcare. BMJ Qual. Saf. bmjqs-2012.

Index

9780128095232